GUIDE TO
VISITING VINEYARDS

GUIDE TO
VISITING VINEYARDS

Anthony Hogg

London
Michael Joseph

First published in Great Britain by Michael Joseph Limited
52 Bedford Square, London WC1B 3EF
1976

ISBN 0 7181 1560 0

Filmset by Filmtype Services Limited, Scarborough and printed
in Great Britain by Ebenezer Baylis & Son Limited, The Trinity
Press, Worcester, and London
and bound by Dorstel Press

Contents

Illustrations

Maps

The maps have been drawn by Chichester Design Associates Ltd.

A SPECIAL
ACKNOWLEDGEMENT

Having completed this book when inflation and taxation (the duties on wines and spirits have never been higher) have put so many of the pleasanter things of life beyond our means, I was told that it could not be a commercial proposition because its price would have to be much higher than people would want to pay for a guide book.

I am, therefore, extremely grateful to Justerini and Brooks, one of the oldest of the companies in International Distillers and Vintners, for a most generous and timely act of sponsorship, which has made publication possible.

Appropriately too, it was also a visit, not to a vineyard or cellar but to London in the days when there may still have been vines in Vine Street, that made Justerini and Brooks possible. What he, Giacomo Justerini, said to her, Margherita Bellino, doesn't matter, but the consequence has been a firm of West End wine merchants holding for over two hundred years all told, Warrants of Appointment to numerous kings and queens of this realm. Let Dick Bridgeman, a Director, summarise the story.

J & B – A Rare Story
Fickle fortune was responsible for the founding of 'J & B' in 1749. Giacomo Justerini, desperately in love with the beautiful Italian opera singer, Margherita Bellino, could bear the parting no longer when her successful appearance in London was extended, and so he followed her to England.

Soon he ran out of money, but with him he had brought the receipts for various liqueurs, which his family had for years been distilling in their native Bologna. He began to distil, setting up shop at No. 2 The Colonnade, Pall Mall, to sell his concoctions.

For the next two hundred and five years Justerini and Brooks conducted their business at this address – through all the Seven Years War, the Peninsula War, the Crimea, the Boer War and both World Wars. No. 2 Pall Mall, as it became, was a favourite haunt of 'the Quality', meeting there to gossip, and to drink a glass of wine whilst placing orders for the replenishment of their cellars.

In the late 1920s Justerini and Brooks discovered that they were gaining acclaim for their own brand of Scotch whisky and the then Managing Director, Eddie Tatham, sensing the end of prohibition in the United States, began to lay the foundation of what could now be called the second J & B success story. It was not immediate; before there was time for sales to boom, the 1939–45 war had begun, bringing the operation to a complete halt.

Peace brought a slow revival; in 1945, when I joined the company, we were exporting 25,000 cases of J & B Rare. Today the brand ranks among the first three in the markets of the world.

'I don't care where the water goes if it doesn't get into the wine', said Chesterton's Noah. We go further; we like to keep our whisky clear of the wine too. Our business as wine merchants is run separately from the whisky 'empire', but it would be a transparent fib to pretend that the whisky profits have not been very helpful in enabling us to build and maintain superb wine stocks, remaining moreover, through difficult days, in our traditional place, London's West End.

Though there are distilleries to visit as well as châteaux in this Guide, wine predominates and it is perhaps because Justerini and Brooks are now, as ever, pre-eminent in the wine business that we have chosen to sponsor a unique Guide not previously attempted.

R. L. O. Bridgeman

PREFACE

*Let us get up early to the vineyards, let us see if the vine
flourish, whether the tender grapes appear
– Song of Solomon 7 v.12.*

When I joined the wine trade in 1956, to visit a wine shipper abroad you wrote to your wine merchant, who wrote to the shipper's London agents, who sent you a letter of introduction. The host firm then needed a further letter, or a telephone call, saying when you intended to arrive.

This courteous, but involved procedure struck me as admirable for the leisurely times of Bertie Wooster (when Jeeves would have attended to everything in impeccable French or Serbo-Croat) but a little unpractical nowadays. There were – and still are – enough formalities in leaving the country for a mere fourteen days as it is. Who wants to add to them by becoming committed to dates, times and the tedium of telephone calls, in a foreign language, in order to spend a pleasant hour with wine at its source?

It seemed to me that if people could just turn up as they felt inclined, far more of them would do so, to the greater benefit of themselves and the wine trade. I set about persuading some of the firms large enough to spare staff as guides to this view. Soon whole packs of invitation cards with how-to-get-there plans on the back began to reach me, cards from this firm and that, designed to be given to members of the public on request.

There was no obligation: John Bull, in his shaky old Dormobile, bearing an invitation to Moët et Chandon, had no need to present himself at those august portals in Épernay if his hangover dictated otherwise, or his wife was tired, or the children scotched the plan by being car sick.

My own employers, International Distillers and Vintners, offered a set of these cards, giving the holder a choice of over a dozen establishments abroad, and when a change was needed, I edited a *Passport to Vineyards and Cellars,* which looked sufficiently like the real one for an absent-minded holder to have his baggage turned inside out after presenting it in error to an outraged official at Dieppe.

This *Passport* had thirty-two 'visa' pages, one each for a variety of 'host firms', spread about France, Germany, Italy, Spain and Portugal, with which International Distillers and Vintners had close associations. Throughout the 1960s the idea of enjoying a little 'grape to glass' instruction when on holiday abroad was growing apace. Vinous travel articles in the Press, including *Wine Mine,* a combination of magazine and illustrated price list which I edited, supported in the Company's shops with various leaflets describing possibilities, led to a free service, in which we prided ourselves on giving anybody an introduction to some sort of wine establishment, wherever wine was made commercially in Western Europe.

If some of my correspondents, trying to say where they were going, were at times a little uncertain as to where the vineyards were ('I shall be on holiday in Paris, can you fix me up with an introduction?'), others – probably just as uncertain ('I would like to see a vineyard on the Gargano peninsula') – made me wonder, as I turned over maps feverishly, whether my own geography was much better.

Having sent the clients forth on their Bacchanalian business, 'East and west and south and north' like Lars Porsena's messengers, I decided in 1972, five years before I expected to retire, that a Guide book to the whole subject was not only needed, but could conceivably occupy me in retirement keeping it up-to-date.

Apart from a little French booklet, none appeared to exist in any language; after all, there had been no need for a Guide, when drinking wine was a pastime largely limited to the gentry and their butlers. But now, when in one year two hundred thousand Europeans go round the Moët et Chandon cellars in Épernay and may want to see a bit of Burgundy or Bordeaux the next, a Guide could fulfil a great want, none the less for being a short-felt want.

I decided my object should be:
1. To provide interesting and instructive visits for those on holiday.
2. To reward those who do open their establishments to the public.
3. To encourage more interest in wine.
4. To amuse myself.

Modelled on Raymond Postgate's original *Good Food Guide 1951–2,* there would be no advertisements and no payments demanded of, or accepted from, those with entries in it.

The firms, châteaux, cellars, distilleries, etc., would have to be limited to those having products on sale in Britain. No attempt should be made to compile an exhaustive list; a reasonable number of 'hosts' in each of the principal wine-growing districts of Western Europe would suffice.

The nearest approach to such a work was a little French booklet, published in 1971 by the Comité National des Vins de France called *ler annuaire des caveaux, celliers, chais et autres centres de dégustation des Vins de France.* Though a useful little book as far as it went (which was France) many of the entries were those of small growers, whose wines are not exported. Visitors in any number moreover could be a serious inconvenience to them.

A satisfactory Guide in English would have to state clearly how to get there, in such a manner that even an Australian from the outback, speaking no French and on his first visit to Europe, could find a château without stopping to ask. *Un peu d'histoire,* as the *Michelin Green Guides* are fond of saying, a word about what the visitor can expect at each place, the visiting hours and the method of introduction would be other essentials.

Draft entries on these lines were sent, either directly or through their U.K. agents, to companies, establishments and proprietors having wines and spirits on sale in this country. The majority replied welcoming the project and kindly supplied me with the details necessary to write the entries in full.

I am glad to emphasise that in no case have I excluded an entry originating in this way. Nor have I had to implore or entreat in order to obtain a sufficient number of entries for my purpose.

The result is this *Guide to Visiting Vineyards* covering Western Europe (with a few more outside it for good measure), virtually an invitation book from a fine mixed company of voluntary hosts some two hundred and fifty strong. And since *Guide to Visiting Vineyards* is such a mouthful (and certainly not the sort of mouthful wine drinkers want), let it be known as *The VV Guide.*

Any omissions – and regrettably there are some well-known names missing – are chiefly due to no replies being received to my original approaches. In some cases, of course, companies have understandably declined as a matter of policy. Some lack the time, some the English-speaking staff and some feel their premises are of insufficient interest. Others, I suspect, have visions of coach loads of Britons, behaving like football club supporters on Saturday night when their team has lost. How are they to know that in reality we British wine drinkers, wearing our National Health spectacles, are middle-aged, middle-class and so respectable that we whisper 'Pardon me' whenever we swallow the samples the wrong way?

With such help from my 'company of volunteers', my task has been simplified; like General Wavell naming his poetry anthology*, which took the minds of my generation briefly off World War II, 'I have gathered a poesie of other men's flowers and nothing but the thread that binds them is my own'. Not that there is anything poetic about a Guide book, particularly where modern wine installations are concerned. Rupert Croft-Cooke† lamented that business firms preserve little of their history and that little is apt to run to a trite pattern – 'The founder of a hundred and fifty years or more ago, a shadowy figure in a wig and knee breeches perhaps, the sons and grandsons of sterling worth and great enterprise, the building of the bodegas, the purchase of the vineyards, the setbacks at the beginning of the century, the prosperity of today'.

Mr. Croft-Cooke's 'today' was 1955. Adding mergers, take-overs, duty increases, V.A.T., inflation, steel-lined tanks for temperature-controlled fermentation and automated bottling lines, my Guide could be made about as romantic as a British Leyland assembly shop on a Midlands wet Tuesday.

By no means all wine and spirit establishments open their doors to the public. Recalling my early active service days in the Royal Navy, when the ship was open to visitors, I cannot say I altogether blame them.

Conversely, some of the very large concerns, like Moët and Martell, Martini and Hennessy, keep 'open house' accepting casual visitors by the coach load as a useful public relations exercise.

Otherwise, hosts like to know whom to expect and when, so that they can put their best foot forward. The hospitable wine trade being no exception, most of the entrants in this Guide still prefer to receive people by appointment.

Some years ago I received a letter from a Mr. Hennessy. He explained he was just an Australian and a wine enthusiast, keen to learn more on a first trip to Europe. Could I give him introductions to some Continental firms without pinning him down to dates and times? I was only too glad to help, sending him some 'To whom it may concern' letters to present and adding that, with a name like Hennessy, he should get along all right.

He did. With a chuckle perhaps, Mr. Hennessy presented himself about half an hour before luncheon at the Reims head office of a big Champagne house. 'Mr. Hennessy is here', announced the girl at the reception desk.

What a commotion! 'Mr. Hennessy?' 'From Cognac obviously'; 'One of the family of course'; 'But who can have asked him to lunch and told nobody?' The Directors, taken unawares, not surprisingly were baffled.

However, with the chef enjoined to special effort and an extra place laid at the boardroom luncheon table, Mr. Hennessy was entertained as befits a magnate of Cognac.

Some weeks later, I was asked if I knew anything about this Mr. Hennessy; they didn't seem too amused at Head Office when I told them the story.

In this Guide, the paragraph entitled 'Introduction' – the last item of each entry – names the agent or representative in Britain from whom a letter of introduction to his 'principal' abroad may be obtained. Unfortunately, the old pattern of the wine trade, in which each grower/producer exporting to us appointed a London shipper as his agent and sole distributor of his liquors, has largely disappeared. Such relationships, where they still exist, do not last for decades as in the old days. In these less stable and more competitive times, the big groups often buy direct at the lowest prices they can find; few agency agreements seem to last long.

When such changes occur, I can only crave the co-operation of my Trade colleagues in forwarding my readers' requests to the new agent, with the consoling thought that the forwarding of a letter still requires no extra postage stamp.

But the great majority of those engaged in wine in

---

*Viscount Wavell *Other Men's Flowers.*

†Rupert Croft-Cooke *Sherry* (page 152).

12

Europe are not employees of companies but grower/farmers, whose vineyards cannot provide them *wholly* with a living. They have little time to spare for visitors and probably speak no English. On the other hand, a local hotel proprietor can often fix up the keen visitor with some sort of contact, particularly if there is no language barrier.

Again it must be remembered that a grower's livelihood (and that of his employees) depends, wholly or in part, on a successful vintage. The grape harvest or 'vintage', starting early in September in southern Europe, continuing into November further north, in Germany for example, is the culmination of a year's work. Everybody is expected to 'pick his or her weight', working full time at this hard, monotonous, back-breaking work, until the harvest is home. The French are not averse to strong, healthy-looking volunteers, with well-kept guitars for the after-supper sing song, but when accompanied by a 'crummy-looking' girl friend, they fear the worst as far as their vintage is concerned.

TASTING ON TOUR

Sooner or later anybody putting my Guide to practical use will, I sincerely hope, be handed a glass of wine in some corner of a foreign cellar and, in the unlikely event of the hosts ever having heard of Rupert Brooke, they could come to regard it as a corner 'that is forever England' if an English visitor just gulps his wine down in the manner of the first-time communicant when the priest said, 'Drink ye all of this'.

In fox hunting – so I'm told – the professionals are more impressed if the novice shouts 'Tally-Ho' instead of 'There the foxy b – goes', so perhaps I should explain that when sampling or drinking fine wine, colour, smell and taste, in that order, are the three characteristics with which to be concerned. First, hold the glass up to what light you can find. The colour should be clear and bright.

Firmly holding the glass by the base between thumb and index finger, it is possible to twirl the wine gently. This increases the bouquet as the wine slides down the sides of the glass making it much easier to appreciate and enjoy.

Next, take a reasonable mouthful and roll it round the mouth in contact with the roof, the tongue and the back of the throat. Repeat if necessary. Make up your own mind about the 'weight' in the mouth and the general flavour. There is no need to swallow. In a tasting room there is always a spittoon or basin; in the cellars the floor. Wine – particularly red wine – from the cask in most cellars is young and rough. Your host or guide is unlikely to swallow it, so watch him for a lead.

Whether presented with two or three glasses in turn in the cellars or finding up to half a dozen wines laid out for a comparative tasting in the tasting room, remember that the practice is to taste in order of merit. The first wine is likely to be fairly ordinary; if in need of a generous draught, the last is the one for the big swallow!

For those wishing to know more about tasting, there is a short practical handbook, *Wine Tasting* by Michael Broadbent MW, from Christie's Wine Publications, 8 King Street, London SW1.

Pamela Vandyke Price, *Times* wine correspondent, has also given the subject a good airing in *The Taste of Wine*.

CARE OF WINE

Having tasted and perhaps bought a few bottles, the following notes may help to get them home in good shape.

Spirits are sufficiently strong in alcohol to remain in normal drinking condition for months *after* the bottle or container has been opened and broached. Fortified wines (e.g. sherry and port) being fortified with brandy, are roughly half as strong as spirits, which is why they remain drinkable for some weeks when the bottle is opened and broached or decanted.

Table wines are about half the strength of fortified wines and, exposed to the air, soon become unpleasant to taste, turning eventually to vinegar. It is no good therefore drinking part of a bottle or jar and expecting the wine to taste good again next day. Kept in a hot car boot, deterioration will be hastened.

In *full and closed* containers, sound wine keeps satisfactorily. Wastage can always be kept to a minimum by transferring any wine remaining to smaller containers (e.g. jars to bottles, bottles to half bottles), filling to just below the top and corking lightly. Make sure containers are clean; washing out wine bottles with clean cold water should suffice.

TIPPING

Don't expect too much of your guides. The Directors haven't the time, employees haven't the languages and university students, sometimes employed in the long summer vacation, haven't the product knowledge. Tipping is not expected, but 1–2 Fr. a visitor would reward a good guide. This too would be appropriate in the Bordeaux region if taken round the chai of a famous château by the Maître de Chai, the Head Cellarman, himself.

HOTELS AND RESTAURANTS

Though some are mentioned in these pages, usually as a result of my personal experience, proprietors, managers and chefs can change frequently; it is not therefore the policy of *The VV Guide* to make recommendations and neither author nor publisher can be responsible for disappointments.

DRINKING AND DRIVING

Lastly a word of caution: in France, Germany, Holland, Switzerland and Spain the regulations are the same as in Britain. In the other countries of *The VV Guide* detention and prosecution is at the discretion of the police with penalties as severe. To risk it anywhere is most foolish.

ACKNOWLEDGEMENTS

Not being a wine buyer, spending much of his working life nowadays* travelling abroad among firms such as will be found in the pages that follow, I cannot claim intimate knowledge of every region about which I have written. Yet, since 1949, there has only been one year when, with my wife, I have not set foot in France, often en route to Italy, Spain or Portugal, and always with our car because, being gluttons, it was unthinkable to miss France by flying to some other destination beyond her *bonnes tables*.

During these travels we have covered practically all the regions of which I write. Speaking French and Italian, my wife's attention to 'duty' has been so great that she declares the words, 'For God's sake, not *another* bottling line' will be engraved on her heart as 'Calais' on that of 'Bloody Mary' Tudor.

Many other people have been indispensable to the writing of this book. I would thank again the many directors, châteaux proprietors, régisseurs, museum curators and others, who have so willingly given me information. I am also grateful to other wine book authors, from whom I have learnt much. Many of them will find their works mentioned under 'Recommended Reading' in the pages that follow.

I record thanks to Peter Dominic Ltd. for permission to draw on material from *Wine Mine*, which I edited from 1959 to 1974. To Sir Guy Salisbury-Jones, GCVO, CMG, CBE, MC, and to Mr. Jack Ward, Chairman of the English Vineyards Association, I am also grateful for valuable help on 'the home front'.

I am much indebted to the following for their help in the preparation of various parts of *The VV Guide:* Dr. Mario Adragna, Mrs. Sarah Ames in Vienna, Mr. Derek Barrow, Miss Patricia Cornforth at Schloss Vollrads, The Hon. Neil Hogg, Mr. Joe Hollander in Provence, Mr. R. J. Horowitz, Mr. Ian Jamieson, MW*, Mr. John Lockwood, MW, Mr. Fred May, Mr. Reggie Peck, Mr. David Peppercorn, MW, Mr. P. G. Psares, M. and Mme. Alain Querre of Château Monbousquet, M. Franck Rochard, Dr. Bruno Roncarati, Mr. Michael Weston and Dr. R. R. Wilson.

Not least, I record my thanks to many of my colleagues in International Distillers and Vintners and particularly to the following, who have read those parts of my manuscript in which they are currently expert, eliminating many, though I fear not all, of my mistakes: Tim Ambler, Martin Bamford, MW, Henry Clark, Garry Grosvenor, Geoffrey Hallowes, James Long, Don Lovell, MW, Robin Reid, Mark Ridgwell, George Robertson, George Shortreed, Simon Smallwood, MW and Christopher Whicher, MW. I am also greatly indebted to Garry Grosvenor, stationed in Germany, for invaluable liaison between German companies and myself.

Finally, special thanks are due, first to Mr. Michael Longhurst, of Gilbey S.A. at Château Loudenne, who has not only given me the benefit of his own knowledge but has been a vital link and interpreter in communications with firms and proprietors throughout the Bordeaux region; secondly, to my secretary, Mrs. Winifred Andrew, whose typing and long wine trade experience have been invaluable.

---

*Before the war, as old members of the retail wine trade in London recall, British wine merchants never went abroad to *buy;* it was the sellers and their representatives who called upon *them.*

*Master of Wine (M.W.) is the British wine trade's highest professional qualification. Examinations, both written and practical, lasting several days, are conducted annually by the Institute of Masters of Wine and the most extensive preparation is required to pass them.

THE VV CLUB

All those who buy this book are invited to consider themselves members of The VV Club, founded herewith by Anthony Hogg, who has accepted his own offer to be President until such time as a more distinguished one can be found. (After forty-five years, half in the Navy and half in the wine trade, his liver is already semi-retired and the sooner he steps down to make way for younger drinkers, the better.)

The period of membership will be until a new edition of *The VV Guide* is published. The Club has no premises, no meeting place, no subscription, no funds, no list of members and no privileges to offer them.

Nevertheless, the President hopes to receive their reports, some recommending new entries worthy of the Guide; some enlarging upon existing entries in it and others – no less valuable – being critical. The suggested form which the report should take is shewn herewith:

VISITORS' REPORT FORM

To The Editor
The VV Guide, c/o Michael Joseph Ltd, 52 Bedford Square, London WC1B 3EF.

I visited

on

A. I do not think it is worth a place in the Guide for reasons given in my report below.

OR

B. The information below is additional to that given in the Guide and may be useful for a future edition.

C. The above establishment is not in the Guide. I think it deserves to be (See my report below).

Report   The following is helpful:
Name, Address and Telephone Number of Establishment

How to get there

History. Enclose brochure if possible

Amenities for visitors

When open

Was call made casually or by appointment?

Signature   Date
Name and address (BLOCK CAPITALS PLEASE)

Profession*

Should you require an answer, please send a *stamped addressed envelope*.

*Members of the catering, wine and spirit and associated trades are requested to state in this space whether their firms or groups have any connection with the establishment(s) to which their completed report form refers.

15

ORGANISED WINE TOURS

The following societies and wine clubs organise tours for their members:

The International Wine and Food Society Ltd
11a Seymour Place
LONDON W1H 5AN
Tel. 01–402 6212

The Wine Mine Club
Vintner House
River Way
Templefields
HARLOW
Essex CM20 2EA
Tel. (0279) 416291

Among British Travel Agents, Cox and Kings offer attractive 8-day fly/drive holidays in Bordeaux and in Burgundy and a cabin cruiser holiday along the Canal du Midi. Ingham's Discovery Tours include four to six-day guided tours of Bordeaux, Burgundy and the Rhine; and one in September to the Munich Beer Festival.

Supertravel (Heritage is a subsidiary at the same address) are more expensive, their seven-day Bordeaux visit in September includes meals at some attractive spots, notably a luncheon at Château Lafite. Other wine regions of France and Italy are included. And, let us hope somebody provides a free dram of Islay malt for those who go bird watching with them on that island in November.

Addresses of these travel agents are:

Cox and Kings
46 Marshall Street
LONDON W1V 2PA
Tel. 01–734 8291

Supertravel
22 Hans Place
LONDON SW1X 0EP
Tel. 01–589 5161

Ingham's Travel
329 Putney Bridge Road
LONDON SW15 2PL
Tel. 01–789 5251

For latest information on this and many other aspects of wine/travel, read *Decanter*, quite the best English magazine yet to appear on wine. It is published monthly by Decanter Magazine Ltd., 44–56 Hercules Road, London, SE1. Tel. 01–928 2774.

The entries that follow were prepared with the co-operation of the companies, proprietors, etc., to which they relate, the accuracy of the final drafts being checked by them or their agents. The information on 'Getting There' was relevant in the summer of 1976. Neither I nor the publishers can accept any responsibility for errors occurring due to subsequent changes.

HOW TO USE THE VV GUIDE

Each wine region (or country in the case of the lesser wine lands) has its own introductory chapter on ways of getting there, its wines, its *Routes des Vins* and other wine/travel information. Descriptions of how wine is made and other technicalities (which appear in many admirable wine books) have been purposely avoided. My object has been to keep to the district, in order to enhance any reader's visit, rather than to write about the product, except occasionally to mention local wines worth drinking, or to explain an interesting process such as the making of champagne or malt whisky.

The Guide begins with France, nearest to Britain and the most important country. The order then moves clockwise to Germany and Italy, completing the clockwise circuit west to Spain and Portugal and then north to England and Scotland. Austria and Hungary follow, with notes on lesser lands being combined as 'Other Countries'.

Throughout the Guide, wherever practicable, a clockwise course is taken from one district to the next, starting with the district a visitor from Britain is likely to go to first.

Each entry consists of:

1. Name, postal address and telephone number of the place to be visited.

If an appointment has been made and you are unable to keep it, due for example to illness or breakdown, please inform the host by a telephone call to this number.

2. Instructions on how to find the establishment.

These have been written in sufficient detail, it is hoped, to allow the reader to get there with the aid of a good local map but without having to ask.

Sometimes these instructions refer to town plans in the *Michelin Guide* of the countries concerned and sometimes to the *Michelin Cartes Régionales* series of maps, which cover France in a Nos. 51–86 series. All these are on sale in bookshops throughout Western Europe.

3. A brief history of the firm or place to be visited.

4. What the visitor can expect.

This paragraph deals mainly with such matters as guides, languages spoken, samples offered, tasting and purchasing possibilities. The pattern of events in wine cellars cannot differ widely, so it should be regarded as guidance not precise information. For example, absence of the words 'Free Sampling' should not be taken to mean that the visitor will come away thirsty! Repetition has been curtailed in the interest of readability.

In Scotland distilleries do not hold off-licences and cannot offer whisky for sale.

5. Visiting days and times.

Unless the entry states otherwise, establishments will be closed to visitors on Saturdays, Sundays and national holidays. Dates of national holidays for each country are included at the end of the introductory chapter of the country concerned.

Long Weekends. When national holidays fall on a Thursday or a Monday, establishments may remain closed on the following day, the Friday, or the Tuesday, in order to make a long weekend break. It is advisable to make local enquiries at your hotel or tourist office to avoid the disappointment of finding premises closed.

Summer Holidays. Many firms in France, more particularly in Burgundy, close down completely for the summer holiday for a month, starting either in the last week of July or the first week in August. Information is included where known.

At distilleries in Scotland, there is usually a 'Silent Season' during July, August and September when the warmer weather is not ideal for distilling. Holidays are taken and machinery refitted. The words 'Non-operational' mean that, though open to visitors, there will be no distilling for them to see.

6. Method of introduction.

Where casual visitors are accepted, no introduction is of course necessary. But – as has already been emphasised – the majority of 'host firms' prefer notice and this paragraph gives the agent, or a principal stockist, through whom a visit can be arranged.

If you are a regular customer of any wine merchant or retailer, make your first enquiries there. The company may like to arrange introductions for its regular customers, or it may be part of its policy to make special offers in connection with such visits.

But should you write to the address given by the Guide in this paragraph and receive no reply, it may be because the agency agreement has changed and the U.K. firm no longer represents that Principal abroad. Indeed, in these difficult times, no reply could be due to economies being made in staff or in postage.

Should this occur, a polite request could be written direct to the host firm. If it can be written in *their* language, so much the better. And do enclose an addressed envelope, accompanied *not* by a British postage stamp, but by an *International Reply Paid Coupon*, which can be bought at any post office. Without one, a reply is unlikely from many companies.

Please state clearly and briefly when you would like to call, making sure from the Guide that the place will not be closed because you have inadvertently chosen a Saturday, Sunday, national holiday, or the Summer Holiday closed period.

U.K. Agent and Devolution. Many Principals have two or more agents, one for England and another for Scotland being a typical arrangement simplifying distribution of the goods. Under the heading 'U.K. agent' the Guide usually gives one – the agent for England and I apologise to any devolutionist who feels affronted.

Distances
In the entry pages, which are intended to be used in conjunction with local signposts, distances are in kilometres for Continental countries and in miles for England and Scotland.

In the introductory chapters, distances are in miles in order to convey more to English-speaking readers. (1 kilometre = 0.625 miles).

Areas
In the entry pages, vineyard areas are in hectares (1 hectare = 2.47 acres). In the introductory chapters the more familiar acre has been retained.

Recommended Reading
The following is a short list for those who want further information in one volume on the wines of all countries in this Guide. Books confined to one country's wines (e.g. *Wines of Germany* or to one type of wine (e.g. *Sherry*) are recommended under this same paragraph heading in the appropriate parts of the Guide.
Alex Lichine's *Encyclopedia of Wines and Spirits* New Edition 1976 Cassell & Co.
Frank Schoonmaker's *Encyclopedia of Wine* British Edition 1975 A & C Black Ltd.

FRANCE

France

(Regions are arranged clockwise from the Loire)

Introduction

GETTING THERE

By road. Whatever closures are threatened for British Rail, car ferries from Britain to the Continent seem to increase each year. Here is a summary of those to France in the summer of 1976, which are expected to be much the same in 1977.

CAR FERRIES

Plymouth–Roscoff	
Brittany Ferries (7 hours)	Daily
Weymouth–Cherbourg	
Sealink (4 hours)	Up to 2 a day
Southampton–Le Havre	
Townsend Thoresen (6½ hours)	Up to 3 a day
Southampton–Cherbourg	
Townsend Thoresen (5 hours)	Up to 2 a day
Southampton–Le Havre	
Normandy Ferries (P & O) (7 hours)	One a day; one a night
Portsmouth–St. Malo	
Brittany Ferries (8½ hours)	Daily
Portsmouth–Cherbourg	
Townsend Thoresen (4 hours)	Up to 2 a day
Newhaven–Dieppe	
Sealink (3¾ hours)	Up to 6 a day
Folkestone–Calais	
Folkestone–Boulogne	
Sealink (1⅔ hours)	Up to 3 a day
Dover–Boulogne	
Dover–Calais	
Sealink	
Normandy Ferries	
Townsend Thoresen (1⅔ hours)	Frequent services
Dover–Dunkirk	
Sealink (3¾ hours)	Up to 4 a day

HOVERCRAFT

Dover–Boulogne	
Seaspeed (35 minutes)	Up to 7 a day
Dover–Calais	
Seaspeed (30 minutes)	Up to 6 a day
Ramsgate–Calais	
Hover-lloyd (40 minutes)	Up to 20 a day

By rail: These car ferries now operate the normal cross-Channel passenger services, trains from London and other cities connecting with them.

By air: There are direct Air France and British Airways flights from London to the following French airports, an * indicating those at which Fly and Drive arrangements can be made in advance: Paris*, Lille, Strasbourg, Bordeaux*, Biarritz, Marseilles*, Nice*, Quimper, Toulouse*, Dinard, Nantes, La Baule, Deauville; and from Manchester to Paris.

The wine regions are also well covered by other companies. Air Champagne – Gatwick–Reims and Gatwick–Dijon; Air Alsace and Air Paris have flights Gatwick to Colmar; T.A.T. from Gatwick to Tours and British Air Ferries still fly cars from Southend to Le Touquet.

In an excellent book for the wine student on holiday, Pamela Vandyke Price wrote: 'There is a lot of nonsense talked and written about both food and wine and also about France'†. Having perhaps written some of it when describing the regions in the pages that follow, my introductory chapter to France herself can be brief.

MAPS AND GUIDES

France, greatest wine country and nearest neighbour, provides about half the entries in *The VV Guide*. Finding some of these establishments should be much easier with the *Michelin Guide to France* and Michelin maps (1 cm pour 2 km; Nos. 51–86 inclusive comprise the set). Though the *Michelin Guide* is published annually, the town plans in it rarely change, thus a current edition is not essential for *this* purpose. It is however recommended so that prices of its

---

†Pamela Vandyke Price *Eating and Drinking in France*

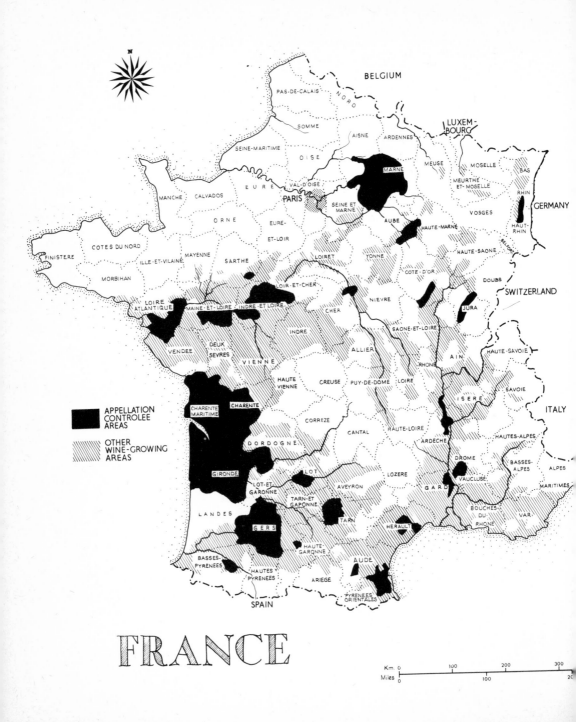

N

BELGIUM

LUXEM-
BOURG

GERMANY

SWITZERLAND

ITALY

SPAIN

PAS-DE-CALAIS

NORD

SOMME

AISNE

ARDENNES

MEUSE

MOSELLE

BAS
RHIN

SEINE-MARITIME

OISE

EURE

VAL-D'OISE

PARIS

SEINE ET
MARNE

MARNE

AUBE

HAUTE-MARNE

MEURTHE
ET-MOSELLE

VOSGES

HAUT-
RHIN

HAUTE-SAONE

MANCHE

CALVADOS

ORNE

EURE-
ET-LOIR

FINISTERE

COTES DU NORD

ILLE-ET-VILAINE

MAYENNE

SARTHE

LOIRET

YONNE

COTE-D'OR

DOUBS

MORBIHAN

LOIRE
ATLANTIQUE

MAINE-ET-LOIRE

INDRE-ET-LOIRE

LOIR-ET-CHER

CHER

NIEVRE

SAONE-ET-LOIRE

JURA

VENDEE

DEUX
SEVRES

VIENNE

INDRE

ALLIER

RHONE

AIN

HAUTE-SAVOIE

HAUTE
VIENNE

CREUSE

PUY-DE-DOME

LOIRE

SAVOIE

ISERE

CHARENTE
MARITIME

CHARENTE

CORREZE

CANTAL

HAUTE-LOIRE

ARDECHE

HAUTES-ALPES

APPELLATION
CONTROLEE
AREAS

OTHER
WINE-GROWING
AREAS

DORDOGNE

LOT

DROME

BASSES-
ALPES

ALPES

GIRONDE

LOT-ET-
GARONNE

AVEYRON

LOZERE

GARD

VAUCLUSE

MARITIMES

LANDES

TARN-ET-
GARONNE

TARN

HERAULT

BOUCHES-
DU-
RHONE

VAR

GERS

HAUTE-
GARONNE

AUDE

BASSES-
PYRENEES

HAUTES
PYRENEES

ARIEGE

PYRENEES
ORIENTALES

FRANCE

Km. 0   100   200   300
Miles 0   100   20

comprehensive list of hotels and restaurants are up-to-date. Select your destination and hotel in advance and get there by 6 p.m., to reserve a room, is a wise counsel for the tourist in France in summer.

The large-scale maps (available singly in most British bookshops) are, I find, indispensable for taking the little roads or for pottering off the beaten track.

CAVES TOURISTIQUES

In addition to the establishments given in *The VV Guide*, there are scores of *Caves Touristiques* and *Salles de Dégustation* by the roadsides of France, enticing the passing tourist with a prominent signboard to stop and taste the local wines. In some the wines will be sound enough, particularly if the place is run by the local Wine Co-operative, a most successful movement widespread in the wine regions of Europe, which makes and markets wines from the grapes sold to it by many small grower/members.

A tasting glass may cost 15–20p and usually wines may be bought in packs of three and six bottles. Whether the price is less than in the shops and supermarkets should be checked; returning home, with V.A.T. and 48p a bottle duty to pay on quantities exceeding the duty-free allowance, the pack may prove to have been no bargain. On many of the popular tourist routes, in the short tourist season, approach warily: large profits quickly gained are likely to be more important to these places than what they sell.

The Cave Touristique is really just a fancy name for a superior Salle de Dégustation, where wine can be bought in bottle, or sometimes in your own jar, filled from a cask to take away. Some of them exercise over-aggressive salesmanship, making the tourist feel compelled to buy at least a few bottles. The Salle de Dégustation is virtually a wine bar.

Some Caves Touristiques and Salles de Dégustation are mentioned under 'Lesser Visits' in appropriate parts of *The VV Guide*.

NATIONAL HOLIDAYS

National Holidays in France are given below. Please remember when planning your trip that, not only will premises be closed, but that it is not uncommon to remain closed for an extra day (e.g. Friday after a Thursday public holiday) to make a long weekend break.

January 1, Easter Monday, May 1, Ascension Day, Whit-Monday, July 14, August 15, November 1 and 11, December 25.

Further Information
The French Government Tourist Office
178 Piccadilly
London W1
Tel. 01–493 3171

THE LOIRE

Introduction

During my wine trade years, I have received so many letters beginning, 'We are going to the Loire and would like some introductions to vineyards and cellars' that it is as well to start here by mentioning that France's longest river does run for six hundred miles and that vines do not line its banks quite like tourists along Buckingham Palace railings. In short, it is as well for people to say to which part they are going.

Fortunately, the regions of great châteaux and of wine largely coincide in the last two hundred miles from

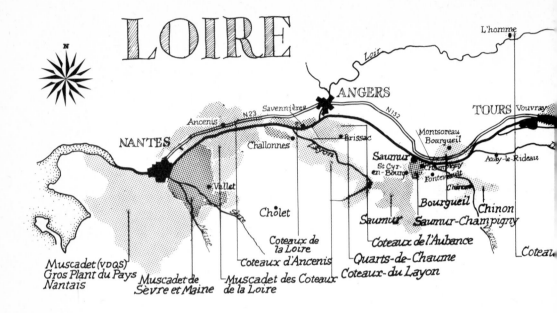

LOIRE

Orleans to the sea, an east to west stretch no more than five hours' drive from the Channel ports of Normandy and Brittany, such as Dieppe, Le Havre and St. Malo. From Paris, the autoroute to Tours is now complete, the journeys by road or rail taking no more than two hours.

BRITTANY

From Nantes, a tour could begin with Brittany's one wine, Muscadet, before working upstream into Anjou and Touraine. This crisp, dry white wine, such a perfect partner for the sea food of the Atlantic, takes its name from a vine brought from Burgundy after a terrible winter in 1709. The vineyards almost surround Nantes, the best being in the south-east sector of Sèvre-et-Maine, particularly in the neighbourhood of La Chapelle-Heulin, Vallet and Clisson. The wine can be drunk straight from the cask starting four months after the vintage, although the best qualities may need a year to show their class. Some Muscadet nowadays is presented on November 15 as Muscadet Primeur, wine of the new vintage for immediate drinking. No Muscadet is likely to improve after two years.

The district also yields the Gros Plant, a lesser dry white wine rated V.D.Q.S. (*Vin Délimité de Qualité Supérieure*, the category below *Appellation Contrôlée*).

Further Information:
Comité Interprofessionnel des Vins d'Origine du Pays Nantais
17 rue des États
44300 Nantes

ANJOU

To the public, the name Anjou is identified with rosé wines and very good they are, particularly if made from the Cabernet grape. The rosé region, about forty miles by twenty, is vaguely a rectangle south of the Loire extending at least from Angers to Saumur. Through this rectangle flows an insignificant Loire tributary, the Layon; indeed it barely flows at all, being little more than a ditch, but the best wines of Anjou are the sweet wines of its slopes, the Coteaux du Layon

Described as 'flowery Sauternes', Quarts de Chaume and Bonnezeaux are separate appellations; Saint-Aubin-de-Luigné, Beaulieu-sur-Layon and Rablay-sur-Layon are labelled Coteaux du Layon, with or without their own names. All are to be found in good restaurants in Angers, notably Le Vert d'Eau, at very fair prices.

To the north, across the river, is Savennières, village of the Coteaux de la Loire, responsible for La Coulée-de-Serrant and La Roche-aux-Moines (q.v.), the best dry white

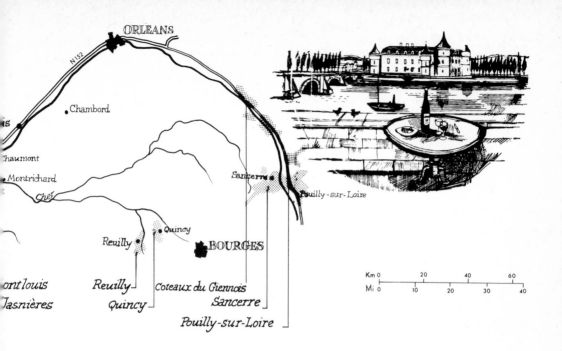

wines of the Loire, certainly comparable with, and longer lived than, many white burgandies.

Still in Anjou, more sound rosé wine is made south of Saumur, a busy town astride the great river. But for over a century Saumur has been the centre of Loire sparkling wine, the best being made by the *Méthode Champenoise* as in champagne itself. The two wines need not be compared; champagne is inimitable, sparkling Saumur (and Vouvray) less expensive, more fruity and less dry. Thus it can be a much more sensible choice for a wedding party in a stuffy tent on a hot summer's day.

Further Information
Comité Interprofessionnel des Vins d'Anjou et Saumur
21 Boulevard Foch
49000 Angers

TOURAINE

By road along the river, Angers to Tours is 65 miles and the dividing line between their two provinces falls about half way at Montsoreau, where the Vienne has flowed sedately past Chinon to merge with the Loire. Nearby, at Fontevrault, a non-vinous visit is recommended to the Abbey to see the tombs of our Plantagenet kings and queens. This stretch, from Angers to Blois and beyond, is the Loire Valley at its best. Great châteaux, like Chenonceaux and Chambord, recall the regal splendour of the Valois. The countryside is benign and even the great trees that rise out of sleepy backwaters look majestic.

In Anjou, the vineyards are mostly to the south of the river; now in Touraine they appear on both banks. Bourgueil looks southward to Chinon, Vouvray to Montlouis, where the Chenin Blanc can rarely achieve quite the quality of Vouvray though the wines are similar. The two red wines too are similar and it is wiser to leave the citizens of Chinon and the burghers of Bourgueil to distinguish between them.

Further Information:
Comité Interprofessionnel des Vins de Touraine
12 rue Berthelot
37000 Tours

POUILLY AND SANCERRE

Here are two more wines, barely possible for the non-native to distinguish. Grown on chalky slopes, the Sauvignon Blanc gives them a dry, flinty smoky flavour and this possibly gave rise to the name Pouilly Fumé. Sancerre is a charming hill town and as this book offers no introductions there, the visitor is recommended to go

25

sampling in the bars and cafés of its main square. There is also a Maison du Sancerre in Paris at 22 avenue Rapp, 7⁰ (Tel. 551.75.91) where wines may be sampled and bought.

Further Information:
Union Viticole Sancerroise
Comité de Propagande des Vins de Sancerre
18300 Sancerre

ST. POURÇAIN-SUR-SIOULE
Far to the south between Moulin and Vichy, small quantities of dry white and rosé wines, rated VDQS, are made, chiefly at the local Cave co-opérative.

BRITTANY
Nantes (near)
Donatien Bahuaud & Co.,
44330 Chapelle-Heulin, Loire-Atlantique
Tel. (40) 78.23.52 and (40) 78.22.14

Chapelle-Heulin is 18 kilometres south-east of Nantes in the direction of Poitiers. The route from Nantes is by N148 bis, forking left after 13 kilometres on to N756 (Michelin map No. 67).

Begun in 1928, here in the heart of Muscadet, this family business has become well-known in export markets in recent years thanks to the energy of Donatien Bahuaud, who has made it one of the largest installations for Muscadet. Visitors see the premises, a vineyard and enjoy a tasting in the old cellars.

Hours: Monday to Friday 0800-1200; 1400-1800. No closed period for summer vacation.
Introduction: None necessary; 48 hours' notice by postcard or telephone call appreciated.

Lusseaud Pierre et Fils,
Château de la Galissonnière, Le Pallet-Vallet 44330, Loire-Atlantique
Tel. (40) 26.42.03

This old walled Muscadet vineyard on N148 bis, though not open to the public, is included here for its historical interest. It is named after an early and celebrated owner, Amiral de la Galissonnière, opponent of the British Admiral Byng at Minorca (1756).

The French ultimately made Galissonnière Governor of Canada and the wine of this vineyard has been supplied to the many French warships named after him. The British shot Byng, 'Pour encourager les autres' as Candide was to put it; or, as written on Byng's monument at Southill,

Bedfordshire, 'To the Perpetual Disgrace of British Justice'. Among the relics, kept by the Lusseauds, is a weighted chest probably used for the confidential fleet signal books and kept on the poop, close to the command at sea, the better to jettison if the ship were captured.

The vineyard makes Muscadet of the highest quality and the Lusseauds, father and son and their families, are pleased to welcome up to twenty at a time in their new reception room. French only is spoken. Arrangements could be made through: Justerini & Brooks Ltd., 1 York Gate, Regent's Park, London, NW1 4PU. Tel. 01–935 4446.

ANJOU
Angers
Cointreau S.A.R.L.,
Carrefour Molière, 49 St. Barthélémy d'Anjou, 49800 Trélazé, Maine-et-Loire
Tel. (41) 88.51.82

Leave Angers by Exit 2 on the *Michelin Guide* Angers plan, which is the Longué road D61. This leads to the Industrial Zone East, five kilometres out at St. Barthélémy d'Anjou where the Cointreau modern factory and headquarters are situated.

One-third each of Cognac, lemon juice and Cointreau! Those of us who shook a pretty cocktail in the twenties have not forgotten the 'Sidecar' recipe. But how much alcohol, sugar and orange peel (bitter and sweet) is needed to make Cointreau itself, only two people know precisely. The two Cointreau brothers made it in Angers in 1849 and now three of their great, great grandsons, all cousins, run this great combination of distillery, bottling lines, administrative offices and people.

Hours: Monday to Friday 0900–1200; 1400–1700.
Introduction: Casual visitors cannot be accepted. All visits must be arranged through the agents, proposals being sent to:
U.K. Agents: J. R. Phillips & Co. Ltd., Avonmouth Way, Avonmouth, Bristol, BS11 9HX. Tel. (02752) 3651.

Angers (near)
Château de la Roche-aux-Moines,
49 Savennières, Maine-et-Loire.
Tel. (41) 41.52.32

The Château, 11 kilometres from Angers, (*Michelin* Map 73) is difficult to find. From Angers take D111, which closely follows the northern bank of the Maine and the Loire downstream. Watch for a sign-post to the left, before Savennières and beyond Epiré. Roche-aux-Moines is then the second château.

The château is a charming private house, occupied by Madame Joly and her husband, who is a leading surgeon in Angers. Her estate comprises two celebrated adjoining vineyards, La Roche-aux-Moines and La Coulée-de-Serrant, which make two of the Loire's finest white wines. This region, on the north bank, is designated Coteaux de la Loire, the village and appellation being Savennières, which has a pleasant camping site on the river. Opposite, on the south bank, is Quarts de Chaume, well known for those charming, flowery, sweet wines from the same grape – the Chenin Blanc.

There is no great chai or extensive cellar here; the charm lies in the walk along the avenue of cypresses, known as the Cemetery of the English, because our Prince John (subsequently King John after his brother Richard I) was defeated here on July 2, 1214, thus losing the province. The vineyards, originally planted by the monks of St. Nicholas in the twelfth century, lie on one side of this avenue; the broad river below on the other. Though production is small (La-Coulée-de-Serrant is only 7 hectares), tasting samples are usually offered and purchases are sometimes possible.

Hours:	Monday to Saturday 0930–1200; 1400–1730.
Introduction:	Madame Joly extends a welcome to all those interested. No introduction is necessary, just ring the front door bell. Little Coulée-de-Serrant is currently shipped to U.K. due to hard times in Britain but any special requirements for visits could be directed to Peter Dominic, who formerly offered it. Peter Dominic Ltd., Vintner House, River-way, Templefields, Harlow, Essex CM20 2EA.

Saumur

Ackerman-Laurance,

49210 Saint-Hilaire-Saint-Florent, Maine-et-Loire.
Tel. (41) 50.25.33.

Saint-Hilaire-Saint-Florent is a suburb of Saumur, on the left bank of the Loire, downstream towards Angers and this establishment is conspicuous on the route.

In 1811, after spending some months in Champagne studying their methods of making sparkling wines, Jean Ackerman settled down to business in Saumur. He found the region pleasant, the wines well suited to the *Méthode Champenoise* and the natural caves ideal for ageing. All he needed was 'a better half' so he married a Mademoiselle Laurance. Ackerman-Laurance is the oldest and the leading house for Saumur sparkling wines (brut and demisec). They also deal in still wines of the Loire.

A notice outside bids visitors welcome. A guide conducts the tour, explaining the various stages in making sparkling wines, which are sampled and may be bought.

Hours:	Easter Sunday and Monday, then daily 1 May to 30 September. 0930–1130; 1500–1700.
Introduction:	Open to all, none necessary.

Bouvet-Ladubay S.A.,

Rue Ackerman, Saint-Hilaire-Saint-Florent, Maine-et-Loi
Tel. (41) 50.11.12.

This well-known sparkling wine house also has its cellars and offices in Saint-Hilaire-Saint-Florent. Étienne Bouvet began it in 1851, being among the first to try the *Méthode Champenoise* in Saumur. Its wines won two Gold Medals and four Diplomas at a London exhibition in 1972.

Visitors are impressed by the immense galleries in the rock. The *Méthode Champenoise* processes of *remuage* and *dégorgement* are demonstrated whenever possible and the company's wines are tasted free.

Hours:	Daily, June to September inclusive 0900–1200; 1500–1700.
Introduction:	All visitors are welcome but the company does prefer advance warning and in particular letters of introduction from:
U.K. Agents:	Jarvis Halliday & Co. Ltd., 102 Bicester Road, Aylesbury, Bucks. Tel. 0296–3456.

Gratien & Meyer,

Château de Beaulieu. Route de Chinon, 49400 Saumur, Maine-et-Loire
Tel. (41) 51.01.54

On Route D947, Saumur to Chinon, along the south bank of the Loire, this establishment is conspicuous three kilometres east of Saumur. It is one of several large local companies, started over a century ago, to make very good sparkling wines by the champagne method from the local Chenin grapes. Gratien & Meyer in fact have a sister house in Champagne and, like all champagne firms, use black Cabernet grapes here as well as the white Chenin.

Production of the region now exceeds seven million bottles a year. Gratien & Meyer were particularly fortunate in establishing themselves on this cliffside with a superlative view across the river and ample space underneath to dig their cellarage. Their sparkling wines reach the British public through Peter Dominic shops and through The Wine Society.

Hours:	Every day 0900–1130; 1400–1730. English-speaking guides and free sampling.
Introduction:	None necessary.

TOURAINE
Bourgueil

Cave Touristique de Bourgueil,

Chevrette, 37140 Bourgueil, Indre-et-Loire
Tel. (47) 58.72.01

Chevrette, marked on good large-scale maps, is a hamlet two kilometres north of the D35, between the village of Saint Nicolas-de-Bourgueil and the small town of Bourgueil. This interesting cellar has been set up in some natural caves in the heart of the vineyards, where the ground has risen to form a hilly ridge. Approaching Bourgueil from any direction, an admirable series of direction signs makes it easy to find.

The Cave is well illuminated and so are the enlarged colour photographs illustrating the eight communes: Saint Patrice, Ingrandes, Restigné, Benais, Saint Nicolas-de-Bourgueil, Chouzé-sur-Loire, La Chapelle-sur-Loire and Bourgueil. The show pieces, however, are the sixteenth, seventeenth and eighteenth century wooden wine presses.

A guide takes visitors round at 15-minute intervals on payment (1976) of 3.50 francs. A glass of the fruity, ruby wine (served at cellar temperature) is given at the end.

| Hours: | Open daily (every day of the year). Summer (Easter to end of September) 1000–1200; 1400–1900. Winter (Oct. 1 to Easter) 1000– 1200; 1400–1800. |
| Introduction: | None necessary. |

Chambord
Etablissements Aimé Boucher S.A.,
41350 Huisseau-sur-Cosson, Loir-et-Cher.
Tel. (39) 46.31.10.

Huisseau-sur-Cosson is a village about a dozen kilometres east of Blois and only three along D33 from the great Château of Chambord. Here, Aimé Boucher and his step-son, Claude Kistner, ply the trade of *négociant-éleveur* in attractive country, on the edge of the huge forest and game reserve.

For many years Monsieur Boucher has been a Government agent, responsible for collecting and accounting for tax payments of wine, for which vignerons and distillers of local liqueurs are liable in France. These duties, taking him into every cellar from Orleans to Nantes, keep him in constant touch with all the best growers, from whom he buys for his own business. Aimé Boucher's Touraine wines are of outstanding excellence and, over the past ten years they have become well-known in Britain through the Peter Dominic shops and from Justerini & Brooks.

M. Kistner speaks English and this busy partnership is always pleased to welcome British enthusiasts for their wines. Arrangements could be made through:
Justerini & Brooks Ltd., 1 York Gate, Regent's Park, London, NW1 4PU.
Tel. 01–935 4446.

Montrichard
Etablissements J. M. Monmousseau,
B.P. 25, 41400 Montrichard, Loir-et-Cher.
Tel. (39) 32.06.70 or 32.07.04.

Montrichard, a charming little town on the right bank of the Cher, south of Blois, is on N76 between Tours and Vierzon. It lies in the midst of some historic châteaux country – Chenonceaux 8 km., Amboise 18 km., Chambord 30 km., Blois 32 km. and Loches 32 km. The Victorian Monmousseau buildings look across the sluggish Cher upon an unchanging pastoral scene.

Visitors can do a 20-minute tour (without covering all 13 kilometres of underground galleries) seeing the *Méthode Champenoise* being applied to make Touraine wines sparkle. Wines are tasted free and may be bought.

Hours:	Monday to Friday 0900–1200; 1400–1900. Also at these times at weekends and on National holidays from 1 April to 31 October.
Introduction:	None necessary. Any special arrangements should be made with agents:
U.K. Agents:	Imported Wines, Monte Cristo House, 24 Gillender Street, Bromley-by-Bow, London, E3 3LD.

Tours
Etablissements Marc Brédif,
Rochecorbon, 37210 Vouvray, Indre-et-Loire.
Tel. (47) 52.50.07.

This company, started by Ernest Brédif in 1893, is easily spotted between Tours and Vouvray going along the north bank of the Loire on the main road.

Marc Brédif succeeded Ernest about 1930, but since 1965 his descendant, Monsieur Jacques Cartier has been in charge, having spent thirty-two years preparing for the post under his predecessor. The cellars extend for some three kilometres in a rocky cliff and are some forty metres below the top of it. Marc Brédif can lay claim to being the leading House of Vouvray; always in spick and span order the cellars are among the finest in Touraine and my reports indicate that much trouble is taken in shewing people round.

Visitors can see a collection of wine presses and bottles from the sixteenth to twentieth centuries and note the interesting process whereby the still wines of Vouvray become *pétillant*. Wines may be purchased.

| Hours: | Monday to Friday 0900–1130; 1400–1730. |
| Introduction: | Open to all – none necessary. Groups (maximum 20 people at a time) should, however, make an appointment. |

Nantes (near)

Caveau des Templiers, Le Pallet. A centre of dégustation in this village of Muscadet wine.

Open all the year, except Thursdays, 1000–1230; 1400–1700. Opening and closing times extended to 0900 and 2000 respectively during the months of July and August.

Bourgueil, Chinon and Champigny

Bourgueil, hardly more than a village, and Chinon, an average size town, make two of the Loire's three Cabernet-Franc red wines.

At Bourgueil, the Cave Touristique (p. 28) is open every day of the year.

In Chinon, the Cave du Syndicat de Défense des Vins de Chinon is open from: Sunday before Easter to 30 October 0930–1200; 1430–1900.

Only very little of the third red wine, Champigny, is made. The village is 7 kilometres south-east of Saumur and the appellation is Saumur-Champigny. All three should be served cool at cellar temperature; at the Hôtel Jeanne de Laval at Les Rosiers they do so to perfection.

Caves Co-opératives

The Cave Co-opérative at Brissac, 16 kilometres south of Angers, welcomes visitors and provides a tasting room for them to sample. A modest charge may now be made. The Cave Co-opérative of Saumur, sited on a hill at Saint-Cyr-en-Bourg, on D93 eleven kilometres south of Saumur, is reminiscent of a London tube station. The grapes go in at the top and are transported as wine at the bottom.

Tours

L'Eglise Saint-Julian, 12 rue Nationale. Clearing a bombed site during World War II revealed these cellars, which were part of a fine mediaeval building that once belonged to the Abbey of Saint-Julian. Now restored, a *Son et Lumière* in praise of the wines of Touraine is produced by the association of Touraine wine merchants (C.I.V.T.) during July and August. Open from: Sunday before Easter to 31 October 0930–1200; 1430–1900.

Vouvray

There is a Co-opérative at Vouvray and it is said there are other houses receiving visitors but more information is wanted.

NORMANDY

Introduction

In terms of the coast, Normandy stretches from Le Tréport (north of Dieppe) to Mont St. Michael, taking in the whole of the Cherbourg peninsular. Inland, the boundary passes by Vernon (between Rouen and Paris), Dreux, Le Mans, curving north to the coast at Mont St. Michel.

None of this is wine country but drink is never far away in France and if no grapes will ripen, the French have the happiest knack of substituting some other fruit that will. Thus, not only is Normandy cider famous, but its distillate, Calvados, is to the apple what Cognac is to the grape.

The great Calvados country is the Pays d'Auge, east of Caen, but virtually the whole of Normandy has made this heart-warming fiery spirit since the Armada galleon, *El Calvador*, was wrecked in 1588 where the invasion fleet was to land in 1944.

Having been off those beaches on 'D' day, I can confidently say that communications with the mainland were better then than they are now. In preparing this book, all letters I sent to establish contacts with Calvados

distillers, let alone one prepared to receive visitors, have only brought forth one reply.

The finest Calvados is distilled twice in a pot still; left to mature in oak casks for a score of years it becomes surprisingly like fine cognac of similar age, yet with its own very delicate bouquet of apple. I used to buy such bottles at two excellent hotels, the Lion d'Or at Bayeux and the Dauphin at L'Aigle, until in the '70s old Calvados prices became prohibitive.

Fortunately rather nearer the Channel ports, the great Bénédictine distillery at Fécamp gives the visiting liquor-lover full value for his small entrance fee. One of the oldest and one of the finest liqueurs, Bénédictine is also the most faked. The museum at Fécamp has a glass case of formidable size in which fake labels from all over the world are displayed. I doubt whether Mother McGinty ever went there. She was the owner of the New York liquor store with the inscription:

'Mother McGinty's New York burgundy –
Beware of Imitation!'

Further Information:
On Calvados: Bureau National Interprofessionnel des Calvados et Eaux-de-Vie de Cidre,
14000 Caen.

Fécamp
The Bénédictine Distillery,
76400 Fécamp, Seine-Maritime.
Tel. (35) 28.00.06.

Fécamp is a Normandy fishing port, 42 kilometres from Le Havre and 64 kilometres from Dieppe. There are frequent car ferry and passenger services from Newhaven to Dieppe and from Southampton to Le Havre, which is France's principal Trans-Atlantic port, under three hours (via motorway A13) from Paris.

In 1066, William the Conqueror's staff invading Britain included advisers from the Fécamp monastery. But this great herb liqueur did not appear before 1510, when Brother Bernardo Vincelli created the recipe that includes the flavours from some twenty-seven herbs and spices.

Thenceforward its fame spread, until about 1790 when the monastery was destroyed in the French Revolution. Forced to disperse, the monks entrusted their secrets to friends – laymen perhaps, who were not persecuted – some very probably to Le Grand, their business agent. Certain it is that M. Alexandre Le Grand, a Fécamp merchant and scholar, was able to recreate the recipe, and by 1863 he was making Bénédictine commercially. By the end of the century he had built a great 'Gothic' edifice, close to the old site, as a Bénédictine distillery and museum, now visited by 100,000 people each year.

For a small fee there is a guided forty-minute tour, seeing the pot-stills and the museum in this imposing building. A recorded commentary in several languages tells the story of the old monastery and that of the liqueur. There is no tasting, but Bénédictine, liqueur chocolates and souvenirs can be bought.

Hours: Summer (Easter Sunday–11 November approx.). Daily, including Sundays and national holidays 0900–1115; 1400–1800.
Winter (12 November approx.– Maundy Thursday) Monday to Friday 0900–1115; 1400–1700 excluding national holidays.

Introduction: None necessary, open to all; any special arrangements (e.g. large parties) should be made with agents:
Rutherford, Osborne & Perkin Ltd., Harlequin Avenue, Brentford, Middx. Tel. 01–560 8351.

Pont-l'Evêque
Calvados Busnel,
30 rue Saint-Melaine, 14130 Pont-l'Evêque, Calvados.
Tel. Pont-l'Evêque 9 and 437.

Eleven districts make Calvados, but the only one where double distillation in a pot still is obligatory is the Pays d'Auge and the small town of Pont-l'Evêque lies in the middle of it. Caen is 46 kilometres to the westward and that pretty yachting harbour of Honfleur 16 kilometres to the north. Coming from Le Havre, via the Tancarville road bridge that spans the Seine, or from Rouen, turn right at the crossroads in the centre of the town.

This is the rue Saint-Melaine, which leads out towards Honfleur and Deauville. The Busnel office is in front of a little chapel built in the eleventh century. Cars can be parked.

The Pays d'Auge grows the best Normandy apples, which make the best Normandy cider. Busnel began distilling them here in 1820. The method is that used in Cognac and when matured in wood for fifteen–twenty years, Calvados is very similar to old Cognac. When young, the spirit is fiery but a tot of it competes with Jamaica rum as the most effective human inner space heater.

Visitors here see the ageing cellars and may buy old Calvados. Commentary is in French.

Hours: Monday to Friday 0900–1200; 1400–1700. Closed the last week of August.

Introduction: None necessary; please call at the office in the rue Saint-Melaine. Any special arrangements, e.g. large party, should be made in advance with the Paris office:
S.E.G.M.(Soc. pour l'Exportation de Grandes Marques), 2 rue de Solférino, 75007 Paris.
Tel. (010.331) 555 92 55.

CHAMPAGNE

L'Aisne

Gueux
Coulommes
Les Ormes
RHEIMS

Ville-Dommange
Sacy

Villers Allerand
Rilly
Chigny-
les Roses
Mailly
Verzenay
Verzy
Villers Marmery

The Mountain of Rheims

Trépail

Valley of the Marne

St Imoges
Louvois

Vincelles
Troissy
Hautvillers
Champillon
Tauxières
Bouzy
Ambonnay

Dormans
La Marne
Damery
Dizy
Avenay

Mardeuil
Ay
Tours-sur-Marne

EPERNAY

Pierry

Ablois
Cuis
Flavigny

Cramant
Avize

Côte des Blancs

Le Mesnil
sur-Oger

Vertus

Bergères

Wine-growing areas

Route A ~ The Marne Valley

Route B ~ Côte des Blancs

Route C ~ Mountain of Rheims

10 Km

Mi

CHAMPAGNE

Introduction

GETTING THERE

Reims and Epernay, the two towns of Champagne, where the shippers – with few exceptions – are based, lie about 90 miles eastwards of Paris.

By road: From Paris allow 3 hours. Recommended route: via autoroute A1, past Le Bourget to exit for Roissy-en-France and Soissons. For Reims take N2 to Villers-Cotterêts and Soissons, thence N31. For Epernay, take D212 to Claye-Souilly, N3 to La Ferté-sous-Jouarre, N33 to Vauchamps, thence D11 via Orbais L'Abbaye.

From Le Touquet or Calais, allow 3–4 hours; recommended route via Arras, Saint-Quentin and Laon.

By rail: There is a good service from Paris (Gare de l'Est), the 108-mile journey taking under two hours.

By air: 'Reims Champagne' is a military airport, which aircraft carrying parties of 60–100 people on day trips from Britain have no difficulty in using by arrangement.

April to October, inclusive, are the best visiting months. The winter can be very cold, with mists, frost and strong, icy winds.

METHODE CHAMPENOISE

Wine is the juice of fermented grapes, but champagne (and any other wine made by the *Méthode Champenoise*) is unique in that it requires *two* fermentations, the second one taking place in every one of those millions of bottles, which we see lying in the natural chalk cellars of Reims and Epernay. Apart from the beauty of these vineyards and of the River Marne, a visit to a big champagne house offers more than is possible with still wines, even in Bordeaux or Burgundy, because the method of making this great sparkling wine is so unusual.

The shipper, or *négociant manipulant* as the French call a

company that makes and markets champagne, is all-important in the story.

These firms buy the grapes, make the wines, blend them the following spring into a cuvée, bottle the cuvée and finally wrap the bottles prettily before shipping them to the ends of the earth. ('The nuisance of the Tropics is the sheer necessity of fizz', as Belloc observed.) The grapes come from three districts: The Mountain of Reims (Black Pinot grapes), The Valley of the Marne (Black Pinot grapes) and the Côtes des Blancs, south of Epernay (White Pinot grapes). Yet champagne is a white wine*, because the grape juice (which is colourless) is separated from the Black Pinot skins before they have time to impart their red colour to the juice.

After pressing, the juice is passed through vats into casks, or special tanks, where the first fermentation continues through the winter. The spring is a time of tasting, cleansing and blending, until in May bottling begins. A machine does the job – 12,000 or more bottles a day – and down they go, extra strongly corked, to the chalk galleries, originally dug out of the natural chalk in days gone by. A little sugar has been added to induce the secondary fermentation in the bottle, the carbonic acid gas so formed dissolving in the wine to produce the famous sparkle.

Thus champagne bubbles naturally in the glass and anybody with a swizzle-ish implement in his pocket, with which to stir those bubbles away, is guilty of carrying an offensive weapon.

Not so welcome as the sparkle is a by-product of fer-

---

*A little rosé or pink champagne is made and rather more *Blanc de Blancs*, solely from the white grapes. *Blanc de Noirs*, solely from black grapes, is also possible.

mentation, an ugly brown sediment that forms, and can be seen, inside the bottle. Before the wine can be drunk, it must somehow be removed. This is done after at least one year of maturing. The bottles are put with their necks inclined downwards, on sloping stands called *pupitres*, where a specialist workman gives each one a deft twist and a further tilt, repeating the operation, which is called *remuage*, at intervals for some weeks. In this way the sediment is made to slide slowly down, inside the glass, collecting in the neck of the bottle, which by then is upside down.

The next operation in this juggling, conjuring act is *dégorgement*, the removal of the sediment. The necks of the bottles, which are still upside down although several years may have elapsed, pass on a moving line through a freezing mixture to a team of *dégorgeurs*. As they remove each cork, out shoots the sediment, now a frozen pellet! The small loss – five to six centilitres – is replaced by a *dosage* of old wine and sugar, the proportion of each depending on the degree of sweetening required, Brut, Sec, Demi-sec, etc. New corks are then fitted and wired; when the bottles have been washed, with their labels and 'frills' added, the champagne is ready for shipment.

All this is a fascinating process, which visitors see during their tour of any champagne cellar. The temperature is invariably 10°C. – cold enough for ladies to need a jersey over summer dresses. Walking round with a guide may take 30 to 40 minutes, making the average visit last about an hour by the time a glass of champagne has been offered, accepted and drunk.

LA CHAMPAGNE

So much for *Le Champagne* – the wine; but there is still *La Champagne* the countryside. A poster map and guide, *The Champagne Road*, available from good wine merchants, describes the Blue, Red and Green routes through the three districts respectively, routes that can be easily followed in a car from the road signs in these colours. Travel then through this hilly countryside of graceful vineyards, through the wine villages, such as Dizy and Bouzy, names the significance of which the Champenois only understood when American and British troops were there during the world wars! There are dramatic views like that from Hautvillers down to the valley of the Marne and some fine twelfth and thirteenth century churches, notably at Sacy, Les Ormes, Villedommange, Coulommes and Avenay, north of the Marne in the Côtes des Noirs; and at Cuis, Flavigny and Le Mesnil-sur-Oger in the Côtes des Blancs, to the south. ·

Looking at Reims across the plain from Heidsieck's Windmill at Verzenay during one of the 'May Days with Monopole' trips we used to do with Peter Dominic's Wine Mine Club, a voice behind me said, 'Somewhere out there they carried me back on a stretcher believed dead in 1916. This windmill was an enemy observation post and the whole plain a sea of mud and trenches. I've never been back since until today.'

In his book (see Recommended Reading) Patrick Forbes calls the champagne winefield a pimple, about the size of Paris but only one-third that of the Burgundy winefield, sitting up in the middle of the province. Sixteen thousand plot-owning vignerons share about 85 per cent. of its 50,000 acres at present under vine,* leaving just 15 per cent. for the 144 shipping firms, some of which just buy the grapes from the plot-owners, without owning any vineyards themselves. Of these shipping firms, eighteen sell two-thirds of the total. Their names are famous and many of those that welcome the public appear in the list that follows.

Though the champagne shippers keep quiet about their poor relations, there are – in the province of Burgundy strange as it may seem – two small regions making white wines, which have the legal appellation right to be called 'champagne'. These regions are Bar-sur-Seine and Bar-sur-Aube. The one village name a few people may know is Riceys, because it also makes a pleasant still rosé, Rosé des Riceys. Curiosity took me there once, stopping in the village to buy sample bottles. They didn't seem at all cheap, nor was the champagne particularly good.

RECOMMENDED DRINKING

The Grande Marque of one's choice is often a matter of sentiment; 'Father always swore by Heidsieck.' Affection for all those dedicated members of the Bolla Club in Evelyn Waugh's *Vile Bodies*; recollection of the Victorian music hall (though this is going back a bit) where George Leyburn's song 'Champagne Charlie' put Moët on the map, so much so that the Great Vance was commissioned to sing another song in praise of Clicquot.

Hitherto not commercial, Still Champagne or *Vin nature de la Champagne*, seems to be coming into favour and, in 1974, the French Ministry of Agriculture gave it an Appellation Contrôlée – *Coteaux champenoise* (wine of the Champagne hills). Laurent-Perrier, Moët et Chandon and Pol Roger are among the companies marketing it.

However, there are also many excellent *marques* which are less famous and less costly. And, if champagne prices are quite beyond the pale, there is always Ratafia, the local aperitif of grape juice and brandy; not forgetting the two brandies, Marc and Fine de la Marne and, above all, the province's excellent table wine, Bouzy Rouge.

RECOMMENDED READING

Champagne, The Wine, the Land and the People by Patrick Forbes, Gollancz, London, 1967.
Bollinger by Cyril Ray, Peter Davies, London, 1971.

Some excellent booklets have also been published in English by the Comité Interprofessionnel du Vin de Champagne (C.I.V.C.), 5 rue Henri Martin, Epernay, the official body responsible for the champagne trade. British wine merchants gave these away at one time.

---

*The legally defined boundaries of the winefield enclose 75,000 acres and those under vine are being increased to meet demand.

Ay

Champagne J. Bollinger S.A.,

B.P. No. 4, 51160 Ay, Marne.
Tel. (26) 50.12.34.

Of all the Grande Marque champagne houses, none is more respected than Bollinger, a comparatively small firm, which has always aimed at perfection, cost being almost a secondary consideration.

It was in 1829 that M. Jacques Bollinger, whose father-in-law owned vineyards at Ay, founded the House of Bollinger. Since then the vineyards have increased to over 100 hectares, located in the six best parts of the champagne district. Madame Bollinger, Président d'honneur, is the widow of the grandson of the founder and has been in full command of the business since 1941, a year when the premises had been damaged by bombing and the Nazis occupied France.

Visitors are shown the cellars and vineyards.

Hours:	Monday to Friday: 1000–1130; 1430–1700.
Introduction:	Strictly by appointment. Letter should be obtained from agents: Mentzendorff & Co. Ltd., Asphalte House, Palace Street, London. Tel. 01-834 9561

Epernay

Champagne Mercier,

73 Avenue de Champagne, 51200 Epernay, Marne.
Tel. (26) 51.74.74.

Champagne Mercier is situated on the north side of the Avenue de Champagne (N3), about half a mile from Epernay's main square, the Place de la République.

The company was founded in 1858 by Eugène Mercier and is run today by his grandson, Jacques Mercier, though it now forms part of the Moët-Hennessy Group. Mercier's vast cellars contain some unusual and attractive carvings in chalk, notably in the Caveau de Bacchus. They also own the largest cask in the world (constructed in 1889 with a capacity of 200,000 bottles); their vat installations are probably the most modern in Champagne.

Visitors are taken round the cellars in a small train. They are then shewn the vat installations, the dégorgement and the packaging. There is also a museum of wine presses. Visitors are offered a tasting and may purchase champagne. There are English-speaking guides.

Hours:	1 March to 31 October only. Every day: 0930–1200; 1400–1700.
Introduction:	None is strictly necessary, but a letter of introduction will be sent on request, in writing, by their agents: Moët & Chandon (London) Ltd. 46 Mount Street, London W1Y 6EJ. Tel. 01–493 3811.

Champagne Moët & Chandon,

18 Avenue de Champagne, 51200 Epernay, Marne.
Tel. (26) 51.71.11.

The Maison Moët & Chandon lies on the south side of the Avenue de Champagne (N3), a hundred metres east of Epernay's main square, the Place de la République. In terms of production it is the largest of the champagne houses and the buses and coaches parked in the forecourt testify to its popularity for excursions.

Claude Moët founded the firm in 1743. It prospered greatly under the direction of his grandson, Jean-Remy Moët, who acquired, after the Revolution, the fine vineyards and monastery buildings of Hautvillers on the Montagne de Reims. At the Bénédictine Abbey of Hautvillers, a hundred years earlier, Dom Pérignon had perfected the methods by which champagne is made. In 1832, Jean-Remy was succeeded by his son, Victor, and his son-in-law, Pierre Gabriel Chandon. Ever since, the firm has been known as Moët & Chandon and today is run by descendants of both families. In 1970 Moët & Chandon merged its interests with those of Jas. Hennessy & Co. of Cognac to form the Moët-Hennessy Group, one of the largest companies in France.

Visitors are shown the cellars (the largest in the world) and the various stages of champagne making. If time permits, they are also shown the famous gardens designed by Isabey, the Abbey and the vineyards of Hautvillers. Visitors are offered a tasting and may purchase bottles in gift packs. There are English-speaking guides.

Hours:	1 April to 31 October: Every day 0900–1200; 1400–1700 (including national holidays). 2 November to 31 March: Monday to Friday 0900–1200; 1400–1630 (except national holidays).
Introduction:	None is strictly necessary, but a letter of introduction and route map will be sent on written application to: Moët & Chandon (London) Ltd. 46 Mount Street, London W1Y 6EJ. Tel. 01–493 3811.

Pol Roger & Cie.,

51200 Epernay, Marne.
Tel. (26) 51.41.95.

Pol Roger was established in Epernay in 1849 by M. Pol Roger. His sons were allowed to join his Christian name to their surname, making Pol-Roger, but nowadays the hyphen is dropped. Proceed from the centre of Epernay towards Châllons-sur-Marne. After passing Moët et Chandon, Perrier-Jouet and de Venoge, turn immediately right (Croix de Bussy).

Perhaps the most respected head of the firm was M. Maurice Pol-Roger, Mayor of Epernay during the German occupation of the first World War and almost continuously Mayor till 1935. The town gave him a bound volume signed by every one of its citizens. His daughter-in-law, Mme Jacques Pol-Roger, has a copy of Sir Winston Churchill's War Memoirs which he inscribed 'Mise en Bouteille au Château Chartwell'. Pol Roger was Sir Winston's favourite champagne; he even named one of his race horses 'Odette Paul Roger' and the filly won four races.

The visit follows the general lines already described.

Hours:	Monday to Friday 0930–1100; 1430–1600. Closed August.
Introduction:	By appointment. Letter of introduction from agents (or from Pol Roger): Dent & Reuss Ltd., Ryelands Street, Hereford. Tel. (0432) 6411.

Reims

Heidsieck & Co. Monopole,
83 rue Coquebert, 51100 Reims, Marne.
Tel. (26) 07.39.34.

These cellars are at the intersection of the Rue du Champ de Mars and the Rue Camille Lenoir, both clearly marked on the Reims (centre) plan in the *Guide Michelin*.

Wool and wine were said to hold a big future for Champagne when Florenz-Louis Heidsieck, son of a Lutheran pastor arrived in 1777. Eight years later at thirty-six, he married a wool merchant's daughter, starting a wine firm, Heidsieck & Co. at the same time. In 1828, Florenz-Louis died and the three nephews who had been needed to run the expanding business went their several ways, creating in due course, three different Heidsieck firms all making champagne in Reims.

Though trading names have changed several times, Monopole has been the name of Heidsieck & Co.'s brand since 1860. Today the firm owns over 120 hectares of champagne vineyards, ships its wines to 139 countries and has over eight million bottles maturing in an 8-mile labyrinth of cellars, carved out of the chalk sub-soil under Reims. They also own the windmill at Verzenay, a national monument worth a stop if only for the view over the plain.

The tour lasts about 45 minutes, no attempt being made to cover all the eight miles! Languages: English, French, German and Dutch. Wines are offered to taste and may be bought by the bottle or case.

Hours:	All year round, Monday to Friday 0900–1100; 1430–1630. From Easter to 31 October, also Saturday and Sunday, by appointment, 1430–1630. Closed August.
Introduction:	Letter of introduction from agents: Bouchard Aîné Ltd., 13 Eccleston Street, London SW1W 9NH. Tel. 01–235 3661.

Champagne Charles Heidsieck S.A.,
46 rue de la Justice, 51100 Reims, Marne.
Tel. (26) 40.16.13.

These premises are at the northern end of the city, close to those of Heidsieck Monopole and Krug. The *Michelin Guide* plan shows the Cemetière du Nord (cemetery). From the Place de la République take the Boulevard Jules César, turning right where the cemetery ends into the rue de la Justice, where Charles Heidsieck will be found on the left.

'You can take a White Horse anywhere' is a Scotch whisky slogan nowadays, but in Napoleonic times Charles-Henri Heidsieck took one to Russia, becoming a celebrity as a champagne salesman on his white steed. In 1851 his son, Charles-Camille, started his own company, becoming known as 'Champagne Charlie' throughout the United States. A fine shot, his shot guns interested the Americans as much as his wine but the Civil War very nearly cost him his life and the business.

Charles is always the first name in this family firm and a succession of Charleses, travelling far afield, have built the company into one of the largest of the privately-owned champagne firms. Jean-Charles, now retired, remains on the Board and his son, Eric, is the great concert impresario. In May 1976, a merger was announced with Henriot, a company with a big holding in the Côte des Blancs.

Visitors are shown the modern cellars, the famous two thousand-year-old Roman caves and, time permitting, the vineyards. They are offered a glass of champagne and may buy wines by the bottle or case. Languages: English, French, German and Italian.

Hours:	Monday to Friday 0900–1100; 1400–1600.
Introduction:	By appointment through distributors, who will send invitation card and route plan on request: Lawlers of London Ltd., 6A Station Road, Redhill, Surrey. Tel. (0737) 61415/7.

Krug, Vins Fins de Champagne S.A.,
5 rue Coquebert, 51100 Reims, Marne.
Tel. (26) 88.24.24 or 88.31.53 or 47.28.15.

The Maison Krug is at the northern end of the city. From the Place de la République take the Boulevard Lundy. The rue Coquebert is the next turning left and Krug is on the corner of Coquebert and the Boulevard.

Krug is one of the two great 'quality' houses for champagne. The Krug family *is* Krug champagne. Every stage of production is supervised by one of them and has been for five generations; they are completely responsible for the Company's own vineyards and for the wine that bears the name. Krug was founded in 1843 by Johann-Joseph Krug of Mainz. His wife was English. The present head is Paul Krug II who is assisted by his two sons, Henri and Rémi.

Hours:	Monday to Friday 1000–1130; 1430–1630. Closed mid-July to mid-August.

By appointment. Letter should be obtained from agents:
Hedges & Butler Ltd., Hedges House, 153 Regent Street, London W1R 8HQ. Tel. 01-734 4444.

G. H. Mumm & Cie.,

34 rue du Champ de Mars, 51100 Reims, Marne.
Tel. (26) 40.22.73 or 88.29.27.

These premises are at the northern end of the city. From the Place de la République, take the Boulevard Jules César and the next turn right, the rue de la Justice. The Mumm establishment is then on the corner, first street left.

The Mumm family were Protestants from Rudesheim where they owned important vineyards. The firm was jointly founded in 1827 by P. A. Mumm and a Frenchman called Gisler, who soon left to start on his own. G. H. Mumm was a grandson. Sold at auction after the 1914–18 war, the company became French. G. H. Mumm, whose Cordon Rouge was first launched in 1876, is a big firm, owning some 200 hectares of vineyards. Seagram Distillers are now important shareholders.

In the 45-minute tour visitors see the whole champagne process and can taste and buy wines afterwards.

Hours: Monday to Friday 0900–1100; 1400–1700. 1 March to 31 October: Saturdays, Sundays and national holidays as above. 1 June to 30 September 0900–1700 daily.

Introduction: None necessary. Large groups however should obtain an appointment through agents:
The House of Seagram Ltd., 17 Dacre Street, London SW1H 0DR.
Tel. 01-222 4343.

Louis Roederer,

21 Boulevard Lundy, B.P. No. 66, 51033 Reims, Marne.
Tel. (26) 47.59.81 and (26) 47.59.82.

Roederer is another house at the northern end of Reims. From the Place de la République, it is a short distance on the right along the Boulevard Lundy.

Run for two hundred years by the family, Roederer sends about half its champagne to export. First associated with his uncle Nicolas Schreider who began the business, Louis Roederer launched his own marque in 1827. Under his son Louis and his son-in-law Jacques Olry until about the end of the nineteenth century, the company prospered. Léon Olry-Roederer continued the direction and his widow, Madame Olry-Roederer, in charge until her death in May 1975, joins the Veuve Clicquot and Madame Bollinger as champagne ladies, who have managed their respective family firms with conspicuous success.

Hours: Monday to Friday 0900–1115; 1400–1615.

Introduction: By appointment, which any stockist of Louis Roederer champagne will arrange through agents:
Rutherford, Osborne & Perkin Ltd., Harlequin Avenue, Great West Road, Brentford, Middlesex.
Tel. 01-560 8351.

Champagne Ruinart Père et Fils,

4 rue des Crayères, 51100 Reims, Marne.
Tel. (26) 40.26.60.

These cellars are in the south-east corner of Reims just off the main road (N44) to Chalons-sur-Marne. Follow the sign posts to the Faculté des Sciences. From the Place des Droits-de-l'Homme take the Boulevard Diancourt to the Rond-Pointe Général Gouraud. The rue des Crayères leads off to the right.

Ruinart Père et Fils, founded in 1729, is the oldest of the existing champagne firms. Now a member of the Moët-Hennessy Group, the firm is still run by a member of the Ruinart family. The House was one of the first to export champagne to the United States, in the early nineteenth century when their wines also achieved great popularity in England. Ruinart's cellars, classified as an Historic Monument, are unique; they embrace a series of chalk pits (*crayères*), 200 ft. tall and pyramidal in shape, which the Romans quarried in 300 A.D. to provide chalk to build Reims (Durocortorum), the capital of Belgian Gaul.

Visitors see the Gallo-roman cellars and various stages in the making of champagne. They are offered a tasting and may purchase bottles.

Hours: Monday to Friday 0830–1100; 1430–1600.

Introduction: By appointment; letter of introduction and map will be sent on request, in writing, by their agents:
Moët & Chandon (London) Ltd., 46 Mount Street, London W1Y 6EJ.
Tel. 01-493 3811.

Champagne Veuve-Clicquot-Ponsardin,

1 Place des Droits de l'Homme, 51100 Reims, Marne.
Tel. (26) 47.33.60.

The above is the address of the cellars in Les Crayères by the Place des Droits de l'Homme in the south-east sector of the city. The office (address below) is in the northern part of the town.

When François Clicquot died tragically at the age of thirty, in 1805, it seemed unlikely that the firm his father had begun in 1772 would celebrate a bi-centenary in 1972. The old man was broken; only François' widow, twenty-seven and simply educated, was left to carry on. It was certainly no time to try; with Napoleon's invasion fleet ready, Britain was as good as finished, yet her blockade had ruined the markets of Germany and Denmark. Russia had placed an embargo on French imports, but somehow 'The Widow' got 10,000 bottles through the blockade in a sailing ship of 75 tons, having already sent her sales manager, Herr Bohne, to St. Petersburg. With the war over and the embargo lifted, Madame Clicquot, way ahead of all competitors, dispatched another 20,000 bottles, sold before they landed. There was a reason why her wine was received with such enthusiasm; it had a clarity, a limpidity, unknown before. She had discovered in her own cellars how to remove the sediment that forms after fermentation in the bottle. It is to her that we owe the process of sliding it to the cork known as *remuage*.

Today the company owns 260 hectares of the finest champagne vineyards and the wine is affectionately known as 'The Widow' throughout the world.

The visit is to the Gallo-Roman cellars, where there are English-speaking guides and a glass of the 'Widow' before it ends.

Hours:	Monday to Friday 0900–1130; 1415–1630. Saturday, Sunday and Public Holidays from Easter to 31 October 1415–1730.
Introduction:	None necessary, but any special requirements should be made with the Veuve-Clicquot office at 12 rue de Temple, 51100 Reims. Tel. (26) 47.67.94. Or through their agents: H. Parrot & Co. Ltd., The Old Customs House, 3 Wapping Pierhead, London E1 9PN. Tel. 01–480 6312.

Tours-sur-Marne
Champagne Laurent-Perrier,
Avenue de Champagne, 51150 Tours-sur-Marne. Tel. (26) 50.67.22.

Tours-sur-Marne, a village on road D1 between Epernay and Chalons-sur-Marne, is 13 kilometres from Epernay and 27 kilometres from Reims. From the latter the route is D9 to Louvois, D34 (Direction Chalons) for 4 kilometres, thence D19.

This house was begun in 1812 by Eugène Laurent who married Emilie Perrier. When he died, Emilie Perrier became yet another widow to take charge successfully. The family fared badly in the first world war, but since the second the export trade has been built to cover ninety countries.

The visit follows the general lines already described.

Hours:	Monday to Friday 0900–1100; 1400–1700. Closed August.
Introduction:	By appointment. Letter of introduction from agents: Findlater Matta Agencies Ltd., Windsor Avenue, Merton Abbey, London SW19 2SN. Tel. 01–540 0601.

ALSACE

Introduction

France's easternmost province lies about 280 miles from Paris, the river Rhine forming her eastern frontier with Germany. There are international airports at Strasbourg, the one city of Alsace, and at Basle Mulhouse. Colmar (55,000 inhabitants), the one fair-sized town of the wine region, has a small airport and makes a suitable headquarters for tourism.

Wine, food, great walking country and a modicum of culture (from Grünewald's famous Retable d'Isenheim in the museum at Colmar to Corbusier's great church at Ronchamp, near Belfort) make Alsace a superb holiday province. From the top of the wooded Vosges, which ward off much of the rain from the west, the road descends sharply into the sunshine of this belt of vineyards, some 60 miles by 10, lying in the plain from Strasbourg to Colmar.

Many of the wine villages had to be rebuilt after destruction in the 1944 campaign of World War II, but Riquewihr happily escaped damage and continues to

ALSACE

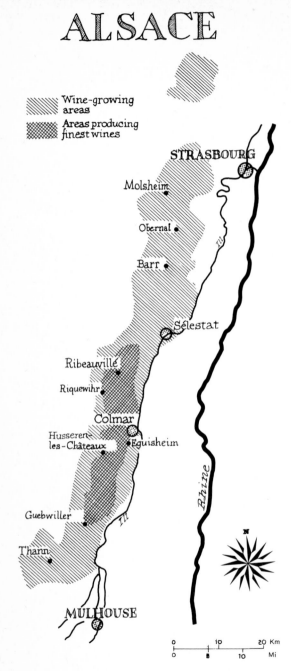

Wine-growing areas

Areas producing finest wines

STRASBOURG

Molsheim

Obernai

Barr

Sélestat

Ribeauvillé

Riquewihr

Colmar

Husseren-
les-Châteaux

Eguisheim

Guebwiller

Thann

MULHOUSE

Ill

Rhine

Ill

0   10   20 Km
0   5   10   Mi

enchant the visitor as a mediaeval town, which might have stepped from the fairy tale pages of Hans Andersen.

By car, sensible tourists follow the well sign-posted *Route des Vins*, which wanders the length of the Alsace vineyards through, or close to, the villages, where *Caves de Dégustation* are so numerous that a teetotal chauffeur might be a wise precaution.

Alsace wines, virtually all white and dry, are sold by the name of the grape – Sylvaner, Riesling and Gewürztraminer being the principal three. Forced to concentrate on quantity without quality when a German province from 1870 to 1918, the Alsace growers saw their land invaded once again after a twenty-year interval. Now, at last, with thirty years of peace, the standard of Alsace wines has come to be recognised and increasingly respected.

No less respected are the fruit brandies *(Eaux de Vie)* of Alsace, the term 'fruit brandies' being used by the trade to denote distilled liquors from fruits *other than* the grape. Kirsch, from certain wild cherries, may be the best known but Alsatians are just as proud of their Mirabelle, their Framboise and their Poires William, distilled from plums, raspberries and pears respectively. No Limousin oak for them; their fruit brandies, matured in glass flagons, are as colourless as water but, at strengths up to 80° proof, a good deal stronger!

This countryside abounds with fruits and berries, wild and cultivated, to such an extent that Cyril Ray, President of Britain's Circle of Wine Writers, has recorded no less than seventeen different fruit brandies made in Alsace. Most of them he tasted at the Dolfi distillery in Strasbourg; others he met 'home made' by local farmers. Finally, on top of some bare Vosges mountain, he discovered *eaux de vie de haie de houx*, (Could it have been *de hoax?*) a holly berry brandy of which not even the fruitiest, 'brandiest' Alsatian had ever heard.

RECOMMENDED READING
Alsace and its Wine Gardens by S. F. Hallgarten
André Deutsch London 1957

Further Information
Comité Interprofessionel du Vin d'Alsace
8 Place de Lattre de Tassigny
68000 Colmar

Strasbourg
Dolfi, Grande Distillerie,
19 Boulevard du Président-Wilson, 67000 Strasbourg, Bas-Rhin.
Tel. (88) 32.64.64

The Boulevard du Président Wilson, clearly shown on the *Michelin Guide* Strasbourg plan, runs north-east from the Place in front of the Central station to the Place de Haguenau. The Dolfi Grande Distillerie is on the left about 500 metres from the station.

Founded in 1888, this great company is under the per-

sonal direction of the Président-Directeur Général, M. Renaud A. Dolfi and his brothers. Remembering what Alsace has suffered in the two world wars, it is remarkable in itself that three generations should have survived to run the business.

Alsace being one of the richest fruit regions of Europe, this business is the distillation and marketing of fruit brandies *(eaux de vie de fruits)*, and of fruit liqueurs. The entire range of both is shipped to Britain and distributed by their U.K. agents.

Dry and colourless, the highly-esteemed fruit brandies are distilled from the fermented mash of fruits other than the grape – cherries and plums, strawberries, raspberries and other soft fruits. Sweet and coloured, the fruit liqueurs are mostly fruit juices with which a spirit has been compounded.

Visitors are likely to see something of these processes with doubtless some palatable instruction.

Hours:	Tuesday to Friday 0900–1100. Closed August.
Introduction:	By appointment, letter obtainable from agents: Rutherford, Osborne & Perkin Ltd., Harlequin Avenue, Great West Road, Brentford, Middlesex, Tel. 01–560 8351.

Ribeauvillé
F. E. Trimbach,
68150 Ribeauvillé, Haut-Rhin.
Tel. (89) 73.60.30.

On the meandering wine road, only 4·5 kilometres from Riquewihr, Ribeauvillé is a small town with 5,000 inhabitants and a good restaurant (Clos St. Vincent) prettily sited among the vineyards. The Trimbachs, after being wine growers in Riquewihr from 1626, settled in Ribeauvillé after World War I. Their establishment will be found on the right, coming from Strasbourg or Colmar, where the two roads (D106 and D416) converge at the entrance to the town.

At that time Frédéric Theodore Trimbach, an outstanding taster and founder member of the Wine Growers' and Merchants' Association, ran the family business. Since his death in 1945, his two grandsons, Bernard and Hubert, have been maintaining the Trimbach reputation, winning prizes for their wines, especially Clos Sainte Hune, a very fine Riesling.

Ribeauvillé has a Wine Fair in September and the Pfifferdag festival on the first Sunday in May when there is minstrelsy in the streets and the town fountain runs with wine.

Hours:	Monday to Friday 0930–1130; 1430–1700.
Introduction:	By appointment; letter obtainable from agents: H. Parrot & Co. Ltd., The Old Cus-

toms House, 3 Wapping Pier head, Wapping High Street, London E1 9PN. Tel. 01–480 6312.

Riquewihr
Dopff 'Au Moulin' S.A.,
68340–Riquewihr, Haut-Rhin.
Tel. (89) 47.92.23.

Though only a small town of 1300 inhabitants, the old walls, towers and houses, dating back to 1291, make Riquewihr a great tourist attraction and the surrounding vineyards are among the best of the region. La Maison Dopff 'Au Moulin' began in 1574 and son has now succeeded father for eight generations. The company is a leading proprietor of well-sited vineyards and much respected for the quality of its wines, notably Riesling Schoenenburg and Gewürztraminer Eichberg. Some *Méthode Champenoise* sparkling wines are also made. Visitors see the cellars, the bottling and the packing. Wines may be bought. In July, August and September, they can sometimes be received, in exceptional circumstances, by appointment on Saturdays and Sundays.

Hours:	Monday to Friday 0800–1100; 1400–1700.
Introduction:	By letter from agents: Mentzendorff & Co. Ltd., Asphalte House, Palace Street, London SW1E 5HG, Tel. 01–834 9561.

Hugel et Fils S.A.,
Rue de la Première Armée, 68340–Riquewihr, Haut-Rhin.
Tel. (89) 47.92.15.

A short walk up the Grande Rue from the Hôtel de Ville leads to the headquarters of Hugel et Fils, another well-known family firm that began here in 1639 and has passed from father to son directly ever since.

At present, Jean Hugel and three sons are in charge and my reports indicate that nowhere in Europe are visitors made more welcome, nor told about the wines in better English. Jean Hugel has in fact written a most useful booklet in English, on Alsatian foods and wines, which is called . . . *And Give it my Blessing*.

Visitors see the pressing room, some of the main cellars, which contain the St. Catherine Vat (made in 1715, capacity 8,800 litres) and the bottling department. Free tastings are given; purchases may be made in the tasting room, which is in the main street.

Hours:	Monday to Friday 0930–1130; 1430–1700.
Introduction:	All visitors are welcome but the company does prefer advance warning and,

in particular, letters of introduction from agents:
Dreyfus Ashby & Co. Ltd., 21–22 Hans Place, London SW1X 0EP.
Tel. 01–589 5433.

Eguisheim
Léon Beyer,
2 rue de la Première Armée, 68420 Eguisheim Haut-Rhin.
Tel. (89) 41.41.05.

Léon Beyer is another wine-growing firm handed from father to son since 1580 when it was begun in the little town of Eguisheim, five kilometres south of Colmar. Visitors are shewn the cellars and winery.

Hours:	Monday to Friday 0930–1130; 1400–1630. Closed last two weeks of August.
Introduction:	By letter from agents: Cock Russell Vintners Ltd., Seagram Distillers House, 17 Dacre Street, London SW1H 0DR. Tel. 01–222 4343.

Husseren-Les-Châteaux
Kuentz-Bas,
Husseren-Les-Châteaux, 68420 Herrlisheim pres Colmar.
Tel. (89) 49.30.24.

Husseren is five kilometres south-west of Colmar, off N83 from Strasbourg to Lyon. This is where the Kuentz family settled from Switzerland in 1795, setting up business in the wines and fruit brandies of Alsace and, as if this were not enough, they also raised a family of thirteen children.

After World War I, a Kuentz daughter married a Bas of Burgundy and, except during World War II, when the family were expelled elsewhere in France, the business has succeeded even though in 1945 it meant starting again from scratch.

Visitors see the cellars and may taste and buy wines in the pleasant reception room.

Hours:	Monday to Saturday 0900–1800.
Introduction:	A brochure welcomes all visitors without reservation but those wishing to meet the management should obtain a letter from agents: Rawlings Voigt Ltd., Waterloo House, 228–232 Waterloo Station Approach, London SE1 7BE. Tel. 01–928 4851.

BURGUNDY (including Beaujolais)

BURGUNDY

Introduction

Burgundy is the name given to red and white wines made within the limits of the *départements* of Yonne (Chablis), Côte d'Or, Saône et Loire (Côte Chalonnaise, Mâconnais, Pouilly-Fuissé and a very small piece of Beaujolais) and Rhône (Beaujolais).

GETTING THERE

By road: From Paris via the A6 autoroute, Chablis (Yonne) is 113 miles, Beaune (Côte d'Or) 194 miles and Mâcon (Saône et Loire) 243 miles.

From Dieppe join the A13 autoroute at Rouen. On the outskirts of Paris keep right on reaching the Boulevard Periphérique, the road that circles the city completely, and follow the signs to the A6 autoroute.

From Calais and Boulogne join the A1 autoroute east of Arras, circle round Paris on the Boulevard Periphérique, leaving it at the A6 exit.

A more leisurely route from any of these three Channel ports is to the east of Paris, via Soissons (permitting a détour to Champagne) and Troyes. South of Troyes, the road, N71, follows the upper reaches of the Seine; the source, in a grotto 6 miles north-west of Saint Seine-l'Abbaye is a place to visit. Beaune by this route is about 360 miles from Calais or Boulogne.

By rail: Beaune, Chagny, Chalon-sur-Saône and Mâcon are all intermediate stations between Dijon and Lyons, principal stops on the Paris-Marseille line. There are frequent express trains with restaurant cars and night sleepers; *Le Mistral,* fastest of these with supplement payable, does the 200 miles from Paris to Dijon by day in two hours twenty minutes. There is also a Paris-Lyons day car train and another (weekly in winter) from Paris to Avignon.

By air: There are scheduled flights London-Paris-Lyons, where self-drive cars can be hired at the airport. Members of the Trade, spending a few days in Burgundy or the Côtes du Rhône, find this route convenient.

By boat: The river and canal systems of France make it perfectly possible for small craft to reach Burgundy from the Channel or the Mediterranean. Canal cruisers can be hired at Baye (south of Tonnerre). One young party made a circuit – Baye, Laroche Migennes, Dijon, Chalons-sur-Marne, Digoin, Decizes, Blaye in fourteen days, discovering in the process that the distance covered was 426 miles and lock gates opened, mainly hand-operated by them, came to 396. 'Locks', said their canal cruising brochure, 'break up the voyage into colourful and sometimes humorous episodes.' What was not a joke was having only fourteen days; three weeks would have been ideal.

Palinarus, a 100-foot diesel-driven barge, converted to accommodate about eighteen people with all mod cons and French chef, covers Burgundy in a series of different weekly cruises from April to October. Details from Conti-

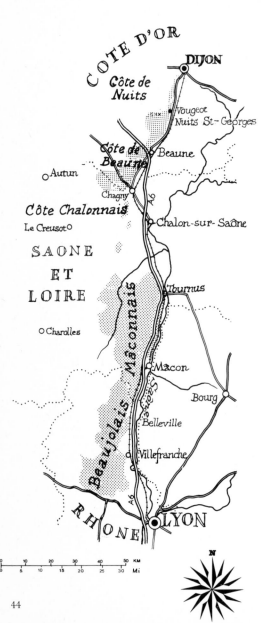

nental Waterways, 22 Hans Place, London, SW1, Tel. 01–584 6773; or other travel agents.

COTE DE NUITS

As a motorist, many wine regions – Champagne and Sherry come to mind – are easy to miss. Not so the Côte d'Or, right alongside the Route Nationale 74 from Dijon almost to Chagny; the original villages – Gevrey, Chambolle, Vosne, Nuits, Aloxe – resenting obscurity a century ago soon became known the world over when they hyphenated their most famous vineyards – Chambertin, Musigny, Romanée, Saint-Georges, Corton – to themselves.

It is pleasant to draw off the main road for a picnic up 'The Golden Slope' trying to fix one's position on the charts of Hugh Johnson's *Wine Atlas*, invaluable in the car when vineyard visiting. A *Route des Vins* beginning at Fixin bisects the great group of Chambertin; Charmes and Mazoyères lie to the left, Latricières, Chambertin and Clos de Bèze to the right, higher up, making wines the better for it. A brief return to the main road occurs at Vougeot, whence another side road circles round the Château du Clos de Vougeot cutting through Echézeaux to the Romanée vineyard, before returning to the main road once more. Similar deviations can be made through the vineyards of most of the Côte d'Or communes nearer Beaune and beyond.

All these named vineyards, making the greatest red burgundies of all, were split into small holdings after the French Revolution; a division which the French laws of inheritance, based on equal shares for all the children of the same parents, have done nothing to heal. In the celebrated walled vineyard, Clos de Vougeot, there are 125 acres divided among about sixty owners, each with his own piece, or *climat*. The seventy acres of Chambertin and Clos de Bèze are divided among some twenty-five owners. Even students of Burgundy may be surprised to know that in any year it is theoretically possible to find twenty-five domaine-bottled Chambertins, no two of which need necessarily taste the same. And, if the owner bottles only some of his wine and sells the rest to *négociants*, there could legitimately be even more different Chambertins on the market.

Throughout the Côte d'Or, the largest acreage in single ownership is Jules Regnier's Clos de la Maréchale (24 acres in Nuits-Saint-Georges); the average holding is nearer three acres, which will make about 500 cases (twelve bottles each). It is much the same story with the generic wines, those from the lesser *unnamed* vineyards in these communes. The plots are rarely more than ten to twenty *pièces*, a *pièce* making twenty-four cases. 240 cases or even 480 are quantities insufficient to interest today's wine buyers needing to cater for chains of perhaps three hundred shops.

In this region of growers making insufficient wine to market themselves, there grew up the firms of *négociants*, buying the growers' wines, or their grapes, and blending them into *cuvées* of their own. The big names are well known – Bouchard Père et Fils, Bouchard Aîné, Calvet, Chauvenet, Drouhin, Geisweiler, Jadot, Louis Latour, Morin, Patriarche. Others, formerly content with their home trade, such as Clair Däu, Henri Gouges, Comte Lafon, Armand Rousseau, have now entered export markets. These firms are growers too, owning some of the finest *climats*.

With burgundy, who shipped it and who bottled it are vital factors. That bottle of Beaune seen in the shop window at 10½p less than expected is tempting, but unlikely to be any more satisfactory than a snail from your back garden masquerading as an *escargot de bourgogne* on your plate.

Red Côte d'Or burgundy must be made from the Pinot Noir grape and white from the Chardonnay. Beaujolais must be made from the Gamay grape. Having made these firm pronouncements, I come inevitably to exceptions. Passe-Tout-Grains is a comparatively inexpensive red wine made from one-third Pinot Noir and two-thirds Gamay. Aligoté, a white counterpart of the Gamay grape, makes light perfumed white wine. Separate decrees permit 'Bourgogne' to precede these two names.

After the august magnificence of Champagne, Burgundy cellars are small and rather bourgeois, which in truth is far more typical of commercial wine cellars in most places. The visitor will enjoy tasting cask samples from a silver or glass Tastevin, comparing the rude purple and rough taste of last year's wine with that of its predecessor. Even the novice will notice how the extra year in cask mellows the hard purple towards ruby and softens the taste.

White wines make up only 20 per cent. of the Côte d'Or total compared with 57 per cent. in Bordeaux. The red wine total in Burgundy as a whole is roughly the same as that of the Bordeaux A.C. region but 78 per cent. of it is Beaujolais and Mâconnais. Red burgundy from the Côte d'Or is only 12 per cent. of Burgundy's total and is approximately one-eighth that of red Bordeaux.

The one to eight ratio makes it a pity that burgundy is as popular, if not more so, than claret. The region is right *on* the beaten track to Italy and the Riviera; the wine, less of an acquired taste than claret, is a shade sweeter and ready to drink sooner. With so great a gap between demand and supply, it is hardly surprising that burgundy is much imitated and faked in other countries.

COTE DE BEAUNE

The Côte de Beaune, second half of the Côte d'Or, begins with the commune of Aloxe-Corton, after which the wine road crosses the A6 autoroute before leading in to Beaune, having come 24 miles from Dijon. A fine old town of churches, walls and cellars, Beaune is the commercial centre of this wine region and a gastronomic centre too. Of its nine hotels and restaurants in the 1976 *Guide Michelin*, though only one may now have a *bonne-table* star, the fact remains that no other wine region, indeed no other part of

France can excel – though some may equal – the cuisine of Burgundy. Rich pastures nourish the Charolais cattle bred for beef, as well as the brown-beige Schwitz cows that grace the rural scene coming to Dijon from Chatillon-sur-Seine. And, contemplating the size of those *escargots de bourgogne,* the pastures do pretty well for snails too!

There is the raw ham of barren Morvan; *écrevisses* from the rivers Yonne, Saône and Seine; Dijon's mustard; Cassis, the blackcurrant liqueur and an immense number of cheeses headed by Bleu de Breese and – if we stretch Burgundy a little south into Haute-Savoie – Reblochon. What a guzzling ground! Even the plump chickens of Bresse have their own Appellation Contrôlée.

Hospices de Beaune. Neither vineyard nor cellar, though renowned for its association with both, the Hospices* or Hôtel-Dieu is an almhouse, unique in the beauty of its design, with a five-hundred-year background of devoted service to relieve the sufferings of the poor.

Walking the cobbled streets of Beaune (so pleasant at night when the cars have all gone) a view of the patterned polychrome yellow, green and black roof of Nicolas Rolin's building appears at unexpected moments when the narrow streets widen into open spaces. Rolin was the

*Hospices (plural) is usual to distinguish it from another charitable foundation in Beaune.

tax-collecting Chancellor of Burgundy when he founded the Hospices in 1443. A hard man it seems; even the King, Louis XI, was alleged to have said that Rolin, having made so many people poor, was wise to appease the Almighty by providing for a few of them.

Being the first to give their vineyards to the Hospices, Nicolas and his wife, Guigone de Salins, set a precedent to be followed over the centuries by some thirty other benefactors, and it is their wines that are sold, by auction after the vintage on the third Sunday of November each year amid a weekend of celebrations. Originally made by private treaty, this Sale was changed to an auction in 1851, the venue being moved to the town's covered market place in 1967, when the Hospices was no longer large enough for the growing host of bidders and others coming from foreign parts. The Sale is in fact a Wine Trade occasion with almost more buyers than barrels, high prices being paid for this reason as much as for the cause.

Hospices wines are known by the name of the donor, e.g. Cuvée Billardet, Cuvée Charlotte Dumay. My advice to readers is to leave them to the restaurateurs, who like to have one or two in their wine lists for prestige purposes. Should a bottle be met *without* the Cuvée name included on the label, it is high time to change your restaurant.

Likewise the banquets, called *Les Trois Glorieuses*! On Saturday night the *Chevaliers du Tastevin* dinner, held in the Château of the Clos de Vougeot (p. 00); on Sunday night the *Diner aux chandelles* in the cellars of the Bastion, near the Hospices. On Monday there is no dinner, just a luncheon, *La Paulée de Meursault*, that lasts till dinner. These parties, celebrating the end of a year's anxious work, which may have been spoilt by a wet vintage, are really for the vignerons and their guests. Tickets are expensive and very difficult to obtain for others. Moreover, well as the French always stage these events, they are not prepared to lavish old vintages on dinners for five hundred to a thousand people. So the wines are likely to be younger than can be bought from a good wine merchant at home.

With money to spare for a good blow-out in Burgundy, far better choose the *menu gastronomique* and a fine old bottle or two at a really good restaurant. The *Michelin Guide's Bonne Table* map shews just how many of these there are in Burgundy. Fifty miles north-west of Lyons, within striking range of Beaune is H. Frères Troisgros at Roanne, which some say is now the best restaurant in all France.

To visit Beaune without seeing the Hôtel-Dieu would be like missing the Louvre or the Eiffel Tower when in Paris (the choice depending on your high-mindedness). Besides, part of the therapy for the patients in the modern hospital facing the Hôtel-Dieu is to watch us tourists assembling in the courtyard between the two buildings.

Conducted tours every three-quarters of an hour from 0900–1115 and from 1400 to 1745 (1615 in winter) take place daily for a few francs. Though my own mind's eye is far from retentive of the visual arts, it does retain those original blue and white pots for mediaeval potions that lie in the old pharmacy, the great Gothic chimney piece in the kitchen and the red curtains which give privacy to the inmates of the cubicles in the vast dormitory ward. It can't however remember any detail of the acclaimed Flemish masterpiece, Roger van der Weyden's Last Judgement, except that it must have frightened poor dying peasants out of such wits as they possessed.

LE MONTRACHET

South of Beaune the vineyard wanderer takes N73, through Pommard and Volnay, and can follow the signs to Meursault and the communes of Puligny-Montrachet and Chassagne-Montrachet. Le Montrachet itself, quintessence of white burgundies, is not difficult to find, under a mile above the village of Puligny-Montrachet, in a spot where it cunningly catches every ray of sunshine for much of the day. After this high spot of the Côte de Beaune, there is the last commune, Santenay, making some good but less distinguished red burgundies, before the wine region of the Côte d'Or ends.

CHALONS AND MACON

This happily is by no means the end of Burgundy. The Côte Chalonnaise has four good communes – Rully making good white and sparkling white burgundies, Mercurey with a reputation for its light reds, Givry, which makes both red and white and Montagny, exclusively white. Thirty-five miles south of Châlons-sur-Saône comes the small town of Mâcon, restored at long last to tranquillity since the A6 autoroute was completed. On the way is Viré, a village with a thriving co-operative that makes Mâcon-Viré, a white burgundy, which British people have come to like in the past ten years.

Five miles west of Mâcon a giant crag called Solutré towers above the village of Solutré, lying at its foot. This is the district of Pouilly-Fuissé, which really should be Pouilly-Solutré; not only do the Solutré vineyards make the best wine, but Pouilly is merely a hamlet in the commune of Solutré. Nevertheless, Pouilly-Fuissé is famous – so famous that much of the wine so labelled must be totally unconnected with Fuissé, Solutré, Vergisson, or Chaintré, which are the five communes of its legitimate birth. The road twists and turns around the great rock of Solutré through two more villages – Vinzelles and Loché, either of which may also be hyphenated with Pouilly. Here, a little Aligoté is allowed with the Chardonnay, both grapes together giving a pale golden white burgundy only surpassed by its northern neighbours at Meursault and the Montrachets, made from Chardonnay grapes alone.

South of Vinzelles lies the Beaujolais; the country wine road enters it at Saint-Veran or Verand, which now has an appellation for its white wine, formerly described as white Beaujolais.

BEAUJOLAIS

Forty-five miles long but never more than ten miles wide, Beaujolais is shaped like the last quarter of the moon, the short curved side being the N6 road that follows the Saône valley from Mâcon to Lyons. From this main road, minor roads at intervals south of Mâcon lead up a few hundred feet to what appears to be an ocean of Gamay vines among the hills and dales as far as the eye can see.

The Bas-Beaujolais, flatter southern half of this gently sloping countryside, gives the plain straightforward 9° Beaujolais and is not of great touristic interest. Better wines are labelled Beaujolais Supérieur and Beaujolais-Villages; both have to be at least 10° and the latter appellation is restricted to some thirty named villages. The best Beaujolais has the appellation Cru Beaujolais, the crus being the eight communes and, if a site name follows that of the commune, e.g. Juliénas Clos des Capitans, the strength must be 11°.

From north to south the communes, named after their principal villages, are Saint-Amour, Juliénas, Chénas, Moulin-à-Vent, Fleurie, Chiroubles, Morgon, Brouilly and Côte de Brouilly. A *Route des Vins*, running along the top for the most part, with a distant view now and then of the Saône glistening in the sunshine, connects most of the villages, which are well organised to refresh the traveller with – I need hardly say – Beaujolais.

Young, light, fruity, alcoholic . . . there is certainly no wine better able to do this job! On Saturday nights organised parties have fun and games in some of the Beaujolais chais, hired to them for the evening. The chais are cool, so too the wine, drawn straight from the cask. This is why wine merchants recommend drinking Beaujolais slightly chilled, which I myself achieve in summer at home in Sussex by leaving a bottle for up to half-an-hour in the refrigerator. There are exceptions; the best crus of Moulin-à-Vent are bigger wines, worth laying down for a few years and drinking, like Côte d'Or burgundy, at about room temperature.

CHABLIS

Ten miles east of Auxerre and separated from the main Côte d'Or burgundy region by a hundred miles, there is a hamlet of a mere 2,000 inhabitants called Chablis. Why this village should have been heavily bombarded twice in the last war is mysterious, unless some simple map-studying staff officer believed that the destruction of the place would put an end to his enemy's Chablis supply. Far from it, Chablis being the most abused name among the world's wines, the barrels of Abyssinian, Afghanistanian and Albanian Chablis would have continued to roll out just the same!

In peacetime the growers have to protect themselves from another enemy – late spring frosts, particularly prevalent here and capable of putting paid to a normal crop during the same year. In the Grand Cru vineyards up the south-facing, chalky slope north-east of the town, gas heating from portable pipes can now be laid on when danger threatens.

Accommodation for the visitor consists of the·Etoile-Bergerand, which the *Michelin Guide* classes as a restaurant with sixteen bedrooms. Robert Speaight (see Recommended Books) recommends it as an admirable headquarters for exploring around Tonnerre and Auxerre. L'Etoile, in the Bergerand family since 1948, has had many distinguished clients.

Briand drank Petit Chablis for breakfast there. Herriot, Doumerge and Jean Cocteau are among others who have contributed more than their marks to the visitors' book.

There are no great names of shippers or *négociants* in Chablis and firms in Britain usually buy the wine through their principals in the Côte d'Or. Arranging visits for the public to individual growers with whom one does not do business is not a highly successful operation, so the best way of seeing Chablis might be to spend a couple of nights at the Etoile, where one doesn't have to be a Premier for M. Bergerand to arrange matters.

None of this belittles the wine. Dry steely Chablis, second only to Meursault and the Montrachets, is perfect with oysters and shellfish. The trouble is that throughout the world only 10 to 25 per cent. of so-called Chablis is genuine, or so the leading growers estimate.

There are seven Grands Crus and twenty-two Premiers Crus. Stick to these classified vineyards – their names are given in every wine book! The others – plain Chablis and Petit Chablis – are not reliable. As to visiting, by all means make a worthwhile détour from the autoroute (A6). There is the Maison du Chablis at 16 rue Émile-Zola, but I doubt if formal introductions by letter from wine merchants are practicable.

RECOMMENDED READING
The Companion Guide to Burgundy by Robert Speaight Collins London 1975
An up-to-date general guide to the whole province, with good wine information largely based – says Mr Speaight – on Mr Yoxall's book below.
The Wines of Burgundy by H. W. Yoxall International Wine and Food Society Michael Joseph London 1968

Further Information
Comité Interprofessionnel de la Côte d'Or et de l'Yonne
Petite Place Camot
21201 – Beaune

Union Interprofessionnelle des Vins de Beaujolais
24 Boulevard Vermorel
69400 Villefranche-sur-Saône
Rhône

Comité Interprofessionnel des Vins de Bourgogne
3 bis rue Gambetta
71000 Mâcon
Seine-et-Loire

Lionel J. Bruck S.A.,
6 Quai Dumorey, 21700 Nuits-Saint-Georges, Côte d'Or.
Tel. (80) 06.07.24.

The fact that Lionel J. Bruck is not well-known to the British wine trade signifies very little. The family have been established in this old house in Nuits-Saint-Georges for a hundred and fifty years and their customers probably consist of faithful clients in France and other continental countries.

Hearing that this Guide was in preparation, he wrote asking to be included and I hope Scottish and West Country readers will try some of his wines through the agencies below. Visitors are shewn their cellars and a tasting is included.

Hours:	Monday to Friday 1000–1100; 1500–1600. Closed August.
Introduction:	By appointment. Letter of introduction from agents but holders should then give the firm 48 hours' notice of arrival by letter or telephone.
U.K. Agents:	Messrs. W. R. Garrad, The Old Howgate Inn, Near Penicuik, Midlothian, Scotland. Tel. (0968) 74244.
	Mr. Leslie Joseph, Wine Investments Ltd., 51 Foxwell Street, Cirencester, Gloucestershire. Tel. (0285) 4239.

F. Hasenklever,
Rue du Moulin, 21700 Nuits-Saint-Georges, Côte d'Or.
Tel. (80) 23.51.61–2.

This company was established in Napoleonic times in the ancient Ursulines' nunnery, which retains its picturesque look in spite of restorations. The original cellars are interesting and ideal as wine cellars. Since 1877, the Bruck family (above) have controlled the business so that the name Hasenklever is best known for their Corton Clos du Roi and Nuits Les Saints Georges, where Lionel J. Bruck owns sites.

Visitors are shewn the cellars and a tasting is included.

Hours:	Monday to Friday 0930–1130; 1430–1630. Closed 4 August–21 August (approx.).
Introduction:	By appointment. Letters of introduction from agents*:

*New agents to be appointed.

Morin Père et Fils,
Quai Fleury, 21700 Nuits-Saint-Georges, Côte d'Or.
Tel. (80) 06.05.11.

Founded in 1822, Morin Père et Fils is a well-known exporter of burgundies, including Chablis and Beaujolais. The family own parts of *premier cru* vineyards in the Côte de Nuits, such as Nuits Cailles, Nuits les Pruliers and the Château de la Tour vineyard, which is one of the many parts that make up the great walled vineyard of Clos de Vougeot.

Visitors are shewn their cellars. There is an English-speaking guide and wines may be tasted and bought.

Hours:	Every day 0830–1130; 1400–1730. Closed Saturdays and Sundays during June.
Introduction:	For numbers up to ten, none necessary; over ten please make appointment, preferably by telephone if in the district, though earlier arrangements would be better made through agents: B. Wood and Son, 44 Bow Lane, London EC4. Tel. 01–236 3725

Vougeot

Château du Clos de Vougeot,
21640 Vougeot, Côte d'Or.
Tel. (80) 06.86.09.

Since French troops passing along the N74 are said to halt and 'Present Arms' on sighting the massive Renaissance château 800 metres up the slope, the passing motorist should have no difficulty in finding it. There are two turnings in the village, which follow the wall of the Clos up to the château.

By the middle of the twelfth century, Cistercian monks had achieved a vineyard and a cellar, their successors a hundred years later installing the huge oak wine presses still on view here. The Renaissance château dates from 1551 and the Abbots of Citeaux were in charge up to the Revolution. Dom Goblet, last of the line and so suitably named, built a magnificent private cellar and when Napoleon required some bottles he replied, 'I have some forty-year-old Clos de Vougeot, if he wants it let him come here.'

Since 1944 the château has been owned by the Confrérie des Chevaliers du Tastevin, the vinous brotherhood of Burgundy now known the world over, which holds investitures and banquets with much pomp, gaiety and pageantry. Visitors see the courtyard, the *cuverie* and the *Grand Cellier* where the galas are held.

Hours:	Every day, all the year 0900–1130; 1400–1730. Guided tours start approximately every half hour. Entrance: 1–12 people 2F each; 13–25 people 25F; over 25 people 1F per person.

Beaune
Bouchard Père et Fils,
Au Château, 21201 Beaune, Côte d'Or.
Tel. (80) 22.14.41.

The *Michelin Guide* plan of Beaune marks the old château against the ramparts on the eastern side of the town, where the Avenue du 8 Septembre runs in from the railway station.

Michel Bouchard of Beaune started a cloth business but in 1731 he thought his customers in northern France might be kept a little warmer with some burgundy too. Obviously they were; only the wine business has been continued for the last 250 years and today Bouchard Père et Fils are principal producer/proprietors in the Côte de Beaune, owning some eighty hectares of priceless vineyards, including a part of the great Montrachet itself. They have also a small site in the Chambertin vineyard at the other end of this wine region.

In summer there is an English, German and French-speaking guide, who takes visitors on a tour of these historic cellars, which have had to be considerably extended in the last ten to fifteen years. Some white and red burgundies are usually sampled.

Hours:	Monday to Friday 0900–1130; 1400–1730. Closed August.
Introduction:	By appointment. Letter or card from agents: Morgan Furze Ltd., City Cellars, Micawber Street, London N1. Tel. 01–253 5263.

Calvet S.A.,
6 Boulevard Perpreuil, B.P. No. 114, 21201 Beaune, Côte d'Or.
Tel. (80) 22.06.32

No French wine company's name is better known than Calvet and their Burgundy headquarters are easily found on the south-east side in the Boulevard Perpreuil, which is part of the road, outside the ramparts, that encircles Beaune, and are visited by many tourists.

Though associated first with the Côte du Rhône, where Calvet began in 1818, and secondly with Bordeaux, there has been a Burgundy branch since 1870, with no less than 1,200,000 bottles lying in its cellars.

The tour is conducted in English; some wines are offered for sampling and purchases may be made by the case or bottle.

Hours:	Open to all without warning every day except Mondays but including Saturdays, Sundays and national holidays 0900–1100; 1400–1700. Open all the year round but not in August.
Introduction:	None necessary.

U.K. Agents:	Sherry House Ltd., 92 New Cavendish Street, London W1M 8LP. Tel. 01–580 0301/8.

Caves-Exposition de la Reine Pédauque,
21201 Beaune, Côte d'Or.
Tel. (80) 22. 23. 11.

La Reine Pédauque is a company owning some vineyards and maturing cellars in Beaune and Aloxe-Corton, a restaurant in Paris and these Exhibition-Cellars, easily seen on entering the walled town of Beaune from Dijon. The origin of La Reine Pédauque is a legendary figure, carved on the portals of old churches.

In their Caves-Exposition, which are charmingly furnished in old Burgundian style, a large variety of wines of Burgundy – red, white and some rosé – are offered free, so that it is not surprising that thousands of visitors come here each year.

Hours:	Every day, including Saturdays and Sundays 0830–1200; 1400–1800. Closed only on 1 January, 1 November and Christmas Day.
Introduction:	Open to all; none necessary.

Joseph Drouhin,
7 rue d'Enfer, 21202 Beaune, Côte d'Or.
Tel. (80) 22.06.80.

The entrance to these cellars is by the west door of the Church of Notre Dame in the centre of Beaune, very close to the Ducal Palace, which contains the Wine Museum.

Founded in 1756, the firm became Joseph Drouhin in 1880. After Joseph came Maurice Drouhin, who distinguished himself in World War I, winning the D.S.C. when a liaison officer on the staff of General MacArthur. Head of the firm from 1918 to 1957, Maurice has been succeeded by Robert Drouhin. The firm owns sites in the great vineyards of Chambertin Clos de Bèze, Musigny and Clos de Vougeot and – almost on the doorstep – in Beaune Clos des Mouches.

The Drouhin cellars, used in their time by the Dukes of Burgundy, the Kings of France and the College of Beaune are of great historical interest and visitors are fascinated to find thousands of casks and at least a million bottles maturing there.

Hours:	Monday to Saturday 0900–1130; 1400–1730.
Introduction:	Casual visitors welcome, but appointments preferred, made through agents: Dreyfus, Ashby & Co. Ltd., 21–22 Hans Place, London, SW1X 0EP. Tel. 01–589 5433.

P. de Marcilly Frères,
22 Avenue du 8 Septembre, 21201 Beaune, Côte d'Or.
Tel. (80) 22.16.21.

This company began in 1849 at Chassagne-Montrachet, one of the three villages (and communes) south of Beaune, known for their white burgundies. The move to Beaune itself took place in 1912. Possessions include fine sites in Gevrey-Chambertin, Beaune and in Chassagne-Montrachet, where they own part of the Clos Saint-Jean and retain the original *cuverie* in operation.

Visitors may see the cellars, cuverie and vineyards.

Hours: Monday to Friday 0900–1100; 1430–1700. Closed August.

Introduction: By appointment. Letter of introduction from agents:
Rutherford Osborne & Perkin Ltd.,
Harlequin Avenue, Great West Road,
Brentford, Middlesex.
Tel. 01–560 8351.

Musée du Vin de Bourgogne,

Logis du Roi 21200 Beaune, Côte d'Or.
Tel. (80) 22.08.19.

Just north of the Hospices and at the centre of the town is the Logis du Roi, once the palace of the Dukes of Burgundy. Now it houses what H. W. Yoxall* describes as 'the best wine museum I know', but apart from the modern tapestries by Lurçat he says nothing of its contents.

In addition to the inevitable collection of ancient wine presses, there are some good collections of *tastevins* and *coupes de mariage* – wine chalices for newly married couples. Perhaps the main attraction is the display shewing every aspect of Burgundy's wine production from early times to the present.

Hours: Open every day 0900–1130; 1400–1730 with guided tours starting as necessary. Entrance fee (1976) 2F a person, 1F a person in a group (size unspecified).

Patriarche Père et Fils,

Ancien Couvent des Visitandines, 5 et 7 rue du Collège, 21204 Beaune, Côte d'Or
Tel. (80) 22.23.20.

Jean Baptiste Patriarche founded this company when a humble grower at Savigny-les-Beaune, a village up the famous slope three kilometres out of the town. The Revolution gave him the chance, in 1796, to buy the former seventeenth century convent of the Nuns of Visitation in Beaune itself. The spacious cellars now hold large stocks of Patriarche wines.

A guide conducts the tour round the 'thirteenth to twentieth century cellars', where millions of bottles of Patriarche burgundies and of the sparkling Kriter are to be seen. The 45-minute tour includes a free tasting of the aforesaid wines, the charges being 3 francs a person, reduced to 2 francs a person for groups of more than ten people. Children under 16 go free.

Hours: Every day 0900–1200; 1400–1800. Guided tours every 20 minutes.

Introduction: Open to all, none necessary.

U.K. Agents: Matthew Clark & Sons Ltd., 185 Central Street, London EC1V 8DR. Tel. Tel. 01–253-7646.

MACON

Piat Père et Fils,

21/23 rue de la République, 71009 Mâcon, Saône-et-Loire.
Tel. (85) 38.21.43.

The rue de la République is just off the Quai des Marans, a kilometre downstream from Mâcon's bridge. Visitors step into a charming old courtyard, with the offices above and the cellars beginning below.

Growers and shippers in Mâcon since 1849, Piat Père et Fils own vineyards in Beaujolais-Villages, Moulin-à-Vent and in the Clos de Vougeot, a hundred kilometres to the north in the Côte de Nuits. In the Beaujolais, exclusive agreements with leading property owners enable them to ship some outstanding crus, notably Château de Saint-Amour, Château de Bellevue and Domaine de Beauvernay. Charles Piat, a respected figure in both wine and golf circles, who has been largely responsible for giving Mâcon a golf course and an attractive club house, retired as head of the company during 1976, having seen his Piat-shaped bottle become well-known in many lands.

Visitors descend below the courtyard into a labyrinth of large ancient casks, concrete vats, maturing cellars and the latest stainless steel tanks, in which the white wines can be kept quite fresh. Le Piat de Beaujolais and Le Piat de Mâcon Viré are usually tasted; wines may be bought.

Hours: Monday to Friday 0830–1030; 1430–1630.

Introduction: Not essential but preferable. Letter or card obtainable from agents:
Gilbey Vintners Ltd., Gilbey House, Harlow, Essex CM20 1DX
Tel. (0279) 26801.

BEAUJOLAIS
Belleville-sur-Saône

Château de Corcelles,

Corcelles-en-Beaujolais, 69220 Belleville-sur-Saône, Rhône.
Tel. (74) 66.00.24.

Going south, leave the motorway at Mâcon-Sud, continuing south on the old main road, N6, to the junction of Thoissy with Villié-Morgon, then follow the signs to this château, one of the most picturesque in the upper part of Beaujolais. (Going north, leave the motorway at Belleville-sur-Saône and continue north on N6.)

Château de Corcelles, built as a fifteenth century for-

---

*H. W. Yoxall, *The Wines of Burgundy*.

tress to defend the local inhabitants, is classified as a French national monument and restoration work has done much to preserve its appearance. The inside court-yard, with Renaissance wainscoting, the old well, the carved panels of the chapel and its grandeur are generally much admired; likewise its viticultural domaine of some 40 hectares.

The outbuildings associated with wine were added in the seventeenth century, notably Le Grand Cuvier, a handsome low building 80 metres long, with the wines maturing in giant casks at ground level and magnificent cellarage below.

Visitors see the property, cellars and vineyards, tasting and buying wines if they wish.

Hours:	Every day 1000–1200; 1430–1900.
Introduction:	Casual visitors may call but a letter is preferable from agents: Jarvis Halliday & Co. Ltd., 102 Bicester Road, Aylesbury HP19 3AT. Tel. (0296) 3456.

Lesser Visits

The following are recommended as other places of interest connected with wine, and in many cases for sampling. They are likely to be open from about 0900–1200 and 1400–1900 in summer, often seven days a week.

Beaujeu
This small Beaujolais town has a tasting room and exhibits of cooperage and other local crafts in the museum.

Belleville-sur-Saône
Maison de Beaujolais combining a tasting room with a regional (Beaujolais) wine information office.

Chénas
This commune's wines may be tasted at the Caveau du Cru Chénas on the Route de Beaujolais.

Domange-Igé
A Caveau de Dégustation.

Fleurie
Tasting cellars for wines of this Commune of Beaujolais.

Juliénas
Tasting room for the wines of this commune in a church, deconsecrated at the time of the Revolution. Murals added since then are 'rather crude . . . and extremely inhibited in subject', declares Pamela Vandyke Price.*

Mâcon
Maison Mâconnaise des Vins, Quai de Lattre-de-Tassigny. Marked in the *Michelin Guide* Mâcon plan,

this is by the Saône at the northern end by the new tourist office. It is said to be a restaurant.

Mercurey
Wines of Mercurey, Côte Chalonnaise, may be tasted at the Caveau on N78.

Moulin-à-Vent
Caveau au Pied du Vieux Moulin and Maison de l'Union des Viticulteurs on N6 – tasting of wines of this commune.

Romanèche
A small wine museum in the Maison Raclet.

Saint-Amour
Caveau on the Route de Beaujolais; shews the wines of this excellent Commune.

Solutré and Vinzelles
So much Pouilly-Fuissé is faked, particularly in the United States, that a visit to Caveau du Pouilly-Fuissé and Caveau du Pouilly-Vinzelles would shew what the white Pouilly burgundies of this district, near Mâcon, truly taste like.

Villié-Morgon
A popular Caveau de Dégustation in the basement of the Town Hall.

Viré
Viré is a Mâconnais commune, less than 20 kilometres north of the town via N6, wholly devoted to producing the excellent dry white Macon Viré. Viré has grown vines since time immemorial. The hub of the community would seem to be the bar at the Co-operative, where all are welcome to drink the wine and smile at the mural that depicts Viré's *jumellage* (twinning) with Montmartre in Paris.

---

*Pamela Vandyke Price, *Eating and Drinking in France Today*.

JURA AND SAVOIE

Introduction

Red, white, pink, yellow . . . Jura wines are a colourful collection. Château-Chalon, the *vin jaune* made from a Jura grape called Savagnin, is the most famous because it is France's longest lived wine. Harvested late, the wine lies in cask without topping up for as much as six years developing *flor* in the manner of Jerez (q.v.). After another fifteen years or so in bottle, this rather strange, sherry-like wine (though not fortified) reaches its peak. A deep golden colour, dry, costly and rare, some say the taste is of hazel nuts and others that it is maderised like old champagne.

Vin de paille is another wine made from grapes late-gathered in the manner of Sauternes. Laying them on straw for two months concentrates the juice and a period of about two years in casks of Russian oak ensues before bottling. Not as like Sauternes as might be expected, the Savagnin grape (with the Melon among others permitted) and the russet-coloured soil give vin de paille a sweet, earthy taste that some people find irresistible. Unfortunately for them production is limited and bottles may not be easy to find in London, but Henri Maire in Arbois have stocks of Vin de Paille and Château-Chalon, which can be tasted and bought there to take away.

The reds, whites and pinks (particularly the last-named, Rosé d'Arbois being one of the best rosés of France) are more straightforward table wines, with good sales in Britain. Sparkling wines are also made in the Jura and Henri Maire is expected to launch his well-known brand, Vin Fou, across the Channel.

The wine region surrounds the unspoilt little town of Arbois where, almost exactly a hundred years ago, Louis Pasteur was conducting experiments destined to shew beyond doubt that micro-organisms from the air, by multiplying in the vessel containing the juice of the grape, brought about the transformation to wine. His own small vineyard can still be seen there.

Attacked late in the nineteenth century by the phylloxera, Jura wine production fell into decline. The revival is almost entirely due to one man, Henri Maire, who finished his studies in Paris and returned to Arbois after the last war to make the family name well-known in the world of wine once more.

There are plenty of Bars Dégustation around Arbois, one worth a call being the Caveau des Jacobins, headquarters of the Co-operative at Poligny, established in a fine old Gothic church now given to the worship of Bacchus.

Whether the second entry in this section should be under Jura or Haute-Savoie I am uncertain. The former has it for convenience. The excellence of Haute-Savoie sparkling wines is not generally known and I am pleased that Varichon and Clerc welcome visitors in their pleasant little town on the Rhône near Annecy.

Further Information
Société de Viticulture du Jura, Maison de l'agriculture, Lons le Saunier, 39000 Jura.
And from the Head Office of Henri Maire at:
Château Montfort, 39600 Arbois, Jura.

JURA

Arbois
Henri Maire,
Les Tonneaux, rue de l'Hôtel de Ville, 39600 Arbois, Jura.
Tel. Arbois 75.

Arbois, still a quiet small town 80 kilometres east of Beaune, was the work-place of Pasteur (1822–1895). The Maire family, also of Arbois, have made Jura wines since 1632 but it has required Henri Maire of the present generation to revive their reputation in France and beyond. Separating administration very wisely from public rela-

tions, Henri Maire has created in Les Tonneaux a real reception centre for visitors, close to the Town Hall in the centre of the town where there are car parks close by.

There are tours of the cellars, vineyard tours by minibus, films and a tasting – all (or at least some of these) being free. However, in Arbois they are keen salesmen too and if you leave with more cartons in your boot than you intended, rest assured the Blanc de Blancs, the red wine called Frédéric Barberousse and the pink Vin Gris are good wines; though a little arithmetic to make sure that by the time Customs Duty has been paid they would not have been cheaper at home is always prudent.

Hours:	Daily including national holidays 0800–2000.
Introduction:	None necessary. Any special arrangements could be made through a U.K. principal stockist: Gilbey Vintners Ltd., Gilbey House, Harlow, Essex CM20 1DX. Tel. (0279) 26801.

Seyssel

Varichon & Clerc,
Château des Séchallets, 01420 Seyssel, Ain.
Tel. Seyssel (Ain) 15.

Seyssel is a small town on the road N92, 45 kilometres south-west of Geneva and half-an-hour's drive from Annecy. The Rhône flows through it, dividing the town into two parts, that on the right bank being in the département of Ain and that on the left bank being in Haute-Savoie. Coming from Geneva on N92, cross the bridge and turn left; the Château des Séchallets is a short way down.

Bugey (Ain) makes some V.D.Q.S. wines and Crépy (Haute-Savoie) is a dry, A.C. white wine; both are worth meeting locally because small production makes an encounter unlikely elsewhere. More important however are the still and sparkling white wines with the Seyssel appellation, first commended by the *Oenologie Française* in 1827.

Varichon & Clerc have made a speciality of these since 1901, particularly a Blanc de Blancs *Méthode Champenoise* sparkler, offered in Britain, amongst others, by Justerini & Brooks, holders of the Royal Warrant. Indeed, with champagne so expensive, this Blanc de Blancs may have passed through many a palatial portal by now.

Visitors see over an interesting installation, taste the wines and may buy what they please.

Hours:	Monday to Friday 1000–1130; 1500–1630. Closed August.
Introduction:	By appointment. Letter of introduction from agents: G. W. Thoman Ltd., 63/5 Crutched Friars, London EC3N 2DT. Tel. 01–381 8866.

SAVOIE

Voiron

Chartreuse Diffusion S.A.,
38503 Voiron, Isère
Tel. (76) 05.19.09 Ext. 19.

Voiron is a small town 31 kilometres north-west of Grenoble; coming from there, N75 passes the Distillerie de la Grande Chartreuse entering the town.

Chartreuse is the only French liqueur which is still made by monks. It all began in 1605 when the silent Order of the Carthusians was presented with a manuscript for a wonderful Elixir requiring one hundred and thirty herbs. Many years passed until this neglected manuscript reached Brother Maubec at the Great Monastry here, and on the eve of achieving the liqueur Maubec died, breaking his vow of silence to pass his discoveries to Brother Antoine. Then, in 1764, Antoine made the first green Chartreuse.

With the Revolution in 1792, Carthusian property was confiscated but the State chemists merely regarded the monks' hieroglyphics as incomprehensible nonsense and the secret came back to them intact. In 1817, with the Order happily re-assembled at Grande Chartreuse, Brother Jacquet discovered yellow Chartreuse and

thenceforward the two liqueurs steadily became world-famous. Today income from sales is about a million pounds a year but there are twenty-six monasteries and eight hundred monks to look after.

Chartreuse is said to be the only liqueur in France made entirely from natural ingredients. Every day during the distilling season three monks come down from the monastery to prepare and distil the 150 different ingredients required for green and yellow Chartreuse. The distilled alcohol is then matured in large oak vats.

English-speaking hostesses are available and a free drink is served to every visitor.

Hours:	From 1 April to 31 October: Every day including Saturdays, Sundays and national holidays: 0800–1130; 1400–1830. From 1 November to 31 March: Monday to Friday, except holidays: 0800–1130; 1400–1730.
U.K. Agents:	J. R. Phillips & Co. Ltd., Avonmouth Way, Avonmouth, Bristol BS11 9HX. Tel. (02752) 3651.

COTES DU RHONE

Introduction

Côtes du Rhône is the appellation of red, white and rosé wines made in delimited parts of the Rhône valley between Lyons and Avignon. The wine region begins about twenty miles south of Lyons at Côte Rôtie, a conspicuous hill with a bald patch on the top, across the Rhône from Vienne. There are vineyards and wines of varying importance at intervals for the next hundred miles, until the Côtes du Rhône comes to an end at Avignon.

The Rhône valley vineyards must be the oldest in France; the Romans, penetrating from the south, encouraged viticulture, but long before them the Greeks had reached Marseille by sea, moving inland up the valley, planting the vines they had brought. Long regarded as making the best red wines after Bordeaux and Burgundy, vinification has been adapted during the sixties to mature them sooner, and with the big rises in Bordeaux and Burgundy prices between 1970 and 1974, exports just about trebled in that period.

GETTING THERE
The Paris-Riviera railway and the A7 autoroute (A6 becomes A7 south of Lyons) run through the region. The London-Paris-Lyons air services are convenient for self-drive hire from Lyons Airport, whether the objective is south to the Côtes du Rhône or north to Burgundy.

WATERWAYS TO COME
In the last ten years the commercial tonnage carried by barge on the Rhône has increased threefold to nearly four million tons. By 1977 canalization plans should make it possible to transport fifteen million tons a year; dredging and the widening of locks make fifty million tons a possibility in due course. Hitherto, for various reasons, the Rhône, compared to the Seine, has hardly been used commercially. By 1977 one of its barge trains should be carrying as much as a hundred 30-ton trucks at a third of the cost a ton-kilometre by road. And by 1982, when the Rhône and Rhine canals become linked at Mulhouse, there will be an international 280-mile waterway from the heart of Europe to the Mediterranean and a vast increase in hydro-electric power associated with it.

It is to be hoped that the cost of transporting wines from a district that is becoming much more important will be reduced – and through water paradoxically.

Before departing altogether from water, mention should be made of the *Frédéric Mistral*, a shallow draught diesel vessel which in summer used to do the 160-mile journey downstream from Lyons to Avignon once a week between dawn and dusk. This was a pleasure trip, with catering by Monsieur Nandron, whose restaurant on the Quai J. Moulin in Lyons is world famous. The luncheon could be slept off till 1930 hours, the arrival time at Avignon. Perhaps when we British can supply our own oil, and if it doesn't cost too much, the *Frédéric Mistral* will sail again.

Rhône wines begin in Switzerland. As the motorist turns south at Brigue for the Simplon Pass, the river is rolling down from an icy source in the Furka Pass high above. Those first vineyards that make crisp dry white wines from the Fendant grape – Fendant de Sion and Fendant de Valais – are lower down in the Valais. The flinty red wine of Dole belongs here too.

GENEVA TO LYONS
After flowing through Lac Leman to Geneva, the Rhône loops its way to Lyons through Haute-Savoie, where excellent sparkling and other wines are made at Seyssel (p.54). Only south of Lyons does the appellation become Côtes du Rhône.

The vineyard wanderer is recommended to use the N86 on the right bank, a fairly pleasant road but for the odd cement factory at intervals. It passes the Blonde and the Brune, the two sides of the hump of Côte Rôtie, which have light and dark soil respectively, and then keeps close

CÔTES du RHONE

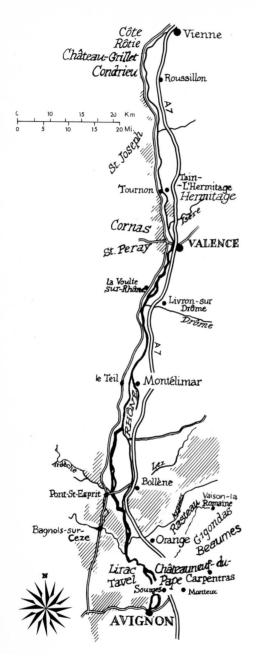

to the river for many miles, almost to Avignon. Condrieu has an appellation for its dry, flinty white wines made from the Viognier grape, and the delightful Hôtel Beau Rivage right on the Rhône is a place in which to try them. South of the village, perched above the river within the communes of Saint Verin and Saint Michel-sous-Condrieu, is Château Grillet, the smallest (3½ acres) vineyard in France to be given its own appellation. Château Grillet is certainly the best wine of Condrieu and many would say the best white wine of the Côtes du Rhône. The output cannot be much more than 200 cases a year, some of which are shipped to Britain and the U.S.A.

HERMITAGE

Some thirty miles further down the road comes Tournon, bright and cheerful in summer as its twin town Tain, across the bridge, is dour. Chapoutier and Jaboulet make their red and white Hermitage wines from the conspicuous hill above their offices in Tain, slipping over the bridge to lunch with their guests under gay coloured awnings in summer.

Red Hermitage wines, even more than Côte Rôtie and Châteauneuf-du-Pape, improve enormously in bottle and given ten, twenty or even more years' bottle age can become superb. Sad to relate, tying up capital in bottles for so long has become increasingly out of the question and the best Côtes du Rhône wines are now made to mature far sooner. The white Hermitage wine too was one of the few dry white table wines that lived long and derived benefit from it. Cheaper red and white wines emanate from the lower slopes of the hill, being sold under the name 'Crozes-Hermitage'.

North of Tournon, Saint Joseph has its own appellation for red and white wines, rather lighter than those of Hermitage. Further downstream, nearly opposite Valence, Cornas makes a robust red wine like Hermitage from the Syrah grape, while its neighbour, Saint Peray, is known for the region's best sparkling wine, for which Wagner is recorded as being a customer in 1877.

This point is virtually half way house. To the north the appellations Côte Rôtie, Condrieu, Château Grillet, Crozes-Hermitage, Hermitage, Saint-Joseph, Cornas and Saint Peray. Ahead to the south, the appellations Côtes du Rhône Villages, Rasteau, Beaumes, Châteauneuf-du-Pape, Lirac, Tavel and Gigondas.

In addition to these, there are another one hundred and thirty-eight communes entitled to the appellation, 'Côtes du Rhône'; no more, no less!

After Valence (still going south on the right bank road N86), the vineyards are absent for about forty miles until the Ardèche flows in to the Rhône at Pont-St. Esprit. The luxurious hotel La Cardinale at Baix is a gastronomic stop, beautifully furnished and right on the Rhône. If its prices are quite beyond your means, there's always a stick of the celebrated nougat at Montélimar.

CHATEAUNEUF-DU-PAPE

In the southern half, one name is known the world over –

Châteauneuf-du-Pape. Branching left from N86 on to N580 at Bagnols-sur-Cèze leads to this charming hill town in a dozen miles, crossing the Rhône on the simplest little car ferry, which moves across from one bank to the other under the sole influence of the river's strong current and its own rudder. (Châteauneuf-du-Pape is eight miles due north of Avignon; from A7 leave by the Orange or the Avignon Nord exit.)

By Côtes du Rhône standards, the Châteauneuf-du-Pape vineyards, which surround the town and descend almost to the river, are not of great antiquity. Côte Rôtie and Hermitage wines were drunk and praised by the Romans, among them Pontius Pilate, who was appointed Governor of Vienne after his time in Jerusalem. When the Papacy moved to Avignon in 1309, the first Pope was a wine man – none other than Clément V, Archbishop of Bordeaux, who had given his name to Château Pape-Clément, still one of the very best *cru classé* clarets of Graves. Clément V, missing the cool breezes of the Atlantic as he sweltered in Avignon, built himself a summer residence on the hill outside and it seems likely that his successor planted the vineyards. Until 1850, the village was called Châteauneuf-Calcernier. It was the change of name and some subtle propaganda by the local owner, the Marquis de Nerthe, ascribing his virility in old age to the wine, that started its popularity. Later, in the twenties, another proprietor, Baron Le Roy, persuaded his fellow growers to start a prototype scheme which was to develop, a decade later, into the national system of *Appellation Contrôlée*.

Much as every Mediterranean fish worth eating goes into the *bouillabaisse*, so every Rhône vine contributes to the soft round purple wine of Châteauneuf, thirteen species in all. By increasing the proportion of Grenache it is now made to mature within two or three years, though doubtless leading growers still make a small proportion in the old way, which transforms a very strong rough beverage into a mellow ruby wine in about ten years, the bouquet so splendid that they used to serve it there in big balloon glasses. Possibly they still do at La Mule-du-Pape, a well-appointed restaurant with lovely views, or at the Château des Fines Roches, where M. Mousset has opened another.

Across the Rhône, Châteauneuf's neighbours, Tavel and Lirac, are justly famed for their rosé. Tavel, a village of only 1,000 people, has long been said to make the world's best dry rosé, but since the war its popularity has declined in the face of rivals from the Loire and from other countries, notably Portugal, selling at lower prices. Discouraged, Lirac is making much more red wine, but Tavel keeps exclusively to rosé.

ROUTE DES VINS

Increasing demand now brings other Rhône wines to British and American tables. Gigondas, some twenty miles north-east of Châteauneuf-du-Pape, makes a similar A.C. red wine; Vaucluse and Rasteau to the north,

Beaumes-de-Venise to the south are becoming known as sources of wines well above ordinary standard. A pleasant run is from Orange along N576 up the valley of the Aygues, crossing the river to Cairanne and Rasteau after ten miles. At Vaison-la-Romaine, on another smaller tributary, the Ouvèze, there is much to see before continuing south-east to Flasson, a lovely village in red and ochre stone at the foot of Mont Ventoux. Another stop can be made, on the way back to Orange, at the *Maison du Tourisme et des Vins* in Carpentras.

Further Information
Comité Interprofessionnel des Vins des Côtes du Rhône, 41 cours J. Jaurès, 84000 Avignon, Vaucluse.

Tain l'Hermitage
M. Chapoutier S.A.,
18 Avenue de la République, 26600 Tain l'Hermitage, Drôme.
Tel. (75) 08.28.65.

Tain l'Hermitage, a small town on the left bank of the Rhône south of Lyons, is close to the Paris-Marseille Autoroute A7, which has an interchange for Tain 83 kilometres from Lyons and 153 kilometres from Avignon. The Avenue de la République leads to the station from the Avenue Jean Jaures.

Founded in 1808, Chapoutier has specialised for five generations in the wines of the Côtes du Rhône – from Côte Rotie to Châteauneuf-du-Pape. The family own vineyards within these appellations, particularly on the conspicuous hill of Hermitage dominating the town, where the vine has been cultivated since pre-Roman times. The red and white wines of Hermitage and Crozes-Hermitage, found in restaurants and cellars in many parts of the world, originate on these slopes.

English is spoken and visitors are shewn the cellars, winery and perhaps some of the vineyards. They are always cordially received but M. Max Chapoutier regrets, in a charmingly expressed letter in English, that their resources do not permit 'a special welcome department' or 'a touristic visit', so they must limit numbers and insist on a letter of introduction.

Hours:	Monday to Friday 0930–1130; 1430–1630. Closed August.
Introduction:	By appointment. Letter of introduction from agents: Aug. Hellmers & Sons Ltd., 76/80 Druid Street, London SE1. Tel. 01–237 1666.

Paul Jaboulet Ainé,
40 Route de Romans, 26600 Tain l'Hermitage, Drôme.
Tel. (75) 08.22.65.

Contemplating a new career in the wine trade shortly before I retired from the Navy in 1955, it was M. Louis Jaboulet, who kindly took me over the Hermitage hill,

pointing out his 'wineyards' as he called them, a term I later adopted for 'Peter Dominic's Wineyard', a weekly piece that appeared in the Epicure columns of *The Times* then on the front page.

Louis, assisted since 1969 by his two sons, Gérard and Jacques, is still in active command of this leading house of Rhône wines, which his grandfather started in 1834. His brother, Jean, supervises the 'wineyards' and between them they own some fifty-five hectares, sharing with Chapoutier the bulk of the export market in Hermitage wines.

Route de Romans is close north of the Avenue J. Nadi (see *Michelin* Tain plan). Visitors see over the cellars where 500,000 bottles lie maturing and thousands of hectolitres are in cask and vat.

Hours:	Monday to Friday 0800–1200; 1330–1830. Closed August.
Introduction:	By appointment. Letter of introduction from agents: O. W. Loeb & Co. Ltd., 15 Jermyn Street, London SW1 6LT. Tel. 01–734 5878.

Châteauneuf-du-Pape
Société des Grands Vins,

Château des Fines-Roches, 84230 Châteauneuf-du-Pape, Vaucluse.
Tel. (90) 83.51.30.

About 1½ km. east of the hill town, Fines-Roches is marked on *Michelin* map No. 81. By road from Avignon, turn westwards at the village of St. Louis and turn left after 3 km. when this ninteenth century 'castle' will come into view.

Although this particular company was only formed by M. Jacques Mousset in 1958, the Mousset family have been in the wine trade as growers and négociants for at least three generations. They may well be the biggest vineyard owners in Châteauneuf-du-Pape and have other vineyards in the districts further north.

Visitors are received at Château Fines-Roches, the family's proudest possession, which makes one of the finest crus of Châteauneuf-du-Pape.

Wines may be tasted, purchases are possible and doubt-less M. Mousset will be delighted if visitors patronise the restaurant here which he opened in 1974.

Hours:	Every day in summer including Sundays and national holidays 0900–1200; 1430–1630.
Introduction:	Not essential but preferable. Letter or card should be obtainable from: Justerini & Brooks Ltd., 1 York Gate, Regent's Park, London NW1 4PU. Tel. 01–935 4446.

Père Anselme S.A.,

B.P. No. 1, 84230 Châteauneuf-du-Pape, Vaucluse.
Tel. (90) 83.51.64.

Châteauneuf-du-Pape is a small but conspicuous hill town thirteen and seventeen kilometres from Orange and Avignon respectively; the Père Anselme headquarters will be found on the approach road from Avignon. Anselme was the Christian name of the ancestor who founded the company in 1681. His age and wisdom prompted the nickname, 'Father Anselme', and the family have continued to specialise in the wines of the Côtes du Rhône ever since.

The business is run by Monsieur J. P. Brotte, who has built up such a fine collection of wine presses and old wine-making implements that they feel it merits the name 'Museum'. Many visitors come here. Wines may be bought and visitors can be shewn the best drive through the vineyards, worth seeing if only to notice the gigantic pebbles characteristic of Châteauneuf vineyards, which play a part in reflecting the heat on to the vines.

The company also runs La Mule-du-Pape the town's excellent restaurant, which does not have its usual *bonne table* star in the 1976 *Michelin Guide,* due one hopes merely to a change of both Manager and Chef, for this is normal practice until a (strictly anonymous) *Guide* inspector has tried out the newcomers' abilities. Whatever the result, there is a superb view of the Rhône valley from the windows.

Hours:	Daily throughout the year 0800–1200; 1400–1800.
Introduction:	Open to all; none necessary.

Lesser Visits

The following are recommended as other places of interest connected with wine, and in many cases for sampling. They are likely to be open from about 0900–1200 and 1400–1900, often seven days a week.

Avignon (by the bridge)
La Maison des Côtes du Rhône, a dégustation centre.

Cairanne
Caveau de Dégustation.

Carpentras
Maison du Tourisme et des Vins in the Place du Théâtre is a dégustation centre for V.D.Q.S. Côtes du Ventoux wines, which being V.D.Q.S. cost less than A.C. wines.

Carpentras is in Vaucluse, 23 kilometres from Avignon by RN542.

Châteauneuf-du-Pape
Caves du Clos des Papes, and Caves 'Reflets du Châteauneuf'.

Die
Caves Co-opératives for Clairette de Die. This semi-sparkling, yellowish gold wine is said to be an acquired taste and this is the place to acquire it (or otherwise).

Gigondas
Cave Co-opérative des Vignerons de Gigondas. Gigondas, 37 kilometres north-east of Châteauneuf-du-Pape, makes similar wines and was awarded its own Appellation Contrôlée for them in recent years.

Rasteau
The Caveau de Dégustation de la Cave des Vignerons is sixteen kilometres from Orange by RN575 and close to Gigondas.

Roquemaure
Caveau des Vins de Lirac.

Tavel
Sixteen kilometres north-west of Avignon, the famous rosé wines of this district may be tried at the Cave des Vignerons.

Vaison-la-Romaine
Maison du Vin.

Vinsobres
Caveau de Dégustation.

PROVENCE

O, for a draught of vintage! that hath been
Cool'd a long age in the deep-delved earth,
Tasting of Flora and the country green,
Dance, and Provençal song, and sunburnt mirth!
– Keats

Introduction

The south-eastern province of France, bounded by the Rhône, the Alps, the Italian frontier and the sea, is loved for many reasons but, in spite of Keats, wine cannot really claim a high place – though it can be excellent – among them. Sea and sun in summer, the mild climate of the Riviera in winter, the gay life of casinos and the luxury of such places as Eden Roc and La Réserve restaurant at Beaulieu provide the glamour. But Provence has always attracted the painter for its light. Cézanne was born at Aix; Van Gogh, Renoir, Picasso, Braque and Matisse are among many who came to live and work here. The Picasso ceramics in the Grimaldi Museum at Antibes draw the art lovers and the summer music festival at Aix is known far afield through broadcasting. For archaeologists the place

60

PROVENCE

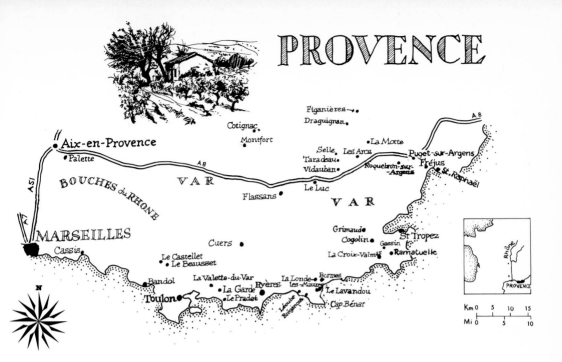

abounds in Roman relics, spread about a beautiful countryside, and over-earnest intellectuals can immerse themselves in the poems of Frédéric Mistral (1830–1914) or
even the Provençal language itself, so beloved of that poet.

A RENAISSANCE

Yet, rather less than a century ago, until that insect the
phylloxera ravished the roots in almost every vineyard in
Europe, Provence – according to that most reliable of wine
writers, Edmund Penning-Rowsell of the London *Financial Times** – was second only to Bordeaux as an exporting
wine region. With growing demand for sound table wines
at moderate prices, the pendulum is already swinging
back, not only towards the rosé wines – made near the
coast and therefore more familiar – but inland to new red
wines being exported from both Provence and
Languedoc-Roussillon.

   With Government financial help and encouragement
there has been a great deal of replanting with better
species of vine over the past ten to fifteen years. These
species are Carignan, Cinsault, Mourvèdre, Grenache
and some Cabernet-Sauvignon. Blended in the right
proportions, the wine is good; moreover, the plants stand
up to the Mistral, the searing north-westerly wind that
can blow across the whole region for days at a time,

*Edmund Penning-Rowsell, *Provence in the Pink*, Financial Times 24
April 1974.

fraying tempers and making the Gulf of Lions a place of
misery for those prone to seasickness.

   Officially, there is no longer a Provence, just the
départements which comprise it – Bouches-du-Rhône,
Var, Vaucluse, Basses-Alpes and Alpes-Maritimes.

CASSIS AND BANDOL

In Bouches-du-Rhône there is a fifteen miles by ten miles
V.D.Q.S. region, Coteaux d'Aix-en-Provence, south-east
of the town, the best part around Palette enjoying A.C.
status. On the coast at Cassis, the rosé of that name from
the vineyards high above the town that trap the southern
sun, is perfectly matched with the fish brought in fresh
daily from the bay. Further along towards Toulon, the
rosé of Bandol is at least as good; indeed, the rosé wines
along the coast to Cannes are seldom disappointing.

MIDDLE AND HIGH PROVENCE

The wines of Var, extending from Toulon to St. Raphael
along the Riviera and inland to Draguignan, mostly qual-
ify as 'V.D.Q.S. Côtes de Provence'. Free maps and
brochures on the *Routes des Vins* can be obtained from the
Syndicat office at Les Arcs (address below; and enclose an
International Reply Coupon if writing). One route, well
mapped and sign-posted, frequently crosses, recrosses
and runs parallel to N7 in the vicinity of Vidauban. 'The
shore line between the Maures mountains and the sea, the
rich central plain, the wooded blue-green hills of middle

and high Provence, the valleys tucked away in the heart of the Maures and behind the porphyry peaks of the Esterel range, old villages with names like Taradeau, Ramatville and Gonfaron are all features making the wine tour a delight to the eye and an experience for the palate', wrote Joe Hollander, an English journalist, who has retired to live there and drink the wine of the country*.

Though there may be few formal Provençal introductions in my pages, rest assured the palate will not lack for experience! Wine being abundant in this hot southern region, there are few formalities. Co-operatives and small growers welcoming the passing motorist abound, particularly if he brings out from his boot a few empty five litre *bombonnes* that need refilling.

Further Information
Comité Interprofessionnel des Vins des Côtes de Provence
3 avenue de la gare
83460 Les Arcs
Vkar

Aix-en-Provence (near)
Château de Fonscolombe,
Proprietor: Marquis de Saporta,
13610 Le-Puy-Sainte-Reparade, Bouches-du-Rhône.
Tel. (91) 28.60.05.

This estate is about 16 kilometres due north of Aix-en-Provence. Leave Aix on the Sisteron road, N96; after about 3 kms., fork left on D13 marked to Le-Puy-Ste.-Reparade. Look out for the estate before Le Puy after the village of Saint Canadet.

The proprietor, the Marquis de Saporta has two properties about 25 kilometres apart – this one and Domaine de la Crèmade. Both have been in the family since 1720 and both are rated V.D.Q.S. Coteaux d'Aix-en-Provence, the best possible rating of the district. Thus there has been much modernisation at Fonscolombe in recent times.

Visitors will see that while the red wines still mature for up to ten years in *foudres* of Hungarian oak, modern techniques have been introduced for the white and rosé wines. They may taste, buy and take a walk in the park, admiring both the swans and the fine façade of the château.

In Britain the wine is in the lists of Justerini & Brooks and Peter Dominic.

| *Hours:* | Monday to Friday 1430-1630. |
| *Introduction:* | Not essential but preferable. Letters if required from: Justerini & Brooks Ltd., 1 York Gate, Regent's Park, London NW1 4PU. Tel. 01–935 4446. |

---

*Joe Hollander 'Riviera Wine Road' Wine Mine Summer 1970.

Bandol (near)
Moulin des Costes,
Proprietors: M. Paul and M. Pierre Bunan,
La Cadière d'Azur, 83740 Var.
Tel. (94) 98.72.76.

Bandol is a pleasant bathing resort 17 kilometres west of Toulon. Eight kilometres inland is the hill village of La Cadière d'Azur. M. Bunan and his brother have some 30 hectares of vineyards here and at Le Castellet, the charming little fortified hill town nearby. The terraced vineyards with stone walls, the views across the plain towards the sea and, in early summer, the splashes of yellow when the abundant broom is in flower, make this a highly pleasing small wine district.

The wine is mainly rosé (A.C. Bandol) of a higher quality than most Provence rosé; a good red is also made by blending the Mourvèdre grape with Grenache and Cinsault. Mas is old Provençal for 'maison' or 'house'; M. Bunan's two wines are Mas de la Rouvière and Clos du Moulin des Costes.

| *Hours:* | Monday to Friday 0900–1200; 1430–1700. |
| *Introduction:* | None necessary but a telephone call in advance is essential to M. Bunan at the number above. |

Cassis
Mas Calendal,
Proprietor: M. Jean Jacques Bodin-Bontoux,
Avenue Emile Bodin, 13260 Cassis, Bouches-du-Rhône.
Tel. (91) 01.71.62.

From the charming little resort of Cassis, on the coast 23 kilometres east of Marseilles, take the road to La Ciotat (N599). At the second crossroads from Cassis (after about 550 metres) take the left turn at signpost 'Jeannots/Super Cassis.' This is the Avenue Emile Bodin. A sign on the left before the road narrows directs to Mas Calendal.

The A.C. Cassis rosé and white wines, which are made from precipitous rocky vineyards above this fishing port, receive high praise in Alexis Lichine's *Encyclopaedia of Wines and Spirits*. Louis XIV liked them too, placing a special order for two *demi-feuillettes* in 1672. Drink them yourself – not too cold – with your *bouillabaisse*.

M. Emile Bodin, grandfather of the present owner M. Jean Jacques Bodin-Bontoux, died at a grand old age in 1969. I remember him as an authority on the Provençal poet Mistral and a horticulturalist who collected plants from all over the world, growing them here in garden and greenhouse.

In earlier days M. Bodin had been the first grower in Provence prepared to risk quality by grafting his vines on to root stock from Texas, which was then the untried safeguard against the devastating phylloxera insect. Later he gave the lead in taking up domaine bottling.

Visitors see the cellars and taste the wines. No English spoken.

Hours: Monday to Friday 1000–1200; 1400–1600.

Introduction: None necessary but it is essential to telephone in advance.

Hyères (near)
Château Montaud Pierrefeu,
Proprietor: M. F. L. Ravel,
83390 Cuers, Var.
Tel. (94) 28.20.30.

Michelin map 84. This is an ancient vineyard, marked on the map 4 kilometres east of Pierrefeu du Var, north of Hyères in the heart of the Maures mountains between Toulon and Saint-Tropez, on the picturesque D14 road linking Hyères, Pierrefeu, Collobrières, Cogolin, La Foux and Saint-Tropez.

Since 1958 the domaine has been reorganised; the present ultra-modern vinification plant is capable of handling, treating, filtering, conditioning and bottling 2½ million litres. Visitors may taste, and buy wines in cartons of 12 or 6 bottles.

Hours: Monday to Friday 0800–1200; 1330–1600. Saturday 0800–1200.

Introduction: Open to all; none necessary.

Saint Tropez (near)
Château Minuty, S.A. Domaines Farnet,
Proprietor: M. Gabriel Farnet,
83580 Gassin, Var.
Tel. (94) 97.12.09.

Château Minuty, near Saint Tropez, is reached via N559 and D89. The present owner, M. Gabriel Farnet, took over in 1936 and runs it with Madame Malton, a married daughter. The château, a fine country mansion in which they live, is not open to visitors, but M. Farnet welcomes anybody interested to his vineyard and cellars, where he makes and keeps red, white and rosés classified Côtes de Provence.

Hours: Monday to Saturday inclusive 0900–1200; 1400–1730.

Introduction: None necessary. Agents are: Capital Wine & Travers Ltd., Central House, 32–66 High Street, London E15. Tel. 01–534 7536.

Vidauban
Domaine de Saint-Martin,
Proprietress: Comtesse de Gasquet,
Taradeau, 83460 Les Arcs, Var.
Tel. (94) 73.02.01.

The property is 50 kilometres from Saint Tropez, 30 kilometres from Fréjus and about five kilometres from Les Arcs railway station on the Paris-Ventimiglia line. From the N7 crossroads at Les Arcs take the road marked Taradeau and Lorgues. Domaine Saint Martin is three kilometres before the entry to Taradeau village.

The vineyards were originally planted by the mediaeval monks belonging to the Cistercian Order on the Lerins Islands off Cannes. Until October 1972, the proprietor was Comte Edme de Rohan-Chabot, President of the Côtes-de-Provence Winegrowers' Association. Since his death, his daughter, Comtesse de Gasquet owns and runs the estate.

Visitors are shewn the vaulted cellar dating from the sixteenth century. The wine is kept in vats carved into the rock, then in great casks of Hungarian oak. Red, white and rosé wines may be tasted and bought in various shapes and sizes of bottle up to 'cubitainers' containing eleven litres.

Hours: Monday to Saturday inclusive 0800–1800.

Introduction: None necessary.

Domaine des Féraud,
Proprietor: M. E. A. Laudon,
83550 Vidauban, Var.
Tel. (94) 73.03.12.

Vidauban is on the N7 between Aix-en-Provence and Nice. To reach it from the A8 motorway (Esterel-Côte d'Azur) from Nice, Antibes or Cannes, leave the motorway at the Cannet-de-Maures exit. From Vidauban take the D48 road to Grimaud. The property is about 4 kilometres from Vidauban.

The vineyard was first planted in the last century by Emile Rival, grandfather of the present owner, M. Laudon. Later it was run by Paul Rival, proprietor of Château Guiraud, Premier cru of Sauternes, who planted better vines – Syrah, Mourvèdre and, above all, Sémillon and Cabernet Sauvignon from Bordeaux. Blanc de Blancs wines are still made from Sémillon grapes planted in 1927.

Red, white and rosé wines may be tasted and bought in the cellars, rebuilt in 1968.

Introduction: The domaine is open to all every day at normal times. No introduction necessary but the proprietor does ask for prior notice of parties, say, over ten people.

Antibes
Domaines Ott,
22 Boulevard d'Aiguillon, 06601 Antibes, Alpes-Maritimes.
Tel. (93) 34.08.91 or 34.38.91.

Marcel Ott, an expatriate of Alsace, founded this company in 1896. Believing that good and natural table wines could be produced in Provence, he has certainly proved his point, with the help of his sons and grandsons, first at Château de Selle, then at Château Clos Mireille and more recently at Château Romassan (Bandol A.C.), properties which can be visited by appointment made through Domaines Ott or their agents. Domaines Ott's rouge, rosé and Blanc de Blancs wines are now exported to some sixty countries.

The offices (address above) are at Antibes, the fashionable resort between Nice and Cannes, which has a charming Picasso ceramic museum and, nearby at Villeneuve-Loubet, a rather different museum – almost a Mecca for gastronomes – in the house where Escoffier, the great chef, was born.

Visitors are received at the following properties, all arrangements being made with the office in Antibes through the London agents.

Bandol (near)
Château Romassan,
Le Castellet, Var.
Tel. (94) 98.71.91.

Michelin map 84. This property is about 8 kilometres inland from Bandol on D66, between La Cadière and Le Beausset. From Bandol take N559, turn left on D82 and right on to D66. Romassan is then a kilometre beyond on the right.

The Domaine makes elegant rosé and full-bodied red wine, A.C. Bandol, and is in the middle of some of the best rosé wine country of Provence. There is a double-vaulted cellar that dates back to the Middle Ages.

Le Lavandou (near)
Château Clos Mireille,
La Londe-les-Maures, Var.
Tel. (94) 66.80.26.

Michelin map 84. This vineyard, making excellent Blanc de Blancs, is almost on the coast, west of Cap Bénat. Take N98 from Hyères for ten kilometres towards Le Lavandou. Turn right. D42 bis, at La Londe-les-Maures. A fork left after a kilometre follows the coast line eastwards to Léoube and Brégancon. Clos Mireille is on the right about 2 kilometres short of Brégancon and 5–6 kilometres from La Londe-les-Maures.

Vidauban
Château de Selle,
Taradeau, Var.
Tel. (94) 73.02.02.

Michelin map 84. From Aix-en-Provence transfer from the autoroute A8 to N7 at the Le Luc exit. Near Vidauban take D73 to Taradeau, continuing on D73 to Selle, about three kilometres beyond Taradeau.

In the eighteenth century, the château was the autumn residence of the Comtes de Provence. The vineyard is planted on rocks broken into fragments and makes worthy rosé and good red wines, which lie in fascinating oak casks.

Tastings and purchases can be arranged at all three domaines. In London the wines are usually on sale at Harrods, Fortnum and Mason, and André Simon Wines Ltd.

Hours: Monday to Friday 0930–1130; 1430–1700.
Introduction: By appointment. Letters from agents: Mentzendorff & Co. Ltd., Asphalte House, Palace Street, London SW1E 5HG.
Tel. 01-834 9561.

LANGUEDOC-ROUSSILLON

Introduction

GETTING THERE
By road: A guide book is no place for a digression but the route to the South of France recalls Leonard Woolf's 'discovery' of the Massif Central.* It was March 1928; he and Virginia, having motored out to Cassis down the Rhône Valley, mapped out a pleasant return route through the centre of France.

*Leonard Woolf *Downhill all the Way: An Autobiography of the Years 1919–1939.*

65

LANGUEDOC/ROUSSILLON

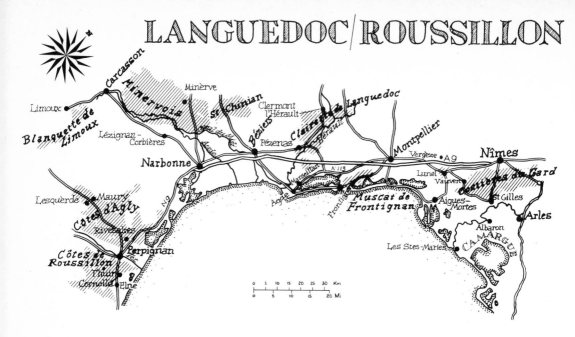

'Being fairly ignorant of geography', he wrote, 'I was completely unaware that the centre of France consisted of the Massif Central, the very formidable mountains of the Cévennes and Auvergne.'

In March 1928 the result was a blacker and blacker sky, as they climbed and climbed into a tremendous snow storm, with punctures every 25 miles for the last 500 miles to Dieppe.

Motoring today in the Massif Central, except in winter, presents no such hazards but it does become really slow going (30 m.p.h. average) south of Châteauroux. Otherwise, with time to spare, a route on the minor roads crossing the Indre, Creuse, Dordogne, Truyère, Lot and Tarn can be chosen to emerge where and when one wishes between Nîmes and Narbonne. And for me, France has few greater pleasures.

The fast route is the autoroute A6, Paris to Narbonne.

By rail: Narbonne and Perpignan are principal stops on the Paris-Toulouse-Barcelona line, with connections from London (Victoria).

Car sleeper trains run direct from Boulogne to Narbonne weekly in summer and sometimes weekly from Paris to Narbonne.

By air: Scheduled flights by Air France include Paris-Toulouse and Paris-Marseille.

By canal: The Canal du Midi runs 190 kilometres (with 43 locks, a small number as they go) from Port de l'Embouchure at Toulouse south-east through the Minervois to the Bassin de Than, close to Marseillan. It is also linked with Bordeaux via the Canal Lateral à la Garonne and the Garonne itself. Cabin cruisers may be hired through Blue Line Cruises, 27a Clifton Road, Rugby, Warwickshire CV21 3PY. Tel. (0788) 75029 or through Cox and Kings, Vulcan House, 46 Marshall Street, London W1V 2PA. Tel. 01–734 8291.

Water Wanderer, a 74-foot British barge, converted to accommodate up to sixteen people, with all 'mod cons' and a station wagon to take them out to dinner locally, covers Languedoc-Roussillon from Sète to Toulouse in a series of different weekly cruises from April to November. Details from Continental Waterways, 22 Hans Place, London SW1. Tel. 01–584 6773 or other travel agents.

CHANGES FOR THE BETTER

All sorts of changes have been taking place quietly in the vast Midi vineyard, where vines seem to be everywhere west of the Rhône, from Nîmes to Banyuls on the Spanish frontier. This has traditionally been the region supplying the French – man, woman and child – with the daily litre of *ordinaire,* essential to the maintenence of national morale. There never was enough and it used to be blended with cheap North African wine. But when the French colonies became independent, many displaced settlers resettled here, persuading the Government to protect them by restricting imports of cheap North African wine.

This in turn revealed just what poor stuff the Midi Vin

Ordinaire was *without* the North African, and led to general French disenchantment with their own low quality of Vin Ordinaire.

Slowly, for two decades now, replanting with better varieties of wine grapes, principally Carignan, Grenache and Cinsault has been going on. Experienced vignerons, resettled from Tunis, Algeria and Morocco, aided by Government support and modern methods of vinification and quality control, now make red, white and some rosé wines far better than any Languedoc-Roussillon produced before. Many of them are rated V.D.Q.S. (*Vins Délimités de Qualité Supérieure*) a class inferior to A.O.C. (*Appellation d'Origine Contrôlée*, usually A.C. for short) but infinitely superior to V.C.C. (*Vins de Consommation Courante*) now generally known in the West (and in Australia where it originated) as 'plonk'. It takes a good five years for new vineyards to bear a good crop and it wasn't until about 1970 that the 'new' Languedoc-Roussillon output became plentiful. With Bordeaux and Burgundy prices rocketing, the door to the European market was opened to the Midi.

One big company, Salins du Midi, has 'Vins des Sables' vineyards in the sandy soil spread about the shores of the Gulf of Lions where the sea has receded. They are good wines – red, white and rosé – justifying their claim that the intruding Bacchus has scored a victory over the sombre regions of Neptune.

SWEET WINES

South of Nîmes, Costières du Gard is a rectangular region, 15 miles by 10, where the red wines are the best. Further west, by the coast, there is nothing new about Muscat de Frontignan, a sweet natural wine that attains the legal minimum of 15° without fortification; the vineyards date back to the returning Crusaders, perhaps – though nobody knows – to the Romans. Frontignan, the town, is an unattractive spot yet hospitable; the concentration of its *bars dégustation* looks even greater than that of public houses along Portsmouth's Hard.

These strong, sweet dessert wines bob up again in several isolated spots; at Lunel, half-way up between Nîmes and Montpellier, at St. Jean-de-Minervois, at Rivesaltes and Maury, north-west of Perpignan and lastly at Banyuls almost into Spain. Old sweet wines, particularly at Banyuls, acquire a pungent burnt taste, known as *rancio*, a term also applied to the heavy white wine, Clairette du Languedoc, after it has been aged for three years.

STANDARD FOR V.D.Q.S.

The name Languedoc dates from the thirteenth century. Everywhere else in France the word for 'Yes' was *Oil* (which became oui). In the Languedoc they said 'Oc' for 'Yes' and so it became the province of the *langue* (tongue or language) of the 'Oc'.

Typical of the new vineyards is that of St. Martin de Toques, lying at the foot of a ruined castle perched on a hill south-west of Narbonne, this dominating position making it a commanding fortress, a Montecassino in miniature, in the countless wars fought over the centuries by Roman, Visigoth, Arab and Berber in turn. The laws permit M. Villeneuve, retired architect and now proprietor, to make 1500 hectolitres of red V.D.Q.S. Corbières from his 65 hectares, newly planted ten years with 55 per cent Carignan, 25 per cent Cinsault, 10 per cent Grenache and 10 per cent divers species. Of more interest to drinkers than the varieties is that samples have to be sent to local tasting commissions, which decide whether wines are up to the strength and general standard decreed for V.D.Q.S..

Being a land of small vineyard proprietors, Wine Co-operatives are intermingled with private owners. The purpose of the Co-operative is to save its members the expense, the time and the risk of making wines individually. The Co-operative buys their grapes, makes all the wine, blends as necessary – the whole business of vinification taking place at a modern central winery staffed by properly trained specialists. These wineries are rather clinical but they are open to all without formalities and some may take (paid) volunteers for picking at the vintage in September and October.

AUTUMN BEAUTY

Turning inland and northwards, almost anywhere from Nîmes to Perpignan, the *étangs* and the flat plain soon give way to a lovely landscape of hills, valleys and dried up rivers. In late October it becomes a patchwork picture of light red, dark red, brilliant orange, yellow and golden shades, with an occasional patch still bright green. The vintage is over; different species of vine are in varying stages of preparation prior to shedding their leaves and taking the brief winter rest that is vital to them all.

Late autumn is a good time for a holiday in these parts, particularly in Roussillon, which begins a dozen miles north of Perpignan and extends to the Pyrenees. Prices are a little less in this south-east corner, the holiday rush to Spain is over, with luck, clear skies and a warm midday sun still linger. Either Perpignan or Narbonne are good centres for exploring this wine land of the rivers Aude and Hérault. The old walled city of Carcassonne will be deserted; on the way there from Narbonne, after tasting at Valar's new centre at Lézignan-Corbières, a deviation can be made through the hilly Minervois.

This rectangular region, about 25 miles by 15, is reputed to make the best red wines of the Midi and good value they certainly are for comparatively little money. Good value too is a look at the little walled town of Minerve, mounted in a setting of vine, cypress and olive that only Tuscany can surpass.

Bordeaux prices have now fallen very considerably, but the red wines all take at least three years before they are attractive to drink. Those of Languedoc-Roussillon improve very little in the bottle and are ready much sooner. Costing appreciably less, the local growers' injunction to 'Drink *our* wines and keep the Bordeaux's for ageing' makes good sense.

Further Information
Conseil Interprofessionnel des Vins de Corbières,
55 Avenue G. Clémenceau
11200 Lézignan-Corbières
Aude

CAMARGUE
Aigues-Mortes
Domaine de Jarras-Listel,
30220 Aigues-Mortes, Gard.
Tel. (66) 38.32.32.

Jarras-Listel is one of several Domaines belonging to Domaines Viticoles des Salins du Midi, a big company dealing in salt and wine. From Aigues-Mortes take N579 towards Le Grau-du-Roi; three kilometres from Aigues-Mortes on the left, a long tree-lined drive with signboard 'Domaine de Jarras-Listel' leads to the chai.

Extensive replantings in the sandy coastal soil around the Gulf of Lions have produced some very pleasant 'Vin des Sables' wines and the chai here at Listel, with sixty-four giant *foudres*, is impressive. Their prices however are a little high, which explains why the wines may be difficult to find in Britain until better times return.

Red, white and rosé wines, tastings and purchases can all be laid on in the tasting room, but the small staff working here are unlikely to speak English and any 'red carpet' treatment would have to be arranged with the head office in Montpellier.

Hours: Summer: Every day 0900–1200; 1400–1800. Winter: Monday to Friday 0900–1200; 1400–1700.
Introduction: Not essential. For any special arrangements (e.g. a large party) communications should be sent in good time to the head office:
Domaine Viticoles des Salins du Midi, 68 Cours Gambetta, 34000 Montpellier.
Tel. (67) 92.48.92.

LANGUEDOC
Lézignan-Corbières
Valar Diffusion S.A.,
Avenue Général de Gaulle, 11200 Lézignan-Corbières, Aude.
Tel. (64) 27.00.07.

Close to the level crossing, east of Lézignan-Corbières, a small town between Carcassonne and Narbonne, there is a new wine centre with a display and tasting of wines of the entire region. Valar Diffusion, the organisers, are a fairly new group, composed of many local growers and shippers, marketing wines from most parts of the Midi.

Visitors may buy at the centre, where an English-speaking hostess will advise them on pleasant places to see in this little known but charming region, stretching northwards through the Minervois.

Hours: Monday to Friday 0900–1200; 1400–1800.
Introduction: Not essential. Firms in Britain shipping Valar wines include Justerini and Brooks, Peter Dominic and Charles Kinloch. Any special arrangements could be made by way of them through Valar's U.K. representative at 52 Prince of Wales Drive, London SW11. Tel. 01–228 3397.

Marseillan
Noilly Prat & Cie.,
34340 Marseillan, Hérault.
Tel. (67) 77.20.15.

A glance at the map (*Michelin* 83) shows Marseillan east of Béziers, on the Bassin de Thau seven kilometres along D51 from Agde. The visit is to the production plant for Noilly Prat vermouth.

Noilly Prat was founded in 1813 by Louis Noilly and Claudius Prat in Lyons. Shortly afterwards they moved to Marseille and were the original *dry* Vermouth manufacturers. Their vermouth was also used to mix the first Dry Martini cocktail in America about 1860. Martini & Rossi were not exporting a dry vermouth to the States then, so Noilly Prat cannot have been too amused when the name of the inventor of the cocktail turned out to be Martinez.

Visitors see the wines maturing in vats and where the maceration with herbs, etc. takes place. Interesting too are the open air enclosures, where these wines mature under the Midi sun, a costly process giving Noilly unique distinction. English-speaking hostesses take parties round throughout the summer.

Hours: Monday to Friday. 1 July–30 September 0930–1130; 1430–1730. 1 October–30 June 1000–1100; 1430–1700.
Introduction: None necessary. Martell own Noilly Prat and their U.K. agents are: Matthew Clark & Sons Ltd., 183 Central Street, London EC1V 8DR. Tel. 01–253 7646.

The following are recommended as other places of interest connected with wine and in many cases for sampling. They are likely to be open from about 0900–1200 and 1400–1900 in summer, often seven days a week.

A Corbières leaflet, obtainable free from Syndicat d'Initiative offices, has an admirable map showing cellars (both those of Co-opératives and private companies) where wines may be sampled without charge.

Béziers
Musée du Vieux Biterrois et du Vin has special exhibits related to wine.

Limoux
23 kilometres south of Carcassonne, Limoux gives its name to the excellent sparkling wine Blanquette de Limoux. The wine is more attractive than the town and may be tried at the Cave Co-opérative.

Montpellier
The Ecole d'Oenologie, the Ecole de Viticulture and the Ecole Nationale Supérieure Agronomique train aspiring vignerons. They are not open to ordinary visitors. Montpellier has a Wine Fair in October.

Thuir
The Company Dubonnet Cinzano Byrrh have vast installations here where visitors are welcome. Thuir is about 15 kilometres south-west of Perpignan.

Vergèze
The position of Source Perrier, off the N113 main road between Nîmes and Montpellier, is clearly marked on *Michelin* map 83. It was during the nineteenth century that a certain Dr. Perrier rediscovered this spring of water that sparkles. He sold it to John Harmsworth, brother of the Lord Northcliffe to be, who promised the doctor that the water would always bear his name. Perrier welcome visitors and get plenty of them.

ARMAGNAC

Introduction

If you wish to lose yourself in France, far from the madding crowd, where the cars are few and the supermarket has barely arrived, proceed south-west from Bordeaux to the country of d'Artagnan. 'Rolling, wooded country, charming small towns and villages, and a plentiful assortment of old buildings, religious and secular', summarised Edmund Penning-Rowsell writing on the district in *Country Life*. Condom, on the river Adour, which carried the local brandy to Bayonne and thence to England in the Black Prince's time, is charming and still the commercial centre of the trade.

This region of France's second brandy is reached in about two hours by car from Bordeaux, Toulouse or Biarritz, all of which have airports, with good services from London and Paris, at which self-drive cars can be arranged. The nearest good rail connection is at Agen, 25 miles from Condom.

The region, demarcated in 1909, is divided into three parts: Haut-Armagnac, the high hilly ground on the eastern side which makes the lowest quality Armagnac; Bas-Armagnac, on the flat low land to the westward extending from the département of Gers into the Landes, which

makes the best Armagnac; and – between the two both geographically and for the quality of the spirit – Ténarèze.

Condom and the comparatively uninteresting town of Vic-Fezensac are in Ténarèze; Eauze, where on Thursdays the buyers and sellers of the Trade form a market in the streets, is in Bas-Armagnac.

Whereas cognac is distilled twice in a pot still, Armagnac is distilled once in an *alambic armagacais,* an apparatus more like the patent still used for large-scale production of grain whisky in Scotland. The strength is lower – only about 80° proof – and it matures in wood faster, the time being from three to twenty-five years. The grapes are mainly Folle Blanche and Picpoule; half the production, which is about a tenth that of Cognac, being the work of the Union of Co-operatives based at Eauze. As in Cognac, brands of Armagnac exported are likely to be blends from the three regions.

Dry, pungent, full of flavour, Armagnac has escaped the attentions of 'developers' so far, but there are signs that the long arms of the big Cognac firms are stretching south-eastwards.

At one time in the Middle Ages, the 'Armagnacs', led

by their Count, were the leading political party in France, but for their greatest heroes we have to wait until the seventeenth century and for Dumas to immortalise them later as *The Three Musketeers*. In 1976 *La Compagnie des Mousquetaires de l'Armagnac*, (the equivalent of Burgundy's *Chevaliers du Tastevin)*, which boasts an authentic of descendant of the d'Artagnan family, celebrates its twenty-fifth birthday. Let us hope the horticulturally minded poet, who described this great Gascon distillation as: 'The scent of wood violets softly exhaling through the mingled aromas of ripe quinces, greengages and burnt hazel nuts' will be in even more dazzling form!

By no means a gastronomic desert this provincial neighbour of Perigord, the foie gras is excellent and the Hotel France at Auch bears two gastronomic stars in the *Michelin*. Darroze at Villeneuve-de-Marsan is another *relais* with an exceptionally fine cellar of old vintages; an Arab gentleman friend may be advisable to foot the bill.

For a couple of nights at Condom I have been very comfortable at the modest Hotel Continental facing the river and eaten magnificently in the tasteful *Salle gothique* of the Table des Cordeliers restaurant given two gastronomic stars by *Michelin* which means 'Excellent cooking, worthy of a détour'.

After the winter rains, before becoming burnt in summer, the undulating countryside has a fresh, southern English look. In the Hundred Years War it was a battle ground, in which some of the bastides, strong points set up by both sides, are still worth seeing. At Fourcès a river winds round the fortified castle as if it were a moat. Montreal has another bastide, while that at Larresingle, associated with the Black Prince, has been restored by an Armagnac merchant and, if close to Condom, should be visited.

The Bureau Interprofessionnel de l'Armagnac at Eauze arranges visits to merchants in Condom, Eauze or Vic-Fezensac. In Britain, Clos des Ducs, Marquis de Caussade, Marquis de Montesquiou and Malliac are the best known brands. I myself recall a pleasant day's instruction on the spot with Monsieur Goux of the *Société Fermière de Malliac*, but I have no doubt visitors will be in equally good hands with Monsieur Michel Janneau.

Condom (near)
Château de Malliac,
32250 Montreal, Gers.
Tel. Montreal 87 (office hours only).

This handsome country mansion is on a hill, off the road (D15) from Condom to Montreal, 2.5 km. from the latter. Until 1973 the *Société Fermière de Malliac* had headquarters in Vic-Fezensac where a M. Corneille had started it in 1855. About 1900, the Bertholon family, who owned this château, succeeded and in 1973, after the sales had increased very rapidly, more ageing and administration space was needed. Very wisely they made their Château de Malliac the new headquarters.

Visitors will learn a lot about Armagnac brandy including the mark of different ages – V.S.O.P., Hors d'Age and Millésimes.

Hours:	Monday to Friday 0900–1200; 1400–1700.	
Introduction:	By appointment. Letter obtainable from agents: Deinhard & Co. Ltd., 29 Addington Street, London SE1 7XT. Tel. 01–261 1111.	

Condom
Janneau Fils S.A.,
B.P. No. 55, 32100 Condom, Gers.
Tel. (62) 28.15.15.

This establishment, among the oldest and prettiest in the region, lies north of Condom on the road to Nérac (on the left going north), facing the river Baise.

Janneau began in 1851 and is now among the first four shippers of Armagnac and beginning to expand into international markets. Visitors will see the maturing and blending warehouses, the bottling plant and a very ancient disused still, which can show how the Armagnac method of distilling differs from others.

Michel Janneau, the only English speaker, is often away on business. Nevertheless, visitors are welcome to take 'tot luck' to coin a phrase.

Hours:	Monday to Friday 1000–1200; 1500–1800.	
Introduction:	Letter of introduction preferred from agents: Matthew Clark & Sons Ltd., 183 Central Street, London EC1V 8DR. Tel. 01–253 7646.	

DORDOGNE

Introduction

GETTING THERE

By road: The N20 runs south-south-west from Paris to Orleans, Châteauroux and Limoges, continuing from there as N21 through Périgueux and Bergerac, ending eventually at the Spanish frontier, near Lourdes. From Châteauroux to Bergerac, on the Dordogne sixty miles east of Bordeaux, is slow going through hilly country.

By rail: Périgueux, thirty miles away, is the nearest rail head, providing connections to and from Paris, via Limoges.

By air: There are good scheduled services from Paris to Bordeaux (Mérignac Airport) and some direct flights from London airports, Heathrow and Gatwick.

Overshadowed by Bordeaux, Bergerac's wines – except for Monbazillac – were largely unknown outside the locality until recently. Attention to their merits was drawn largely by the absurd increases in price of fine Bordeaux wines between 1970 and 1974; and although Bordeaux prices have now fallen, Bergerac wines have probably come to stay, so long as their prices remain much lower.

Monbazillac, named from a village five miles south of the town, is a sweet white wine made in the manner of Sauternes. And because true Sauternes is scarce (the yield is restricted to about 300 bottles to the acre to ensure quality and the district is only 7 miles by 4), Monbazillac is a useful substitute at a much lower price. The output may approximate to double that of true Sauternes and the price may be half. Nobody claims that Monbazillac is as good as Sauternes but it is a very good sweet wine, with its own flavour from the Dordogne soil.

Montravel is a white wine, usually dry; Rosette is another white wine, usually dry; Pécharmant is the best red wine. Most of the production of the region (some fifty million bottles) is 'Appellation Bergerac', made mainly from the same vine species as are found in the Bordeaux vineyards.

The *Michelin Green Guide* of the Dordogne (English edition) recommends a drive through the vineyards leaving the town on D13. This is the way to the Château de Monbazillac. From there take D14 bis westwards, then D107, then a local road (V.O.) leading to Château de Bridoire, returning to Bergerac on N133, which gives a good view of the vineyards.

TOBACCO MUSEUM

Bergerac, centre of the Dordogne's appreciable tobacco industry, has the only experimental Tobacco Institute and, on the second floor of the Town Hall, the only tobacco museum in France. The exhibition there covers the history of tobacco, its influence on society, the cultivation of the plant and other aspects.

Hours:	Monday to Saturday 1400–1900. Sunday 1500–1800.
Admission:	0.60F.

Bergerac (near)

Château de Monbazillac, administered by:
Unidor – Union des Co-opératives Vinicoles de la Dordogne, B.P. No. 1, 24240 Monbazillac, Sigoules. Tel. (53) 57.34.36.

Unidor is the central organisation of ten large co-operative wineries about the Bergerac region, 90 kilometres east of Bordeaux, the largest of them being that at Monbazillac. From Bergerac, taking N133 towards Eymet, the Co-operative is five kilometres and the château six. Unidor handles about 40 per cent. of the region's A.C. wine production – the sweet white Monbazillac, a dry white Sauvignon and a Bergerac Rouge, all sound medium-priced wines.

Visitors should not, however, go to the Co-operative but to the sixteenth century Château of Monbazillac, which Unidor bought and restored. In an attractive setting, the Château is a real regional wine tasting centre, with museums (one of wine of course!) and a restaurant.

Hours:	Daily 0800–1200; 1400–1730.
Admission:	3.00F.
Introduction:	None necessary. One good customer of the Union des Co-opératives is Peter Dominic and any further information could be obtained from them at Vintner House, River Way, Templefields, Harlow, Essex CM20 2EA.

BORDEAUX

> Sou'-West by South – and South by West –
> On every vine appear
> Those four first cautious leaves that test
> The temper of the year;
> The dust is white at Angoulême,
> The sun is warm at Blaye;
> And Twenty takes to Bourg-Madame
> And Ten is for Hendaye
>
>   – Kipling

Introduction

GETTING THERE

By road: The 350 miles along N10 from Paris to Bordeaux, via Tours, are not exactly those to choose for a quiet journey through the beloved land. Fortunately, completion of the A10 autoroute from Paris to Tours has provided a safer and quieter way for over half the journey, and on the old road there are now fewer cars and camions approaching each other at high speeds on three lanes.

By leaving the autoroute at Orleans and joining N675 to Le Blanc, or to Saint-Junien further on, the N10 can be re-joined at Poitiers or at Angoulême, a leisurely route through lovely country, avoiding the heaviest traffic.

From the Channel ports, if the Loire is crossed at Saumur, Angers or Nantes, an excellent plan is to make for Royan at the mouth of the Gironde estuary, where a large car ferry makes frequent crossings in 30 minutes to Le Verdon, at Pointe de Grave, the northern tip of the Médoc.

From June to mid-September the service in each direction is hourly from about 0600 to 2100. During the winter it becomes approximately two-hourly. Single fare is about 30.00F, for a medium-sized car and 5.00F. for each passenger.

From Le Verdon, Bordeaux is a fast seventy miles, but the object of taking the ferry is to proceed slowly to Bordeaux, visiting the châteaux along the D2 *Route des Châteaux*, as described in this section.

Since there are no hotels suitable for overnight stops in the Médoc, it is as well to catch an 0800, 0900, or 1000 car ferry from Royan in order to see the châteaux and make sure of accommodation in Bordeaux (or Arcachon) that

evening. For the previous night, La Rochelle, Saintes, Cognac – or Royan itself – are recommended.

Another way to the Médoc, avoiding Bordeaux, is to take the same company's other car ferry from Blaye to Lamarque, which is between Saint-Julien and Margaux and cuts off about fifty miles. This ten-minute crossing costs about 25 francs for a medium-sized car and driver.

Ferry timetables *(Horaire officiel)* and information are available from the company, Trans Gironde, 9 Place du Parlement, 33000 Bordeaux (56) 52.63.76, or from the ferry embarkation stations:

Royan	(46)	05.29.03
Verdon	(K56)	59.60.84
Blaye	(56)	42.13.99

Reservations – normally not necessary – may be made in advance by telephone to the appropriate embarkation station. Though fares appear high, the cost of the alternative, motoring via and through Bordeaux, must be offset in assessing them.

By rail: Bordeaux is on the main Paris-Madrid line with frequent fast trains from Paris (Gare Austerlitz) taking four hours upwards. Cars may be sent by freight train daily, driver and passengers following by passenger train.

From early July to early September a car sleeper train runs each way between Lyons and Bordeaux nightly.

A New York correspondent of the British *Wine* magazine recommended 'doing' the whole region by train. For Saint-Emilion, he travelled from Bordeaux to Libourne, on the main line to Paris, in twenty minutes, taking a fifteen minute bus ride from there. For Pomerol he again alighted at Libourne, walking three kilometres to enjoy a tour and tasting of the wines at Château Petit-Village (p.92).

For Graves, he took a seven minute journey by train, walking on for twenty minutes to Château Pape-Clément. The Barsac train took thirty minutes, giving him a walk of 1½ kilometres to Château Coutet.

By air: Regular services to Bordeaux Mérignac airport from Paris daily and from London's airports several times weekly.

By sea: I still receive enquiries as to passages in wine boats. Ten years ago they sailed regularly between the Port of London and Bordeaux taking up to half a dozen passengers. The extra staff required to look after passengers proved too costly and with Bordeaux wines now largely carried by road (either to the French Channel ports or, via the numerous Channel car ferries, to their final destinations without unloading), sea transport from Bordeaux to the United Kingdom becomes obsolete.

THE BORDEAUX TRADE

Thanks to the sea, Bordeaux wines have been shipped to Britain – to Plymouth, London and Leith and to the Irish ports of Limerick and Cork – for nine hundred years. In 1152, Henry II of England married Eleanor of Aquitaine, bringing Gascony and most of western France under English rule for the next three hundred years. English ships were built to take out wool and bring Gascon wine back from the port of Bordeaux, a wine fleet in fact which began our long mercantile story.

After 1453, when the defeat of our General Talbot by the French at Castillon, near Saint-Emilion, ended our occupation, the trade continued. Clairette, or claret as it became, was far too good to be given up.

This trade, which had grown apace in the eighteenth and nineteenth centuries, was never wholly interrupted until Hitler occupied France in 1940. Expanding again since the war, the Bordeaux region now makes more fine wine than any other district on earth and half of it is claret.

Claret, or red Bordeaux, comes only from certain districts of the Gironde, which the French laws of Appellation Contrôlée carefully define. The châteaux or properties making it must exceed two thousand. It is undoubtedly the greatest wine in the world; Burgundy may match it for quality, but not for quantity. In Sauternes, Bordeaux can also claim the world's best sweet white table wine, as well as many less sweet from Graves, Entre-Deux-Mers and lesser districts, wines that make the dreariest fish course more palatable.

For claret, the four principal districts (in clock-wise order once more) are: Médoc, Pomerol, Saint-Emilion and Graves, their different soils giving different characteristics to their wines, although the species of grapes are (to a large extent) the same. To taste a glass of claret and identify its age and which district it comes from is an amusing exercise, requiring years of wine-drinking practice, which should be exercised with discretion. 'Not a bad wine but a bit hard – still needs another few years in bottle' is just a conversational opening gambit when one's host is a fellow wine merchant. But when the scene is a private dinner party with the hostess Ottolie, Lady Lapper-Litre, lavishing her last two bottles of a precious 1959 on her guests, such frank appraisal is apt to be misunderstood.

For visitors, April to October are the best months, excepting mid-September to mid-October when the grapes are being gathered and the vignerons have no time for anything of lesser importance. Rain usually comes in November and December, which is when the vine needs it; a great vintage requires only light summer rain, with no storms and above all no hail, no late spring frosts and a generous measure of summer sunshine.

TIME IN CASK

The grapes are fermented in the chai (above-ground cellar) of each château, the new wine remaining in cask for two years, being racked (moved) into new casks and fined (made finer) at intervals to eliminate impurities. In these chais the wines of the last two vintages lie, their casks in serried ranks from which samples are drawn to taste. The wine is rough and purple; in the trade we sample and spit out on the earth floor, though this is almost desecration in these places, arranged and kept with such good taste and in such good order that they might be termed 'Temples of Bacchus'.

Bottling usually takes place in the spring and early summer of the second year (i.e. the 1976 vintage in 1978), some of the wine being château-bottled on the spot, the rest having been sold in cask for the buyer to bottle. At the greatest properties, Château Mouton-Rothschild for example, all the wine good enough to bear the name is château-bottled and then each bottle is allotted its own serial number, which is printed on its label.

A VARIED STORY

The claret story is one of variations. Though only certain species of grapes may be grown, the proportions of each may vary from one property to another, just as the owners' skills as vignerons may vary. Variations of site and soil cause wines – even of adjoining vineyards – to differ, one being markedly superior to the other. The weather brings even greater variations. There are wonderful years like 1970, when sun and rain in the right proportions at the right times bring about an abundant vintage of the highest quality. And there are disappointing years, like 1975, when hail (the vigneron's worst enemy) destroyed much of the crop in sudden storms and rain finally pelted down at the vintage in some places.

As to the properties, these vary from great classical edifices – stately homes of big farming and agricultural estates of which the vineyard is but a part – to small houses, almost cottages, surrounded by an acre or two of vines in which, as like as not, a figure in blue overalls and a beret can be seen, hoe in hand, indicating that the proprietor is present and at work. Few of these will be found

74

BORDEAUX

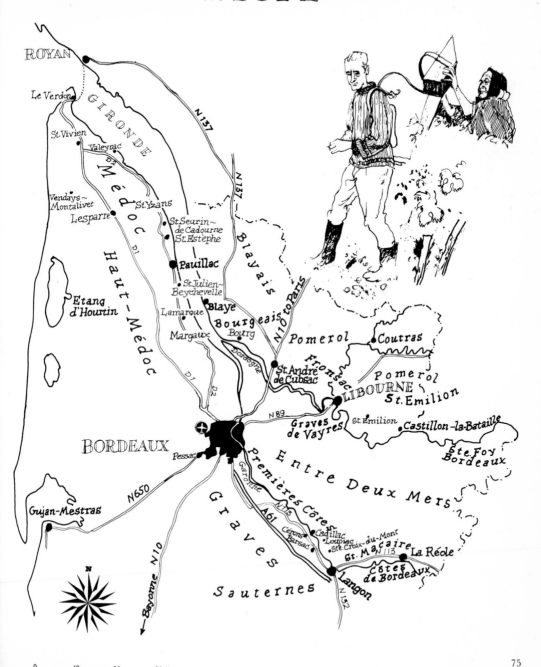

ROYAN

Le Verdon

G I R O N D E

St.Vivien

Valeyrac

Médoc

N137

N137 to Paris

Vendays~
Montalivet

Lesparre

St.Yzans

St.Seurin~
de Cadourne
St.Estèphe

B l a y a i s

Haut-Médoc

Etang
d'Hourtin

Pauillac

St.Julien-
Beychevelle

Lamarque

Blaye

Margaux

Bourg

Bourgeais

N10 to Paris

Pomerol

Coutras

D1

D2

Dordogne

St.André
de Cubsac

Fronsac

Pomerol

LIBOURNE

St.Emilion

St.Emilion

Castillon-la-Bataille

N89

Graves
de Vayres

BORDEAUX

Pessac

Garonne

E n t r e   D e u x   M e r s

Ste.Foy
Bordeaux

Gujan-Mestras

N650

N10

Bayonne

Premières Côtes

A61

N113

G r a v e s

Cérons
Barsac

Cadillac

Loupiac

Ste.Croix-du-Mont

Gr. Maçaire

La Réole

Côtes
de Bordeaux

Langon

N132

S a u t e r n e s

N

0   10   20   30 Km

0   5   10   20   Mi

75

in this Guide; their work is their livelihood and I would not thrust visitors upon them, though they are usually good for a chat about wine with any passing enthusiast who can speak French.

The owners of the big properties are now mainly companies; some, like Barton and Guestier, of long standing in the Bordeaux wine trade. Individual owners have of course formed limited companies to run their estates but it is remarkable in these times how so many family connections still survive.

THE CLASSIFICATIONS

Médoc: Of Bordeaux's two thousand claret-making properties, some have consistently made finer wines than others. Consequently, there have been attempts at compiling orders of merit from time to time. The first of these was in 1855 when the sixty-two best wines of the Médoc were put into five classes, giving rise to the terms *1er, 2me, 3me, 4me* and *5me cru classé.* Though there would be changes if a new classification were ever agreed, these terms are still in general use. At the same time, in 1855, a classification of Sauternes, resulted in nine *premiers crus* and eleven *deuxièmes* crus, Château Yquem being classed as *'cru supérieur'* above them all.

One other red wine came into the 1855 reckoning – Château Haut-Brion, the *premier cru* of Graves, was included with the three *premiers crus* of the Médoc – a number increased in 1973, amid general acclaim, with the promotion of Château Mouton-Rothschild.

Saint-Emilion: The occasion in 1855 was a great Exhibition in Paris. Since the wines of Saint-Emilion and Pomerol were unfashionable then and comparatively unknown, they were not represented. Not until 1955 were those of Saint-Emilion officially classified into a dozen *premiers grands crus classés* and about sixty-five *grands crus classés,* with about five additions made since.

Graves: Two years earlier, in 1953, Graves had been given official attention, thirteen châteaux (including Château Haut-Brion) being given premier status for red wines. A white wine classification, with eight premier properties, followed in 1959.

Pomerol: The district of Pomerol still remains 'without the law' but alas, lack of official recognition by no means keeps prices lower; Château Petrus, the leading *cru,* is virtually a premier in all but title.

The Bourgeoisie: Below all these in status, Bordeaux abounds with *crus bourgeois* – lesser wines at lesser prices, whose distinct merit is that they reach their best far sooner than their classified superiors. They are not *officially* classed at all.* In fact, according to an article in the *Revue Vinicole* some years ago, these lower orders are in reality just as much aristocrats as the *crus classés.* The term is derived from 'Droit de bourgeoisie', which in the Médoc gave the wines certain tax advantages over those of other districts.

The segregation in 1855 of some sixty châteaux into superior classes seems a shocking step in a country which could not have forgotten the *Egalité* of the Revolution. One's reaction is similar to that of Bernard Shaw when taken to the Moscow races in early Communist days. 'Surely the horses should all have come in *equal* first,' he said.

Below the Bourgeoisie, A.C. Bordeaux Supérieur and A.C. Bordeaux complete the claret society. They are not to be despised; crossing to New York in 1974 on almost in the last voyage of the *France,* I found the cheapest 1970 claret was quite the best of the 1970 selection; it was just a Bordeaux Supérieur, which had matured fully, whereas the 'better' wines needed more time to reach *their* best.

THE GRAPES

Médoc vineyards can only be planted with certain grapes and a typical proportion might be: Cabernet Sauvignon 55 per cent. Cabernet Franc 20 per cent., Merlot 20 per cent. and Malbec 5 per cent. Petit Verdot and Carmenère may also be found in small quantities. In the premier cru vineyards, Latour and Mouton-Rothschild for example, the proportion of Cabernet-Sauvignon can be as high as 80 per cent., possibly omitting Cabernet-Franc altogether.

The Cabernet-Sauvignon, Edmund Penning-Rowsell† describes as 'the backbone' of an elegant Médoc claret. In Saint-Emilion the Merlot is much more in evidence. For example, Château Cheval Blanc is planted with one-third Merlot, one-third Bouchet (which is a Cabernet-Franc) and one-third Malbec.

Over the years, each property has come to the proportions of whichever varieties are best suited to its particular soil and site. The life of a vine being at least twenty-five years, this is no matter for annual experiment. A proprietor cannot extend his vineyard as he fancies; for example the boundaries of the commune making claret good enough to be called even plain 'Margaux' have long since been defined and the *Institut National des Appellations d'Origine* would require to make soil tests before giving permission to plant a vineyard outside them, from which the wine could even be described as 'Margaux'.

What would claret be like made wholly from one or other of these grape species? Monsieur Dousson at Château de Pez answers this question in a practical way. For demonstration purposes he makes a little wine from each of them. On a visit there I was fascinated to taste myself his 'Cabernet-Sauvignon', 'Cabernet Franc', 'Merlot' and 'Malbec' wines. Though doubtless they

---

*Classifications are given in Cocks and Feret's *Bordeaux et ses Vins,* but none has received the approval of a national, as opposed to a local, body.

---

†Edmund Penning-Rowsell *The Wines of Bordeaux.*

fairly represented their respective grapes, each was undistinguished, so it is all the more remarkable that, when vatted in the right proportions, the result is fine claret – as great a red wine as exists on our planet.

RECOMMENDED READING
The Great Wine Châteaux of Bordeaux Hubrecht Duijker Times Books 1975.
Translated from the Dutch, details of some hundred and sixty châteaux are included in this beautiful book with numerous colour photographs and maps of the highest standard.
The Wines of Bordeaux Edmund Penning-Rowsell.
International Wine and Food Society Michael Joseph 1969.
Lafite Cyril Ray Peter Davies 1968.

Mouton-Rothschild Cyril Ray Christie's Wine Publications 1974.

Further Information
Maison du Vin, *Conseil Interprofessionel du Vin de Bordeaux*, 1 Cours du XXX Juillet, Tel. (56) 44.37.82, on the corner of the Allées de Tourny facing the Theatre welcomes visitors. Visits can be arranged and wines can be tasted there.

This official body, composed of growers, merchants, brokers and representatives of the Ministry of Agriculture, regulates the Bordeaux wine trade and is responsible for seeing that the regulations made by I.N.A.O. *(Institut National des Appellations d'Origine)* are carried out. It is also responsible for promoting the reputation of, and demand for, the A.C. wines of Bordeaux.

THE MEDOC

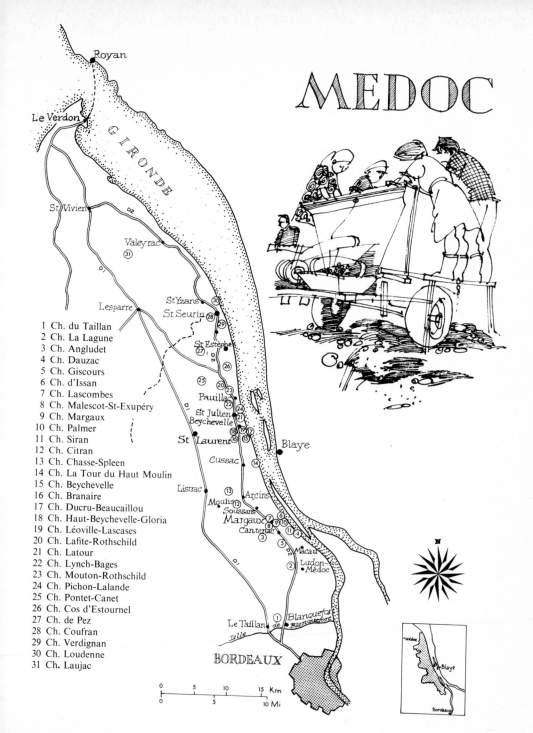

MEDOC

Royan

Le Verdon

G I R O N D E

St Vivien

Valeyrac
㉛

Lesparre

St Yzans ㉚
St Seurin ㉘
㉙

St Estèphe
㉗
㉖

㉕
⑳ ㉓
Pauillac
㉒
St Julien ㉔
Beychevelle ㉑
⑲ ⑰
⑱ ⑯ ⑮

St Laurent

Cussac

Listrac

⑬
Moulis ⑫
Soussans
Margaux ⑦ ⑥
Cantenac ⑧ ⑨ ⑩
③ ⑪ ④
⑤
Macau
②
Ludon
Médoc

Le Taillan ① Blanquefort
Jalle de Blanquefort

BORDEAUX

Blaye

Arcins

N

1 Ch. du Taillan
2 Ch. La Lagune
3 Ch. Angludet
4 Ch. Dauzac
5 Ch. Giscours
6 Ch. d'Issan
7 Ch. Lascombes
8 Ch. Malescot-St-Exupéry
9 Ch. Margaux
10 Ch. Palmer
11 Ch. Siran
12 Ch. Citran
13 Ch. Chasse-Spleen
14 Ch. La Tour du Haut Moulin
15 Ch. Beychevelle
16 Ch. Branaire
17 Ch. Ducru-Beaucaillou
18 Ch. Haut-Beychevelle-Gloria
19 Ch. Léoville-Lascases
20 Ch. Lafite-Rothschild
21 Ch. Latour
22 Ch. Lynch-Bages
23 Ch. Mouton-Rothschild
24 Ch. Pichon-Lalande
25 Ch. Pontet-Canet
26 Ch. Cos d'Estournel
27 Ch. de Pez
28 Ch. Coufran
29 Ch. Verdignan
30 Ch. Loudenne
31 Ch. Laujac

0   5   10   15 Km
0   5   10 Mi

Médoc
Blaye
Bordeaux

The Gironde – estuary of the Garonne and Dordogne – runs roughly north and south; the Médoc is the western strip, about 80 miles by 15, alongside it between Bordeaux and the Atlantic. The Médoc wine district as defined, is narrower – less than three miles wide, vineyards outside its bundaries being only entitled to call their wines 'Bordeaux'. The name Médoc is derived from *medio aquae*, 'almost in the middle of the water', but it is the vineyards *above* the water that make the best wine. Margaux, Saint-Julien, Pauillac and Saint-Estèphe, the four principal communes, are each sited above the general marshy level, where the ground is drier.

When I made my first visit in 1954, there were cheap hotels in Margaux, Pauillac and Saint-Yzans, deemed worthy of inclusion in the *Michelin Guide*. A friend recommended one of two in Margaux. My inadequate French, making a telephone call for a reservation, led me to believe that a carnival awaited us on arrival. To an Active Service naval officer, who had become devoted to wine, forming too romantic an attitude towards it, this was a really exciting prospect. What disillusionment! We found a dirty inn (so rare in France), with the 'bouquet' of a singularly unpleasant earth closet being emitted from one side of our bedroom and the noise beginning about 11 p.m. from what passsd for the local cinema on the other. It was the film that was called 'Carnaval'.

THE D2 ROUTE DES CHATEAUX
In truth, there is little romance about this flat scrubby wine land, except its general air of peace and timelessness. The villages are uninteresting, barring the small Romanesque churches in some of them. The Gironde is neither Mediterranean blue nor even Atlantic green; it is usually churned-up football pitch brown. Yet we English did own Bordeaux for three hundred years (1152–1453) and, even earlier, the building of ships to carry Bordeaux wine to London constituted the beginnings of Britain's merchant navy.

Today the traffic is still two-way – the wines mostly go overland by road and the drinkers come each summer by road, pilgrims to the source of claret, the vineyards and the fine châteaux, which they have only known hitherto from labels.

The Guide follows the D2 *Route des Châteaux* from Bordeaux. The *Green Michelin Guide* (Côte de l'Atlantique, no English edition) also recommends this, the route plan returning to Bordeaux at Saint-Estèphe, via Vertheuil (D104) and the D1 fast road through Castelnau-de-Médoc.

Those proceeding *to* Bordeaux from the Royan-Verdon car ferry, wishing to visit châteaux en route, turn hard left at Saint-Vivien, following the D2 to road to Valeyrac and Saint-Yzans, where there are signposts to Château

Loudenne. (Entry p. 89). The pink, château, flying the Union Jack and the Tricolour, makes a colourful starting point for members of the English-speaking world and the staff are delighted to arrange visits to other châteaux, advise on hotels, etc.

Further Information
Maison du Vin
Quai Ferchaud
Pauillac 33250.
Tel. (56) 59.01.91.

HAUT-MEDOC
Blanquefort
Château du Taillan,
Proprietors: Cruse et Fils Frères
33320 Le Taillan, Médoc.
Tel. (56) 35.02.29.

From Bordeaux this is the first château of the Haut-Médoc, a dozen kilometres from the city between the villages of Blanquefort and Le Taillan. The best route to the château, which can be difficult to find, is to take the D1 road out towards Lesparre. On reaching Le Taillan, turn sharp right immediately after the first traffic lights. This road, following the wall of the park, reaches the gates in about a kilometre and from there the imposing white, early eighteenth-century mansion is easily recognised.

Alternatively, from Blanquefort take the D107 to Le Taillan; when vineyards appear on the right, look for an entrance to the park on the left.

The 200-hectare estate, belonging to the Cruse family since 1896, has a racing stable as well as a vineyard. Two wines are produced: Château du Taillan, a cru bourgeois claret and Château La Dame Blanche, a white Bordeaux Blanc Supérieur.

The château is classed as a national historic monument; in the Médoc they say the back is La Dame Blanche and the front du Taillan.

Hours:	Monday to Friday 0830–1200; 1330–1700.
Introduction:	Visits should be arranged by telephone to the office of Cruse et Fils Frères, (56) 29.35.83, asking for the hostess.

Château La Lagune,
Proprietress: Mme Chayoux,
Ludon-Médoc, 33290 Blanquefort.
Tel. (56) 30.44.07.

Coming from Bordeaux, the D2 road is straight for a

dozen kilometres after Blanquefort. After about six of them, this vineyard is seen on the right and a right turn up a wooded drive leads to the château, with the village of Ludon-Médoc a kilometre further on.

La Lagune, once called Grand-La-Lagune, is the first cru classé on the Médoc *route des grands crus* when coming from Bordeaux. The château itself dates from 1730 and is finely proportioned with handsome railings; the 55-hectare vineyard is almost one piece of land with the buildings around it.

In 1961, M. Chayoux, a champagne man, took over Lagune, where the wine has always had a reputation for great staying power. Since his death in 1969 his widow has kept up the property, with M. Jean-Michel Ducellier as adviser and Madame Boyrie as régisseur on the spot. The wine deserves its high reputation and this is a well-kept (3me cru) property of unusual interest.

Hours:	Monday to Friday 1400–1800.
Introduction:	Not essential. Prior notice required for groups and from those who wish to taste.

Margaux
Château d'Angludet,
Proprietors: Mr. and Mrs. Peter Sichel,
Cantenac, 33460 Margaux.
Tel. (56) 30.30.41.

From Bordeaux, take the D2 road as far as Cantenac (25 kilometres), turn left just before the church and follow the signs.

One of the oldest and prettiest places in the Médoc, the records go back to a Chevalier d'Angludet of 1150. Changing hands frequently, the vineyard became split up and the wine did not even make the 5th class in 1855. Now owned by Mr. and Mrs. Peter Sichel, who have been living here since 1961, the grounds, with their wide expanse of paddock, have a look of English parkland. The vineyard has been extended and improved, so much so that Edmund Penning-Rowsell in *The Wines of Bordeaux* considers the wine worthy of the classed growth status it did once enjoy. At present it is rated one of the six *crus exceptionnels*, which are very superior *crus bourgeois*.

Hours:	Monday to Friday 0900–1130; 1400–1630.
Introduction:	Having obtained a letter of introduction from the U.K. agents, visitors must telephone the Sichel head office in Bordeaux (p. 106) at least 24 hours in advance to fix a convenient time when a member of the staff can meet them at the property.
U.K. Agents:	H. Sichel & Sons Ltd., 4 York Buildings, Adelphi, London WC2. Tel. 01–930 9292.

Château Dauzac,
Proprietors: Les Héritiers de M. Edouard, F. Miailhe,
Labarde, 33460 Margaux.
Tel. (56) 30.34.03.

Twenty-two kilometres from Bordeaux off the D2 road between Macau and Labarde, Dauzac adjoins the cru bourgeois, Siran. Both vineyards are within a kilometre of the estuary and the Miailhe family bought Dauzac, a 5me cru in poor shape, in 1966.

During the eighteenth century, Dauzac was the home of the Galway-exiled Lynch family, associated with Lynch-Bages. Count Lynch, Mayor of Bordeaux when Napoleon was first exiled, presented the keys of the city to the English General Beresford and was instrumental in the French restoration of the Bourbons. One plot of vines here is dated, 12 March 1814, the date of his rousing call of 'Vive Le Roi!' to the citizens. After the Lynches, the place became a vast experimental estate under the Johnston family, who owned Ducru-Beaucaillou. It had its own railway and, here too, Bordeaux Mixture, familiar to gardeners everywhere, is said to have been developed.

The Dauzac estate comprises over 120 hectares and as new plantings begin to bear well during the seventies, a big advance in the wine's quality is expected. The château, painted white, with very light blue shutters, is attractively restored.

Hours:	Monday to Friday 0930–1200; 1400–1630.
Introduction:	By appointment obtained through the administrative office: Les Héritiers de M. Edouard F. Miailhe, 6 Quai Louis XVIII, 33000 Bordeaux. Tel. (56) 48.35.01.

Château Giscours,
Proprietor: M. Nicolas Tari,
Labarde, 33460 Margaux.
Tel. (56) 30.34.02.

Labarde has two crus classés, Giscours and Dauzac, separated by about two kilometres, Giscours being inland and Dauzac close to the estuary. Coming from Bordeaux on the D2 road, the château becomes visible to the left about 1.4 kilometres (say a mile) before the village of Labarde and a Giscours signpost points clearly to the left turn.

The first records of Giscours go back to 1330 and its wine was drunk by Louis XIV in 1656. Thereafter, there have been too many owners – good and bad. The good include a banker Comte de Pescatore (1847–1875), who built the present Renaissance-style château, Edouard Cruse (1875–1913) and Nicolas Tari, the experienced Bordeaux vigneron, who took over a sadly neglected 3me cru in 1952.

Nicolas Tari and his son Pierre, presently Mayor of Labarde, have extended the vineyard to 75 hectares, converted some of the outbuildings into a reception centre, with kitchens and banqueting rooms, and modernised the chai. The result has been a delightful 1962 and a better 1966.

These and subsequent vintages have impressed Gilbey S.A. to the extent of buying the entire 1972 vintage for purposes of world distribution.

Visitors see the vineyard and associated buildings, taste from the cask and can buy some of the wine.

Hours:	Monday to Friday 0900–1200; 1400–1600.
Introduction:	Casual visitors can be accepted but 24/48 hours' warning is preferred. For groups, a week's warning is essential. Arrangement direct with Château or through Gilbey S.A., Château Loudenne, Saint-Yzans-de-Médoc, 33340. Lesparre-Médoc. Tel. (56) 41.15.03.

Château d'Issan,

Proprietors: Cruse et Fils Frères,
Cantenac, 33460 Margaux.
Tel. (56) 30.30.72.

'A seventeenth century fortified manor within the complete moat of a château fort,' says Hugh Johnson*, declaring Issan to be about the most beautiful house in the Médoc. It lies in the flat marshy land, close to Palmer, on the right off the D2 road approaching Margaux village from Bordeaux. The vineyard of this 3me cru, being a little higher, slopes gently and is well sited.

Among the oldest wine-making properties in the Médoc, there was fighting here in 1451, the English troops retreating for home (with the wine, they say) in the general rout of the English from the region. *Regum mensis arisque deorum* 'For the table of Kings and the altar of Gods' is engraved in stone above the entrance, a well worded piece of P.R.

For the property and the wine there was a bad period under changing ownership early this century until M. Emmanuel Cruse bought Issan in 1945, subsequently restoring it with no expense spared. One of the chais shewn to visitors has timber rafters and is an historic monument.

Hours:	Monday to Friday 0930–1200; 1400–1730.
Introduction:	Visits should be arranged by telephone to the office of Cruse et Fils Frères, (56) 29.35.83, asking for the hostess.

---

*Hugh Johnson, *The World Atlas of Wine*.

Château Lascombes,

Proprietors: Bass Charrington Vintners Ltd.,
33460 Margaux.
Tel. (56) 30.30.66.

This 2me cru occupies a well-favoured gravelly site on the north-western corner of Margaux; driving the 27 kilometres from Bordeaux, one turns left in the centre of the village.

Once, it belonged to the Chevalier de Lascombes but the names of the numerous owners until 1952 need no longer concern us. In 1952, it was bought by an American Syndicate headed by Alexis Lichine, whose book *Wines of France* broke new ground in the fifties, to be followed by his *Encyclopaedia of Wines and Spirits* in 1967. Lichine also instituted a series of art exhibitions at the château.

In 1971, the property was bought by the British brewing, wine and spirit group Bass Charrington Ltd., who have effected considerable improvements to the vineyards and chais. In particular, there is a new *chai des vins nouveaux* capable of holding almost 900 hogsheads at floor level. The vineyard consists of almost 100 hectares and, in addition to the claret, a quantity of rosé wine Chevalier de Lascombes is made from vines near the river.

The present château, a romantic house built at the turn of the century, is currently being restored.

Hours:	Monday to Friday 0930–1200; 1430–1730.
Introduction:	By appointment. Write to: Hedges & Butler, Hedges House, 153 Regent Street, London W1R 8HQ. Tel. 01–734 4444.

Château Malescot-Saint Exupéry,

Proprietor: M. Roger Zuger,
33460 Margaux.
Tel. (56) 30.30.68.

Château Malescot is in the middle of Margaux village, 27 kilometres from the centre of Bordeaux, the journey taking about 35 minutes by car. The name is that of two past owners, Simon Malescot in 1697 and the Comte de Saint-Exupéry in 1827. After World War I, W. H. Chaplin, the London shippers, bought it and after World War II came Seager Evans, the British distillers, Paul Zuger running it and its neighbour, Château Marquis d'Alesme-Becker, another 3me cru classé, for them with considerable skill.

Here I must declare a sentimental interest – it was the Zugers who kindly shewed me and my wife the Médoc for the first time. I still recall that hot afternoon in 1954 ending about 8 p.m., having 'done' Malescot, Margaux, Beychevelle, Latour, Grand-Puy-Ducasse (which is the Pauillac Maison du Vin), Mouton-Rothschild, Lafite and then back to Rausan-Ségla. It was quite a marathon. I remember too the magnum of Malescot's magnificent 1948, which we were given to take home.

Today, their son Roger is installed at Malescot, while

the parents live in the Château Marquis d'Alesme-Becker, which will not be open to the public for some years until structural alterations are completed and the wine is again being made there and not at Malescot as at present.

Visitors see the chais and cuvier; M. Roger Zuger speaks English and is often there for a chat. The usual tasting of the last two vintages lying in cask is offered; older vintages can be tasted on payment by arrangement in advance. The Château Malescot brochure is a model of its kind – well-designed and written in sensible English. Being so close to Château Margaux, both properties could be fitted in, given an hour or so to spare.

Hours: Monday to Friday 0900–1200; 1430–1800. Closed 15 August to 10 September.

Introduction: Not essential. Prior notice required for groups of more than six people and from those who wish to taste older vintages in bottle.

Château Margaux,

Proprietors: Ginestet S.A.,
33460 Margaux.
Tel. (56) 30.30.28.

Approaching Margaux from Cantenac, a right turn, clearly marked, leads to the château, which was called La Mothe in the fourteenth century. Edward III is said to have been among the many owners of this famous vineyard. Another was a M. Fumel, who probably first planted it with Cabernet Sauvignon and the other great vine species, for which his countrymen rewarded him with the guillotine in the Revolution. With the State in possession, it is curious that it should have gone back to another Marquis (de la Colonilla) in 1802, for half the market price, except that the buildings seem to have been destroyed. The new château, with its fine classical colonnades, so familiar because it is part of the label, was thus begun then.

Since 1949, the owner has been M. Pierre Ginestet, whose father, founder of the Ginestet firm of négociants in 1899, had been a member of a syndicate that bought the château in 1925.

Remarkable at Château Margaux is the long gallery adjoining the chai, which is the private cellar holding 60,000 bottles. Even this is not enough; another will probably have been completed before this appears in print. Being the 1er cru of the Médoc nearest to Bordeaux brings the visitors. During August at least five hundred gallons go in samples from the cask and throughout the year there are heavy demands on the older wines in bottles for celebrations, many at the château itself.

The property is the only Médoc 1er cru that makes – what is rare anywhere in the Médoc – a white wine, Pavillon du Château Margaux. Both the red and white wines are distributed exclusively by the Maison Ginestet.

Hours: Monday to Friday 0930–1200; 1500–1700.

Introduction: By appointment. Write or telephone to: Ginestet S.A., 83 Cours Saint-Louis, 33027 Bordeaux. Tel. (56) 29.78.62.

Château Palmer,

Proprietors: Société Civile de Château Palmer,
Cantenac, 33460 Margaux.
Tel. (56) 30.30.02 or 30.30.87.

From Bordeaux, the château is conspicuous on the right of the D2 road shortly before coming to Margaux village.

Originally Château de Gasq, the English general Palmer renamed it soon after 1814, extending the vineyard and improving the wine until he went broke in 1840. The next owners the Pereires, being bankers, were rich enough to build the present 'pepper-potted' château. Since 1938, the Union Jack, the Tricolour and the Dutch flag have flown over the roof, representing the three nationalities of a shared ownership, Sichel being the English interest and Mähler-Besse of Bordeaux, the Dutch.

Today, the third growth award of the 1855 classification, definitely under-rates the wine from this vineyard that adjoins Château Margaux. I particularly remember the 1960 wine, which was wonderful in what was generally only a fair year for quality.

The Cellar Managers shew visitors the cuvier and the chai. There is tasting from the cask and purchase by the bottle.

Hours: Monday to Friday 0900–1145; 1400–1700.

Introduction: None strictly necessary but Maison Sichel much prefer prospective visitors to give them an advance telephone call, (56) 29.52.20, or to arrange the visit through the Conseil Interprofessionnel du Vin de Bordeaux (C.I.V.B.).

Labarde

Château Siran,

Proprietor: M. William-Alain Miailhe,
33460 Margaux.
Tel. (56) 30.34.04.

Twenty-four kilometres from Bordeaux off the D2 road in Labarde, the Siran vineyard occupies the central plateau between two classed growths, Dauzac and Giscours. Siran in fact includes vines from Dauzac bought by the Miailhe heirs in 1966; both vineyards are within a kilometre of the estuary.

The origins of Siran go back to the fifteenth century when it was sometimes called Saint-Siran. In the early nineteenth century, the daughter of the house married a

Count de Toulouse-Lautrec-Monfa; the painter, born 1864, was their grandson. The charming château – or rather villa in the Italian sense – looks as if the second storey over the entrance were almost added as an afterthought.

Behind it in autumn when the cyclamens are in bloom, the park becomes a blaze of colour. Any re-classification would undoubtedly promote Siran to cru classé.

Hours:	Monday to Friday 0930–1000; 1400–1630.
Introduction:	By appointment, obtained through the administrative office: William-Alain Miailhe, 6 Quai Louis XVIII, 33000 Bordeaux. Tel. (56) 48.35.01.

Avensan
Château Citran,
Proprietor: S.A. de Château Coufran,
Administrator: M. Jean Miailhe,
Avensan, 33480 Castelnau-de-Médoc.
Tel. (56) 30.21.01.

Going north from Margaux, the turning to Château Citran is marked on the left, about one kilometre after Soussans and the château lies about 2.5 kilometres beyond. The estate comprises 550 hectares of woods and pasture, with a comparatively small proportion under vine. The present château is an imposing edifice built about 1861. The vineyard is in two parts, that in front being on soil similar to Margaux, that on the other side being on soil typical of Moulis. The wine is an excellent cru bourgeois.

Hours:	Monday to Friday 0900–1200; 1500–1800. Closed from a few days before each vintage till a month after its completion (i.e. usually from about 20 September till 1 November). Given sufficient notice, an English-speaking guide will be found for important groups.
Introduction:	By appointment, obtained through the administrative office: Miailhe Frères, 24 Cours de Verdun, 33000 Bordeaux. Tel. (56) 52.25.85.

Moulis
Château Chasse-Spleen,
Proprietors: M. Frank Lahary, M. Michel Bezian,
Moulis, 33480 Castelnau-de-Médoc.
Tel. (56) 30.32.54.

Going north leave the D2 road at the marked turning left at the crossroads in Arcins. After crossing the railway (level crossing) turn right after 500 metres at the next crossroads; Chasse-Spleen is then immediately on the right. From Bordeaux the faster route would be via the main D1 road, turning right to Moulis at Bouqueyran, two kilometres after Castlenau.

Being one of six crus exceptionnels in a classification of crus bourgeois in 1932, the wine of this fine 48-hectare vineyard always fetches a good price. A pleasant property to visit, given half-an-hour to spare, with peacocks an extra attraction. English spoken; parties of up to 100 can be accepted given notice.

Hours:	Monday to Friday 0900–1100; 1430–1700.
Introduction:	Communications to Monsieur Charles Bouilleau, Manager at the château.

Cussac
Château La Tour du Haut-Moulin,
Proprietor: M. L. Poitou,
33152 Cussac-Fort Médoc.
Tel. (56) 20.91.10.

This cru bourgeois is close to the estuary in the commune of Vieux Cussac. Going north on the D2 road, a kilometre after the village of Lamarque, look out for a sign on the right to Fort Médoc. The vineyard and the low white buildings of the property are a few hundred metres down this minor road, which ends at the old dilapidated fort by Vauban.

Cultivated by four generations of the Cezeaux and Poitou families, La Tour du Haut-Moulin is a model estate and M. Laurent Poitou made a superb 1966, which had matured to perfection by 1975.

Hours:	Monday to Friday 0900–1200.
Introduction:	French only spoken. M. Poitou welcomes individuals interested in fine Bordeaux wine. Appointments could be arranged by Gilbey S.A., Château Loudenne, Saint-Yzans-de-Médoc, 33340 Lesparre, Médoc. Tel. (56) 41.15.03.

Saint-Julien
Château Beychevelle,
Proprietor: M. Achille-Fould,
Saint-Julien-Beychevelle, 33250 Pauillac.
Tel. (56) 59.05.14.

Worth at least a stop where the D2 road enters the Saint-Julien commune from Margaux, the front of this large one-storey *chartreuse* house can be admired from the road. But to see the other side, where the garden and lawns slope for over a kilometre down to the estuary, a visit is necessary.

The terrace, with steps at the centre leading down to the gardens, reminds me of Dartmouth when admirals took the salute at Passing Out parades at the Royal Naval

College. This was precisely what Beychevelle's illustrious owner, Grand Amiral of France, Duc d'Epernon must have done as the passing sailing vessels lowered their sails (*baisse-voile*) momentarily in salute. It is no myth to me, having spent an active thirty years in the Royal Navy, that 'Beychevelle' is derived from *baisse-voile*.

Since 1874, ownership has been in the Achille-Fould family, whose members have been prominent in French public life. The former head of the family was an Oxford graduate who became Minister of Agriculture in a pre-war French government. The present owner, M. Aymar Achille-Fould, has been Minister of Posts and Telecommunications.

If Beychevelle's 4me cru, commanding a price approaching that of the first growths, is beyond our means, at least a visit is free and it certainly should be included in the visiting list, along with those first growths, when in the Médoc for the first time.

Hours:	Monday to Friday 0900–1200; 1400–1800. Saturday 0900–1200.
Introduction:	By appointment in advance from Monsieur J. Simon, Exploitation et Présentations Internationales, 164 Bld. Haussman, 75008 Paris. Tel. 227.93.34. This office handles the business of the château, where resident staff are unlikely to speak English. In Bordeaux, any French wine authority (The Maison du Vin, for example) can make appointments direct to the château by telephone.

Château Branaire,

Proprietor: M. Jean Tapie,
Saint-Julien-Beychevelle, 33250 Pauillac.
Tel. (56) 59.08.08.

Branaire and Beychevelle sit astride the D2 road at the southern end of the Saint-Julien commune, their vineyards being less than two kilometres from the estuary; it is worth drawing up for a moment to admire the façades of both these fine buildings. Previously the name was Branaire-Ducru, a M. Ducru having been the proprietor a century ago. The post-war owner, the late M. Jean Tapie, sensibly preferred plain straight-forward Branaire. M. Tapie, who had previous experience in Algeria where methods were often highly enterprising, introduced many technical improvements at Branaire. Among them was a non-vintage blended wine of different years, which was very much better than the sort of wine usually made from just one indifferent vintage. When he died in 1969, his son Jean was well able to continue in his place.

The château is a fine two-storey edifice built about 1795, with a pleasant garden and orangery. The wine, a 4me cru, is not on sale but visitors taste recent vintages in cask. English is spoken.

Hours:	Monday to Friday 0900–1200; 1500–1800.

Introduction:	By appointment. Letter of introduction obtainable through any reputable wine merchant. A week's notice of arrival is required for groups.

Château Ducru-Beaucaillou,

Proprietor: M. Jean-Eugène Borie,
Saint-Julien-Beychevelle, 33250 Pauillac.
Tel. (56) 59.05.20.

The 'beautiful pebble' is in fact a grandiose mansion, approached along a marked turning, to the right off the D2 road when going north towards Pauillac from its next-door neighbour, Beychevelle. The vineyard runs along the Gironde, only a few hundred yards away through a park with fine trees. There have been various owners since M. Ducru tacked on his name, but since 1941 it has belonged to a branch of the Borie family, the present owner, occupier and manager being Jean-Eugène Borie.

Under his direction the wine, after a decline in the early part of the century, has been steadily returning to the high standard of a 2me cru.

Hours:	Monday to Friday 0900–1100; 1400–1700. Saturday 0900–1100. Closed 15–31 August.
Introduction:	By appointment. Groups of up to 10–15 people acceptable. Arrangements best made through a wine merchant, who should contact M. Borie or the Maître de chai at the château.

Château Haut-Beychevelle-Gloria,

Proprietor: M. Henri Martin,
Saint-Julien-Beychevelle, 33250 Pauillac.
Tel. (56) 59.08.18.

Usually referred to as Château Gloria, this is considered the best cru bourgeois of Saint-Julien. The vineyard now includes part of Saint-Pierre Bontemps (Saint-Pierre, subsequently divided, was a 4me cru in the 1855 classification) and covers 31 hectares. The little château is under a kilometre up the road from Château Beychevelle on the left going towards Pauillac.

Gloria is owned by M. Henri Martin, manager of Château Latour and an important figure in the Bordeaux wine trade, which explains its fast-growing reputation.

Hours:	Monday to Friday 0900–1200; 1400–1700.
Introduction:	By appointment, which could be arranged direct or through Gilbey S.A., Château Loudenne, Saint-Yzans-de-Médoc, 33340 Lesparre, Médoc. Tel. (56) 41.15.03.

Château Léoville-Las-Cases,

Proprietor: Société Civile du Château Léoville-Las-Cases,
Administrator: M. Paul Delon,
Saint-Julien Beychevelle, 33250 Pauillac.
Tel. (56) 59.05.19.

Along the D2 road from Bordeaux, which has signs 'Route
des Châteaux' in places, the big stone arch surmounted by
a lion, on the right after the village of Saint-Julien-
Beychevelle, is hard to miss.

In the eighteenth century there was one very large
Léoville vineyard stretching almost the length of the
Saint-Julien commune from Château Beychevelle to
Latour. M. Léoville, the owner, died before the Revolution and the Marquis de Las-Cases, a part owner, saved
himself during it by emigrating. His share became
Léoville-Barton by purchase in 1821 (see Barton & Guestier, p. 104). Another part went to a Monsieur Poyferré,
but by far the largest piece went back to M. Léoville's
family and remained 'Léoville-Las-Cases'.

There was only one château, which is now divided
between Poyferré and Las-Cases, M. Paul Delon the
Las-Cases administrator owning the northern part. To go
with *their* piece, the Bartons managed to buy the delightful
one-storey Château Langoa in 1821, and there they have
lived and still do. All three Léovilles are 2me crus.

At Léoville-Las-Cases, although the château is divided,
there are the chais for making and maturing the wine in
cask. Visitors are invited to see them, calling first at the
office nearby to arrange tasting and possible purchases.
An English-speaking guide is present on Wednesdays
(except in August) and at other times given notice.

Monsieur Paul Delon managing the estate has made
great wines since 1959. Clos du Marquis is the second
string of Las-Cases.

Hours: Monday to Friday 0800–1200; 1400–
       1800. Saturday 0800–1200.
Introduction: Visitors who write or telephone, giving
       at least 24 hours' notice and stating
       their expected time of arrival, are wel-
       come.

Pauillac

Château Lafite-Rothschild,

Proprietor: Société Civile de Château Lafite-Rothschild,
Managed by Baron Elie de Rothschild,
33250 Pauillac.
Tel. (56) 59.01.04.

From Bordeaux on the D2 road north of Pauillac, Lafite is
seen to the left, on the lower ground after passing
Mouton-Rothschild. The archives of Christie's, the London auctioneers, record its wines being sold for 66/- a
dozen at their Sale in 1788, the Latour 1785 having provided them with their first sales entry of fine claret a year
or so earlier.

'The history of Château Lafite', I once wrote, 'could be
woven into a best-selling historical novel'. Since then, Mr.

Cyril Ray has told it in detail in his book *The Story of
Château Lafite-Rothschild,* commended on p. 78.

Briefly, in the Revolution the owner was guillotined and
the estate sold by the government to a Dutch company.
Control passed to a M. Vanlerberghe and a M. Ouvrard,
a smart financier. Together they went bankrupt, Ouvrard
ending in a debtors' prison, while Vanlerberghe crossed
into Holland and died in 1819, a year after he had cleverly
made over Lafite to his ex-wife, Barbe-Rosalie Lemaire.
But in 1821 the lady sold it again – to a Mr. Samuel Scott,
an eminent London banker, acting with complete secrecy
for Vanlerberghe's son. It was not until after the son's
death in 1866 that Scott's bank revealed for whom they
had first bought, and then administered, the estate those
past forty-five years.

In 1868 Lafite came up for auction in Paris, Baron
James de Rothschild buying it for 4,440,000 francs. With
fees and taxes added, Lafite must have cost him just under
£200,000.

The English Baron Nathaniel having taken over
Mouton in 1853, the stage was set for English and French
Rothschild rivalry and they have been competing, one
might say, in Oxford and Cambridge style, for 'Head of
the Vintage' ever since.

There is an endearing, unassuming charm about Lafite;
the house is so modest for an 80-hectare vineyard that
makes a wine consistently fetching the highest price of
them all. And the story of how five brothers set out from
the Frankfurt ghetto in Napoleon's time, each to a different country, to make a lasting fortune, is recalled by the
five arrows joined at the middle with heads fanning out
(the heraldic emblem of the Rothschilds), which form the
weathercock on the little pepper-pot clock tower. As to the
rivals, if the English cousins up the road at Mouton now
have their museum of vinous art treasures, Lafite has had
its Vinothèque since 1798! A few years ago it still contained a few bottles of the 1797 – the year Nelson was
upsetting Napoleon at the Battle of the Nile.

Hours: Monday to Friday 0830–1130; 1400–
       1730.
Introduction: By appointment, obtainable by writing
       to:
       Monsieur Némes, Direction des
       Domaines, 21 rue Laffitte, 75428 Paris
       CEDEX 09,
       or to
       Monsieur Crété at the Château itself.

Château Latour,

Proprietors: Société Civile de Château Latour,
33250 Pauillac.
Tel. (56) 59.00.51.

About forty minutes on the D2 road from Bordeaux,
Château Latour, its isolated tower sticking up out of the
vines, can hardly be missed on the right hand (Gironde)
side just before reaching Pauillac. As at Mouton, the

proportion of Cabernet Sauvignon grapes is said to approach 90 per cent., giving both their wines a staying power that even the other two 1er crus, Margaux and Lafite, rarely match. Latour also has an unmatched capacity for making the greatest wine in poor years.

In Plantagenet times, a line of towers hereabouts formed part of the local defences against sea-borne attack by pirates. There was a succession of owners then, quite a few of them pro-English, so it is not inappropriate that, since 1963, a British (Cowdray) company has held the controlling interest, with John Harvey of Bristol fame and the French Beaumonts, the previous owners, having minority holdings. Some Bordeaux vignerons, resenting the English return having kicked us out five hundred years ago, are said to have referred the matter to General de Gaulle, whose consoling counsel was that even the English were unlikely to remove the soil.

The English have, however, completed a big programme of modernisation – not forgetting the tower. Never part of the old fortifications, though it may have been made from the original stones, this was put near the chai in the seventeenth century. A good landmark and emblem for Latour, it was repaired and restored in 1972.

Visitors see the cuvier and the major cellars in the chai. If invited to taste the young wines, it is as well to remember that they are the slowest of all to develop and are likely to seem even harder than those of their neighbours.

Hours:	Monday to Friday 0930–1200; 1400–1630. Closed during the vintage, normally three weeks from about 20 September.
Introduction:	By appointment. Letter of introduction obtainable direct from the General Manager, M. Jean-Paul Gardère at Château Latour.

Château Lynch-Bages,

Proprietor: M. André Cazes,
33250 Pauillac.
Tel. Château (56) 59.01.84.

The hamlet of Bages is well signed to the left of the D2 road about a kilometre south of Pauillac town and about 50 minutes' drive (50 kilometres) from Bordeaux. Its chief properties, in ascending order of importance, are Haut-Bages-Avérous, Haut-Bages-Libéral, Croizet-Bages and Lynch-Bages.

Classed as a 5me cru in 1855, Lynch-Bages has since become a big sturdy wine that fetches second growth prices. A certain Mr. Lynch, escaping from some religious battle going on in Ireland, bought the place in 1740. A grandson became Mayor of Bordeaux in 1804 and in 1936 it was bought by Jean C. Cazes, whose son André Cazes, the present owner, is now Mayor of Pauillac. Mayors are usually sturdy burghers, so it would seem that this big wine must owe something to its mayoral masters.

Hours:	Monday to Friday 0900–1200; 1400–1800.
Introduction:	All visitors are welcome but it is best for arrangements to be made through a wine merchant, who should contact M. André Cazes at his office: 17 rue Jean-Jaurès, 33250 Pauillac. Tel. (56) 59.00.64.

Château Mouton-Rothschild,

Proprietor: Baron Philippe de Rothschild,
B.P.32, 33250 Pauillac.
Tel. (56) 59.01.15.

Baron Philippe has such a flare for showmanship in the best taste that every visitor wants to see the superb chai and the museum at Mouton, so much so that for the latter, separate appointments and a charge of 2 francs now have to be made. The village of Mouton is two kilometres north of Pauillac on the D2 road leading on to Saint-Estèphe.

For decades there has been general agreement that an injustice was being done in continuing to regard the wine by the 1855 classification of 2me cru and amid cheers, as glasses were raised, it was promoted at long last in 1973. But the property is also the headquarters of Château Mouton-Baron-Philippe (5me cru) and Château Clerc-Milon (5me cru). The company, La Bergerie Baron Philippe de Rothschild S.A., also produces and markets the well-known brand Mouton-Cadet, which is a selection from Bordeaux vineyards.

Mouton is a contraction of *motte de terre* – an eminence, the vineyard being on a slight hill that slopes down to the estuary.

Mouton was bought by Baron Nathaniel in 1853 for less than the previous owner had paid, due chiefly to the disease, oïdium, having made an appearance in the Médoc. His son James added the house, Petit-Mouton, used for guests since the family converted some stables for themselves. After Baron James came Henri and then Philippe, responsible, with his wife Pauline, for so many innovations. She died in 1975.

Visitors are shewn the cellars and the winery, both of the château and of La Bergerie. In the château cellars there are 150,000 bottles, going back at least to the 1870 vintage, by no means all of the property, for a Médoc custom to buy the wines of one's neighbours too. Indeed, another proprietor dining as a guest could easily find himself drinking one of his own wines that he himself no longer had left in his own reserve.

Hours:	Monday to Friday 0930–1130; 1400–1700. Closed August.
Introduction:	By appointment. Letter of introduction obtainable from agents.
U.K. Agents:	Hedges & Butler Ltd., Hedges House, 153 Regent Street, London W1R 8HQ. Tel. 01–734 4444.

Additionally, the museum, which is a collection of

works of art associated with wine, is open to all who care to apply in advance to the secretary of the château. It is best to do this through the agents above.

Château Pichon-Lalande,

Proprietor: Société Civile du Chateau de Pichon-Longueville, Comtesse de Lalande
33250 Pauillac.
Tel. (56) 59.00.02.

The Pichon pair – the Baron and the Countess – facing each other across the D2 road about three kilometres south of Pauillac, evidently titillate the wine writers' imaginations.'A pair of old Duchesses in party clothes', declares Hugh Johnson,* yet the Dutchman, Hubrecht Duijker, regards the Baron as 'the more feminine'. 'The gentleman should be served before the lady' he says. 'The slightly more refined Baron with the main dish and the Comtesse with the cheese.'†

Far from being some 'two-ton Tessie' of a wine, Comtesse Lalande (as is to be expected from such a close neighbour of Latour) is a lady of breed, charm and delicacy. Very good well-balanced wines were made here in 1964, 1966 and 1970; the last two being by no means ready yet.

Originally, of course Pichon, (another second growth like Léoville) was one very large vineyard stretching into the Saint-Julien commune and owned by successive Barons Longueville. The split came in 1855 in the proportion Lalande 55 hectares, Baron 18.

Hours:	Monday to Friday 0930–1200; 1430–1700.
Introduction:	By appointment, obtained from the administrator of the château at the above address.

Château Pontet-Canet,

Proprietor: M. Guy Tesseron,
33250 Pauillac.
Tel. (56) 59.00.79.

Two kilometres north-west of Pauillac and very close to Mouton-Rothschild, there are few better known wine château names in England than Pontet-Canet. The reason is partly that the vineyard of 75 hectares is the largest in the whole Médoc, producing more wine than any other and partly that the late M. Christian Cruse, when head of his firm and earlier, spent fifty years of his life concentrating on the British market. Moreover, Pontet-Canet and Lafon-Rochet (the Saint Estèphe 4me cru classé also owned by M. Tesseron) always shipped all

their wine in cask, (none was château-bottled) which meant less duty to pay in London. In recent years, when demand for fine wine has grown so greatly, growers are far less disposed to entrust bottling to unseen hands, lest their reputation should suffer. Thus, proportions of Pontet-Canet and Lafon-Rochet are now bottled in their châteaux.

Bought by Maison Cruse in 1865 and sold in 1975 to M. Guy Tesseron (whose wife was the daughter of M. Emmanuel Cruse), the 5me cru rating of 1855 under-rates a wine that, in the great year of 1929, was on a par with Mouton-Rothschild itself and usually fetches a 2me cru price.

A good property to see; its trees, lawns and flowerbeds should be approved by our nation of gardeners. There are good views too of its neighbour and – quite a rarity in the Bordeaux districts – real cellars underground.

Hours:	Monday to Friday 0930–1200; 1330–1700.
Introduction:	Visits would best be arranged through: Maison du Vin, Conseil Interprofessionnel du Vin de Bordeaux (C.I.V.B.), 1 cours du 30 Juillet, Bordeaux. Tel. (56) 44.37.82.

Saint-Estèphe

Château Cos d'Estournel,

Proprietor: Domaines Prats,
Administrator: M. Bruno Prats,
Saint-Estèphe, 33250 Pauillac.
Tel. (56) 59.00.33.

Pronounced 'Cos' like the lettuce but looking like a Chinese pagoda, this strange edifice, appearing up the hill on the D2 road north of Château Lafite, is an amusing folly, giving the Médoc about its only touch of architectural eccentricity.

'Cos' was built between 1815 and 1840 by Louis Gaspard d'Estournel, who traded horses with the Far East. The building is not used as a dwelling but as a chai. Today, 'Cos' belongs to Fernand Ginestet's daughter and her sons. The fine brass-studded wooden main entrance door came from the Palace of the Sultan of Zanzibar, adding a touch of arabesque to the oriental.

The vineyard adjoins that of Lafite but Cos is higher up, on the plateau, and though a 2me cru the two wines are not really comparable.

Hours:	Monday to Friday 0900–1200; 1400–1700
Introduction:	By appointment. Letter obtainable from agents: Mentzendorff & Co. Ltd., Asphalte House, Palace Street, London SW1E 5HG.

---

*Hugh Johnson *World Atlas of Wine.*
†Hubrecht Duijker *The Great Wine Châteaux of Bordeaux.*

Château de Pez,

Proprietor: M. Robert Dousson,
Saint-Estèphe, 33250 Pauillac.
Tel. (56) 59.30.07.

A little difficult to find from some directions, it is best to continue north on the D2 road from Pauillac as far as the village of Pez and then take the turning left that is marked to the Château de Pez. Do not confuse with Les Ormes de Pez, the château in the village itself.

The château and chais were built in 1749; the records shew that until 1789 the wines were sold on a par with those of Châteaux Margaux and Haut-Brion. The vineyard, now planted with a very high percentage of Cabernet Sauvignon, is extremely well exposed to the sun on the extensive plateau of Saint-Estèphe.

Since the war, de Pez wines have been remarkably consistent under the energetic owner-manager, M. Robert Dousson. I remember a luncheon when Cos d'Estournel and Montrose 1964, the two 2me crus of Saint-Estèphe, were served first but were outclassed by the de Pez 1964 which followed. And in 1975 I recall a bottle of the 1953 still in its prime, though only rated a Cru bourgeois.

Hours: Monday to Friday 0900–1200; 1430–1800.

Introduction: Groups of up to fifty can be accepted but French only is spoken. Gilbey S.A., Château Loudenne, Saint-Yzans-de-Médoc, Tel. (56) 41.15.03 have a close liaison and can fix appointments by telephone.

Saint-Seurin-de-Cadourne
Château Coufran,

Proprietor: S.A. du Château Coufran,
Administrator: M. Jean Miailhe,
Saint-Seurin-de-Cadourne, 33250 Pauillac.
Tel. Château (56) 59.31.02.

Château Coufran, clearly signed, is on the left of the D2 road, two kilometres north of Saint-Seurin-de-Cadourne, a position giving it the distinction of being the last château entitled to the appellation Haut-Médoc. Beyond lies the Bas-Médoc – not half such a good address – though the term is hardly used, the appellation usually being just plain Médoc. The terms Haut and Bas, referring to high and low ground, are hardly of immense import either, in this territory which is flat as the proverbial pancake almost everywhere. Nevertheless, Coufran is on a hill sufficiently high for it to look *down* on Loudenne in the Bas-Médoc a mile beyond.

Making a good cru bourgeois, Coufran was bought by M. Louis Miailhe fifty years ago and is now run by M. Jean Miailhe. At present the vineyard covers about 66 hectares.

Hours: Monday to Friday 0900–1200; 1500–1800. Closed from a few days before each vintage till a month after its completion (i.e. usually from about 20 September till 1 November).

Introduction: By appointment obtained through the administrative office: Miailhe Frères, 24 Cours de Verdun, 33000 Bordeaux. Tel. (56) 52.25.85.

Château Verdignan,

Proprietor: S. C. du Château Verdignan.
Administrator: M. Jean Miailhe,
Saint-Seurin-de-Cadourne, 33250 Pauillac.
Tel. (56) 59.31.13 or 59.31.34.

Leaving Saint-Seurin, Bel-Orme comes first on the right with Verdignan just beyond it.

A good deal of regrouping has gone on among the vineyards of this locality so that Verdignan now embraces another called Plantey-de-la-Croix.

Hours: Monday to Friday 0900–1200; 1500–1800. Closed from a few days before each vintage till a month after its completion (i.e. usually from about 20 September till 1 November).

Introduction: By appointment obtained through the administrative office: Miailhe Frères, 24 Cours de Verdun, 33000 Bordeaux. Tel. (56) 52.25.85.

Given sufficient notice, an English-speaking guide will be found for important groups.

MEDOC
Saint-Yzans-de-Médoc
Château Loudenne,

Proprietors: W. A. Gilbey Ltd.,
Saint-Yzans-de-Médoc, 33340 Lesparre-Médoc.
Tel. (56) 41.15.03.

From Bordeaux, Loudenne is at the far end of the D2 *Route des Châteaux* and much the faster road is the D1, turning right at Lesparre for Saint-Yzans. There are signposts to Loudenne thereafter. From the airport (60 kilometres) it takes about an hour.

From the Royan-Pointe de Grave ferry, Loudenne is the *nearest* château of consequence, the 50-kilometre journey along the Gironde road (turn left in Saint-Vivien) taking about 50 minutes.

Flying the Union Jack, this charming eighteenth-century pink château is the first to catch the mariner's eye sailing up the Gironde to Bordeaux. Bought by Sir Walter Gilbey in 1875, the 34-hectare vineyard provides – that rarity in the Médoc – a very good dry white wine as well as its well-known cru bourgeois claret. The marque, La Cour Pavillon, is from neighbouring vineyards.

In 1876, Sir Walter added a chai, now a mellow building of Victorian charm but still the largest anywhere in the Médoc. It holds 10,000 barrels and 250,000 bottles. The capacity is needed; Gilbey S.A. (see p. 106) have to supply all manner of Bordeaux wines, not only to Peter Dominic and Westminster Wine in Britain but to other retailers and wholesalers in most countries of the world. For many of these, the chai is an assembly, a maturing and a distribution point.

Visitors enjoy the views of the estuary, a short tour of the chai, an exhibition and a little tasting. Purchases by case and bottle are possible and English-speaking guides (if obtainable) operate from June to September inclusive. An important extra is that the staff will arrange visits to other châteaux in the Bordeaux region, a telephone call there and then often enabling visitors going on to Bordeaux to see them on the way.

Hours:	Monday to Friday 0930–1200; 1400–1630.
Introduction:	None necessary but for any special arrangements (e.g. parties over 10) write to: Justerini & Brooks Ltd., 1 York Gate, Regent's Park, London NW1 4PU. Tel. 01–935 4446.

Bégadan
Château Laujac,
Proprietors: Cruse et Fils Frères, Bégadan, 33340 Lesparre, Médoc. Tel. (56) 41.50.12.

The hamlet of Laujac is best approached on the D103E road from Lesparre-Médoc, turning left after seven kilometres. Proceed through a pair of entrance gates in the village, past farm buildings and through another pair of entrance gates with white railings beyond. The château is on the left.

In 1852, Laujac became the first Bordeaux château to be bought by Maison Cruse. At Blanquefort they claim the Médoc property (see Château du Taillan p. 80) nearest to Bordeaux; this one, nine kilometres north-east of Lesparre and seventy from Bordeaux, is about the furthest. The vineyard overlooks marshes reclaimed from the estuary and surrounded by a dyke built by the Dutch in the seventeenth century. Bégadan has a Romanesque church – one of many in the Médoc – with a notable apse.

Lafitte-Laujac and La Tour Cordouan are other names of parts of this property used by Cruse. The wines are rated cru bourgeois Médoc, the commune of Bégadan being well to the north of where the Haut-Médoc ends.

Hours:	Monday to Friday 0830–1200; 1330–1700.
Introduction:	Visits should be arranged by telephone to the office of Cruse et Fils Frères, Tel. (56) 29.35.83, asking for the Hostess.

POMEROL

Introduction

Moving clockwise from the Médoc, the districts of Blaye, Bourg and Côtes de Fronsac, all providing sound claret, should follow in that order. But since there are better châteaux to visit elsewhere the guide moves on, first to Pomerol and then to Saint-Emilion.

Headquarters of the Pomerol and Saint-Emilion wine trade is Libourne on the river Dordogne, twenty miles east of Bordeaux. The vineyards of Pomerol cluster together in a sector, a mile or so north-east of the town, becoming even thicker around Saint-Emilion, which is only four

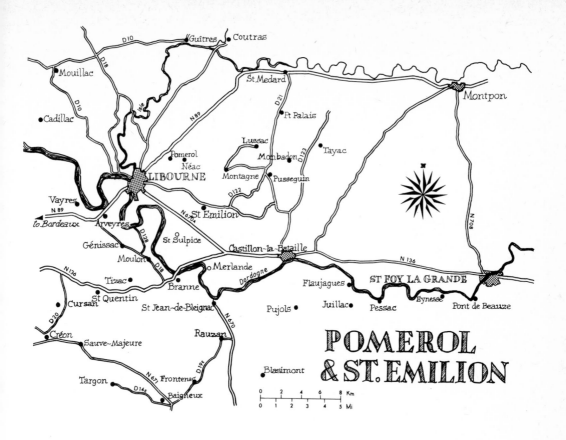

POMEROL & ST. EMILION

miles away. Edward I was responsible for Libourne; in 1268, as Prince Edward, he required a fortified town to be built in this good defensive position on the river. The job was done by Roger de Leyburn from whom the name is derived.

Until motor transport became firmly established, sea-going craft loaded the wines at the quays here, where shippers like J. P. Moueix still have their offices. Even in the present century, the Gilbey family would come out from London for the vintage, disembarking from the Paris night train at Libourne to take a special steamer the sixty miles or so downstream to their Médoc Château Loudenne on the shores of the Gironde estuary.

The slightly hilly country (about 200 feet compared with 50 for the Médoc) makes the Pomerol and Saint-Emilion plateau a more attractive region than the Médoc,

and the undulating terrain continues through Entre-Deux-Mers and Premières Côtes de Bordeaux until the river Garonne is reached at Cadillac, almost opposite Sauternes. There is no official classification of Pomerol wines. Lichine's *Encyclopaedia* lists nearly a hundred and fifty properties, many of them as small as 2·5 acres.

Much of their wine will go to individuals, for the Frenchman has a considerable mistrust of wine firms and likes to buy direct. The châteaux are nearly all modest, pleasant dwellings, of little architectural interest, lived in by their owners. Château Pétrus is the outstanding wine; Vieux-Château-Certan is a fine edifice in the *chartreuse* style, met also in the Médoc notably at Beychevelle, Langoa-Barton and Loudenne; while Château Beauregard so intrigued a visiting American architect that he built a replica of it on Long Island.

SHORT TOUR

Leave Libourne on D21 for Saint-Georges and Montagne. Turn right and go south to Saint-Emilion on D122 thence back to Libourne on D17.

Total about 12 miles (say 20 kms.).

NORTH-WEST

From Libourne go north-west on N670 to Fronsac, Saint-Michel-de-Fronsac and Cadillac-en-Fronsac. After Cadillac turn right on D137 to Mouillac. Turn right at Mouillac on to D10 to Coutras, a possibility for lunch. Continue on D10 for about 6 kilometres bearing left for Saint-Médard-de-Guizières. This wine route now turns south on to D21 to Petit-Palais, whence a left turn will make a circular deviation to Tayac and Monbadon, rejoining the D21 at Lussac. From Lussac, D121 leads back through the Pomerol vineyards and D122 goes through Montagne to Saint-Emilion and thence to Libourne.

Total distance about 60 miles (say 100 kms.).

EAST

From Libourne take D21E and D122 to Montagne and Lussac. Join D17 two kilometres beyond Lussac, turning right towards Puisseguin and Castillon-la-Bataille*, which can be reached in prettier, if more devious fashion, through Saint-Philippe d'Aiguille and Bèlves. From Castillon take the main Bergerac road, N136 to Ste. Foy-la-Grande, a suitable lunch stop.

Having crossed the Dordogne on entering the town, remain on the left bank leaving Ste Foy on N672 (see *Michelin Guide* Town Plan) and stick to D130 along the river through Pont de Beauze, Eynesse, Pessac-sur-Dordogne, Flaujagues, Juillac to Pujols.

From Pujols take D17 to Blasimon, and from there take D127E to Rauzan and Branne, where a bridge re-crosses the Dordogne to Saint-Sulpice-de-Faleyrens, close south of Saint-Emilion town.

Total distance about 70 miles (say 110 kms.).

ENTRE-DEUX-MERS AND GRAVES

Leave Libourne on N670A (the Bergerac road) forking right after five miles through Merlande and crossing the Dordogne to Saint-Jean-de-Blaignac, where the Auberge Saint-Jean has a *Michelin bonne table* star. D123 to Rauzan,

D19E to Frontenac and Baigneaux, D14E to Targon should emerge on N671, heading for Sauve-Majeure and Créon (lunch stop).

From Créon take D20 north to Cursan, and then a V.O. fork right to Saint-Quentin-de-Baron on N136. Two kilometres east along N136 reaches Tizac-de-Curton. From Tizac bear left on D128 to Moulon, left on D18 to Génissac and straight on to Vayres on the Dordogne. Return to Libourne via Arveyres on N.89.

Total distance about 45 miles (say 75 kms.).

Further Information

Maison du Vin,
Place Clocher,
Saint-Emilion.
Tel. (56) 51.72.17.

Libourne

Château Beauregard,

Proprietor: Les Héritiers Clauzel,
Pomerol, 33500 Libourne.
Tel. (56) 51.14.09 or (office) (56) 51.13.36.

From Libourne take the road D21E in the direction of Montagne. Do not take the fork left beyond the village of Catusseau (3 kilometres) but continue on D21E. Beauregard is the next château on the right.

The vineyard is believed to be on a site where, in the twelfth century, the Knights of Malta built an hospice to care for the pilgrims on their way to the relic of Saint James at Santiago de Compostella in the north-west corner of Spain. M. Clauzel's ancestors traced their ownership back to 1610. After the First World War, visiting Pomerol, an American architect, called Coffin, was so struck with the château that he proceeded to erect an exact replica, called 'Mille-Fleurs', on Long Island. It belongs now to Mrs. Daniel Gugenheim.

M. Clauzel makes a very good grand cru, likes to speak English, rather badly he says modestly and, provided he is not away, can take up to twenty visitors at a time.

Hours:	Monday to Friday 1000–1200; 1500–1700. Saturday 1000–1200.
Introduction:	By appointment; please write or telephone M. Clauzel at least 48 hours in advance.

Château Petit-Village,

Proprietor: Société des Vignobles Prats,
Pomerol, 33500 Libourne.
Tel. (56) 51.21.08.

From Libourne take the road D21E in the direction of Montagne. Do not take the fork left beyond the village of Catusseau (3 kilometres) but continue on D21E.

---

*The battle was that on 17 July 1453 which ended The Hundred Years War, General Talbot and the English being defeated by the French. The site of the battle was by the Dordogne, east of the town, where a monument erected eighty years ago now stands between the river and the busy N136 Bergerac road.

Beauregard will be seen immediately on the right; Petit-Village, lying back from the road, a few hundred metres beyond on the left.

Petit-Village is likely to be included in any dozen of Pomerol. The vineyard is planted with an unusually large proportion (relative to normal in Pomerol) of Cabernet Sauvignon; not surprising perhaps, recalling that the Prats family own Cos d'Estournel in the Médoc (p.88), where over 50 per cent. Cabernet Sauvignon is normal.

Hours: Monday   to   Friday   0800–1200; 1500–1700.

Introduction: By appointment. Please write or telephone at least 48 hours in advance to: M. Bruno Prats, 84 rue de Turenne, Bordeaux. Tel. (56) 44.11.37.

Château Pétrus,

Proprietress: Mme L. P. Lacoste-Loubat,
Pomerol, 33500 Libourne.
Tel. (56) 51.17.96.

From Libourne take the road D21E in the direction of Montagne. Just beyond the village of Catusseau (3 kilometres) fork left towards Néac and pass Vieux-Château-Certan in about seven hundred metres and Château Pétrus in another six hundred.

Pomerol wines have never been officially classified so that every proprietor can call his wine a *'grand cru'*. All agree however that Pétrus comes first and the unofficial title *Premier des Grands Crus* is appropriate. My first acquaintance with Pétrus was in Bordeaux in 1955 at a luncheon, when the Calvets kindly gave us their last bottle of the 1929 vintage. In 1925 part of the property had been bought by M. Edmond Loubat and after his death, his widow succeeded in obtaining the rest by the end of the war. Madame Loubat then insisted that just because there was no official classification in Pomerol, her Pétrus was in no way inferior to the other top wines, such as Lafite and Latour, and indeed it now commands much the same price. On her death, Madame L. P. Lacoste-Loubat, a niece, inherited the celebrated 7-hectare vineyard.

Few of the châteaux of Saint-Emilion and Pomerol are imposing places and Pétrus is no exception, just a medium-sized house with a small chai.

Hours: Monday   to   Friday   0900–1200; 1400–1700.

Introduction: By appointment, arranged either directly with: J. P. Moueix, Quai de Priourat, 33350 Libourne. Tel. (56) 51.08.15. or through: Gilbey S.A. Château Loudenne, Saint-Yzans-de-Médoc, 33340 Lesparre, Médoc. Tel. (56) 41.15.03.

Vieux Château Certan,

Proprietors: Héritiers de Georges Thienpont,
Pomerol, 33500 Libourne.
Tel. (56) 51.17.33.

From Libourne take the road D21E in the direction of Montagne. Just beyond the village of Catusseau (3 kilometres), a left fork in the direction of Néac passes Vieux Château Certan in about seven hundred metres. The house itself is *chartreuse* style, not unlike Château Loudenne where the vineyards also come almost up to the windows.

Its wine is in the first dozen of Pomerol and has been known to excel that of its great neighbour, Pétrus. The vineyard is about a third larger than Pétrus, much of the wine going to Belgium since 1924, the year when M. Thienpont, a Belgian wine merchant, became the owner. Among his heirs inheriting on his death, the youngest was M. Leon Thienpont, who lives here, runs the property and can tell visitors all about it in Flemish, Dutch or German.

Hours: Monday   to   Friday   0930–1200; 1400–1700.

Introduction: By appointment, please write or telephone M. Thienpont at least 48 hours in advance.

SAINT EMILION

Introduction

Saint-Emilion wines are hill wines, some grown on chalk and clay around the old town, others on gravelly soil towards Pomerol. The district is twenty miles east of Bordeaux. The town with its moat, old churches and traces of the Romanesque can lay claim to being the most charming of wine towns. Its wine is soaked in history and its history in wine.

A larger district than Pomerol, there are no less than three hundred châteaux belonging to Saint-Emilion and its associated communes, about seventy of them making wine sufficiently good to catch the selector's eye in the classification first made in 1954–5.

In this classification, eleven properties, headed by Ausone and Cheval Blanc, were made *Premier Grand Cru classé*. Nearly sixty others (one or two have been added since) were made *Grand Cru classé*, forming the second category. After every vintage these 'seconds' have to submit samples of their wine to the *Institut National des Appellations d'Origine* (I.N.A.O.) for tasting and may lose their place for that year if their wine is below average.

A third category of about a hundred properties – those not classed above – can call themselves *Grand Cru* (without the *classé*). A fourth comprises the generic wines, which are labelled Saint-Emilion. The following secondary communes have their own separate appellations – Lussac-Saint-Emilion, Montagne-Saint-Emilion, Puisseguin-Saint-Emilion, Saint-Georges-Saint-Emilion and Sables-Saint-Emilion.

Saint-Emilion vineyards divide into two parts – those on the gravelly soil, almost across the western boundary into Pomerol are Graves Saint-Emilion: the remainder, on the hilly slopes nearer the town, are Côtes Saint-Emilion.

A matter of metres outside Libourne, the Pomerol vineyards begin and the road winds by names such as Pétrus, Nenin, Trotanoy and Gazin. Curving west imperceptibly, the wine topographer will suddenly realise that he is in Graves Saint-Emilion when he meets equally familiar names on the gates – Cheval Blanc, Figeac, Ripeau. Here the country is flat but in a kilometre or two the mild slopes on all sides become studded with the vine, the châteaux of Côtes Saint-Emilion rising to a peak in the towers of the town itself.

Within the mediaeval walls there is much to see. The thirteenth-century Château du Roi, a fine parish church, the well in the school garden where the owner once hid her friends from the henchmen of Robespierre, all receive full mention in the guide books. Likewise the monolithic abbey hewn out of rock, its belfry tower on the terrace above; a strange, far from beautiful place, mined for centuries by the monks until its was transformed from a cave to a church supported by ten pillars of original rock.

Every year in the spring, members of *La Jurade de Saint-Emilion*, dressed in their white ermine-lined scarlet robes proceed in procession to the monolithic church for Mass. The *Jurade* is the oldest society for the control of wine in the world, the original *Jurats* being elected when Richard, Coeur de Lion, ruled in Gascony. Their Charter, conferring powers on both magistrates and local councillors, was signed by King John in 1199. Their seal was affixed to every cask of Saint-Emilion wine exported by sea from Libourne to English, Flemish and North German ports. For six hundred years, until the French Revolution, the *Jurade* maintained the Saint-Emilion reputation abroad.

After the First World War chaos gave birth to the system of 'Appellation Contrôlée', which improved standards by defining boundaries and controlling methods, but could not guarantee quality. After the Second World War there was neglect and despair until, in 1948, the *Jurade* was revived.

Once more wines are submitted to it for tasting and approval, their special *Jurade* label being affixed to Saint-Emilion wines of high standard. What cannot of course be wholly prevented is unscrupulous breaking of the law by labelling inferior wine from elsewhere as Saint-Emilion, though recent prosecutions in Bordeaux indicate Government determination to curb abuses.

In September, the *Jurade* are again on view when, from one of the thirteenth-century towers, their *Procureur-Syndic* issues the call for the *Ban de vendanges*, the banquet that heralds the start of the grape harvest.

For the foot-weary, or just plain hungry, the town has good restaurants, notably the Logis de la Cadène and the Hostellerie de Plaisance, each with about twelve bedrooms.

From July to September at the Maison du Vin, Tel. (56) 51.72.17, close to the Plaisance in the Place Clocher, there is likely to be a hostess, who will arrange visits to Saint-Emilion châteaux for members of the public.

Further Information
Syndicat d'Initiative de Saint-Emilion et du Relais Touristique Départemental,
Place des Créneaux,
33330 Saint-Emilion.
Tel. (56) 51.72.03.

Graves Saint-Emilion
Château Cheval Blanc,
Proprietors: Héritiers Fourcaud-Laussac.
33330 Saint-Emilion.
Tel. (56) 51.70.70.

Cheval Blanc is barely ten minutes by car from Libourne and the best route is to leave that pleasant town by D21E through the Pomerol district. Where the road forks left to Néac, keep right toward Montagne Saint-Emilion; a turning right at the next crossroads leads to this 1er grand cru classé in about 800 metres.

Being just on the Saint-Emilion side of the Pomerol border, this celebrated wine ranks as a Graves Saint-Émilion, Graves being the western side which has a gravelly soil. In 1921 Cheval Blanc's wine astonished the wine world and has continued to do so in great years; thus its merit was recognised when an official classification of Saint-Emilion wines came at last to be made in 1955. A modest house, with so small a chai, until modernised and enlarged recently, that château-bottling had also to be permitted in an officially approved cellar in Libourne. It has been owned by the Fourcaud-Laussac family for four generations. The vineyard of 34 hectares makes an average of 12,000 cases a year.

Hours:	Monday to Friday 1000–1200; 1500–1700. Saturday 1000–1200.
Introduction:	By appointment. Please write or telephone in advance to the château.

Côtes Saint-Emilion
Château Ausone,
Proprietors: Vve. C. Vauthier and J. Dubois-Challon,
33330 Saint-Emilion.
Tel. (56) 51.70.94.

Château Ausone, named after the fourth century A.D. Roman poet Ausonius, thought to have lived here, enjoys an elevated position on a southern slope just below the southern gate of the town. The seven-hectare vineyard is a small one so that there is only about one bottle of Ausone for every five made by Cheval Blanc, Ausone's rival of equal top rank, at the other end of the commune.

Some people say that Ausone, which made the outstanding wine of Saint-Emilion fifty to a hundred years ago, has lost ground. M. Dubois-Challon, an expert vigneron, who owned and ran Belair next door in conjunction with Ausone, installed new equipment and increased the quantity of Cabernet Sauvignon grapes in the vineyard. Everybody hopes for an exciting improvement soon. M. Dubois-Challon died in 1974 and Ausone now belongs to his widow and to two Vauthier nephews. Not everybody however believes that Ausone's wine *has* lost ground. One Saint-Emilion proprietor declares that it needs ten to twenty years in bottle and is drunk far too soon.

Meanwhile, Ausone, where the cellar is a deep rocky cave hewn from the soft stone and the view across the vineyard is magnificent, remains an historic château to see.

Hours:	Monday to Friday 0900–1200; 1400–1800.
Introduction:	By appointment. Please write or telephone M. Chaudet at the château or, failing that, try the Maisons du Vin, either in Bordeaux or in Saint-Emilion.

Château Cadet-Piola,
Proprietor: M. Maurice Jabiol,
33330 Saint-Emilion.
Tel. (56) 51.70.67.

From the northern exit of the town take the road D122 north to Montagne. Cadet-Piola is only about 500 metres on the left at the top of the first slope. M. Maurice Jabiol is also the proprietor of Château-Faurie-de-Souchard, another Grand cru classé close by to the north-east where he lives* and of two smaller crus, Château Cadet-Peychez and Domaine de Pasquette.

He speaks 'a little English' and is pleased to have visitors interested in his wines, up to five at a time. They can taste from the cask and wine is sometimes available to buy.

Hours:	Monday to Friday 0900–1100; 1400–1700.
Introduction:	By appointment. Please write or telephone M. Jabiol in advance.

*Tel (56) 51.72.55.

Château La Clotte,
Proprietor: Héritiers de Monsieur G. Chailleau,
33330 Saint-Emilion.
Tel. (56) 51.72.52.

This vineyard is even closer to the town than Ausone. Approaching on D122 from the south, turn right just before entering. The property makes a good wine most years and much of it is sold in Britain. There is, however, an earlier English connection – a cave dug out by English soldiers some time before 1453, the date when General Talbot lost the last battle at Castillon-la-Bataille, ten kilometres down the road. Otherwise there is not a great deal to see here and no tasting because the wine is matured in Libourne.

Since the death of M. Chailleau in 1973, the property has remained in the hands of the family. His daughter, Madame Moulierac, and her husband own the Logis de la Cadène, the restaurant in the Place Marché-au-Bois, which has twelve rooms classed in the *Michelin* as 'plain but adequate'.

Hours: Monday to Friday 0800–1200; 1400–1800.

Introduction: Contact M. Christian Moueix, c/o J. P. Moueix, Quai de Priourat 33350 Libourne. Tel. (56) 51.08.15.

Château La Gaffelière,

Proprietor: Comte de Malet Roquefort, 33330 Saint-Emilion.
Tel. (56) 51.72.15.

From Libourne take N670A, the Bergerac road, turning left after 7 kilometres towards the town of Saint-Emilion, at the little village called Les Bigaroux. The Château is 400 metres beyond the railway.

In the classification of Saint-Emilion, made in 1955, there are only a dozen Premier Grands crus classés and four of them – Ausone, Bel-Air, Magdelaine and La Gaffelière – lie very close to one another on the south-facing slope within a kilometre of the southern gate of the town.

The name Gaffelière is derived from a mediaeval word for a leper colony, which existed here in the Middle Ages. Now there is a 25-hectare vineyard comprising 60 per cent. Merlot, 20 per cent. Cabernet Sauvignon and 20 per cent. Cabernet Franc, which combine to make some 10,000 cases a year.

Visitors see the installation and the park. The proprietor, Comte de Malet Roquefort speaks English.

Hours: Monday to Friday 0900–1200; 1400–1800.

Introduction: By appointment; please write or telephone in advance to the château.

Château Magdelaine,

Proprietor: M. Jean-Pierre Moueix, 33330 Saint-Emilion.
Tel. See below.

Sited just to the westward of Bel-Air and Ausone, off the D122 road at the south of the town, Château Magdelaine

was bought by the Libourne firm of négociants, Jean-Pierre Moueix in the 1950s after being in the hands of the Chatonnet family for three hundred years. The 12-hectare vineyard is south-facing and well placed to make a consistently fine wine, one of the dozen 1er crus classés.

Hours: Monday to Friday 0900–1200; 1400–1800.

Introduction: By appointment which has to be arranged with: J. P. Moueix, Quai de Priourat, 33350 Libourne. Tel. (56) 51.08.15.

Château Monbousquet,

Proprietor: M. Alain Querre, 33330 Saint-Emilion.
Tel. Office in Libourne (56) 51.00.18. Château (56) 51.75.24.

From Libourne take N670A to Castillon for 7/8 kilometre, turning right at the little village called Les Bigaroux; then take the second turning on the left. This should bring you to the handsome eighteenth-century Château of Monbousquet, every man's dream of the perfect retreat. Elegance, a cool informal garden, a lake well-stocked – such is the combination of classic virtues that some past owners have almost forgotten to look after the 30-hectare vineyard. Not so the young and energetic, Alain and Sheila Querre, who took over when his father, Daniel, died in 1973.

On the flat ground some three kilometres below the town of Saint-Emilion the soil becomes more gravelly. The gravel belt in fact extends from Pomerol and Graves-Saint-Emilion so that Monbousquet's wine is nearer in style to Cheval Blanc than Ausone, and under M. Querre, a dedicated vigneron, it should improve steadily. His wife, whose father was an English journalist, described in Peter Dominic's *Wine Mine* magazine how young volunteers came from many countries to pick their grapes each year. Though the vintage is hard monotonous work, the 'United Nations' at Monbousquet sounds a happy party.

Visitors see the chai and cuvier, taste from the cask and can buy in cases or bottle. There is a pamphlet in English.

Hours: Monday to Friday 0930–1200; 1400–1700.

Introduction: Invitation card with route plan is available from agents, but with or without it visitors are welcome, provided they give 24 hours' notice by telephoning to the Maison Querre office number above.

U.K. Agents: Gilbert and John Greenall Ltd., P.O. Box No. 3, Causeway Distillery, Warrington WA4 6RY.

Château Pavie,
Proprietors: Consorts Valette,
33330 Saint-Emilion.
Tel. (56) 51.72.02 or 51.71.71.

Château Pavie, with one of the largest vineyards of
Saint-Emilion, lies a little over a kilometre south-east of
the town, the road to Saint-Laurent-des-Combes being
the one to take. The owner, M. Valette, having two other
vineyards – La Clusière and Pavie-Decesse – formed a
company. The cellars were hewn out of an old quarry and
the wine, well-known in Britain, fully deserves its premier
status.

Hours:	Monday to Friday 0900–1200; 1400–1700.
Introduction:	By appointment. Please write or telephone in advance to the château.

Château Soutard,
Proprietor: M. Le Comte des Ligneris,
33330 Saint-Emilion.
Tel. (56) 51.72.23.

This fine estate, which has been in the Ligneris family
since 1785, is about a kilometre from the northern gate of
the town. From there take the road D17E towards Saint-
Christophe-des-Bardes. A fork left in less than a kilometre
leads to the château.

The Count and Countess des Ligneris are resident
proprietors in this splendid eighteenth-century manor
house. The vineyard is 22 hectares making about 10,000
cases a year of grand cru classé wine.

Hours:	Monday to Friday 0900–1200; 1400–1700.
Introduction:	By appointment. Letter can be arranged through: Gilbey S.A., Château Loudenne, Saint-Yzans-de-Médoc, 33340 Lesparre-Médoc, a company which buys and sells much of the wine of Château Soutard.

GRAVES

Introduction

The Graves region becomes easier to visualise if regarded
as an extension of the Médoc south-east along the
Gironde. In the direction of Bordeaux, after disembarking
from the car ferry at Le Verdon, as has been suggested on
p.73, there are three contiguous regions, each clinging to
the western bank of the Garonne and of much the same
narrow width – 5 to 10 miles. First, there are the twenty
miles through the Bas-Médoc, of little interest until things
begin at Loudenne. Then come 25 miles of celebrated
châteaux through the Haut-Médoc and lastly the 35-mile
length of Graves, that includes the city of Bordeaux at 'the
top end' and the district of Sauternes at 'the bottom'.

Everybody has heard of 'Graves', associating the name
most probably with a fairly sweet white wine, so that this
is what exporters and imitators are apt to provide. But, in
fact, the white wines of Graves vary from fairly sweet to
dry, the latter being the better of the two, partly because
the slightly cloying, earthy taste derived from the gravelly
soil is less marked in them. Better than these – in so far as
white can be compared with red – are the red wines of
Graves, the best being as great as those of the Médoc.

These wines come primarily from two communes –
Pessac, a Bordeaux suburb, barely fifteen minutes from
the city centre by public transport, and Léognan, which is
only seven or eight miles out. Château Haut-Brion is in
Pessac and La Mission Haut Brion, in Talence, as close as
makes no matter to it; each has a white wine vineyard as
well as a red. Getting out of Bordeaux into this wooded

GRAVES & SAUTERNES

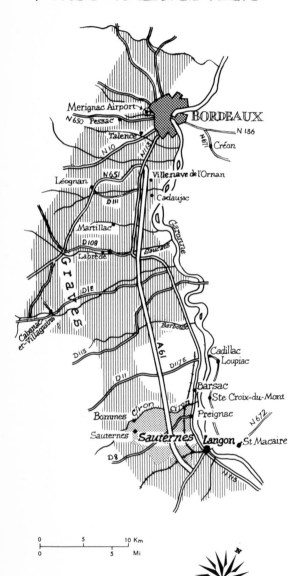

country, intersected by small streams, is easier now that the A61 autoroute to Langon, where Graves ends, has eased the traffic on the main road N113. The village of Léognan is reached by taking a left fork on to N651 and a return to N113 by D111 makes a loop which takes in Châteaux Haut-Bailly, La Louvière and Carbonnieux. (The story of how the monks, who were the owners in the Middle Ages, exported Carbonnieux to teetotal Mohammedans in Turkey under the label 'Mineral Water of Carbonnieux' has many variations.)

The official classification, given below, of the best Graves châteaux dates from 1953 for red wines and from 1959 for white. Properties are widely separated; finding them requires a large-scale wine map, the best solution being a copy of Hugh Johnson's *World Atlas of Wine* in the car.

Since few people seem to realise the excellence of the Graves Grands Crus, the classification is given below; note how the reds outnumber the whites.

CRUS CLASSES DES GRAVES

	Red	White
PESSAC (Suburb of Bordeaux)		
Haut-Brion (classé ler cru en 1855)	*	
Pape-Clément	*	
TALENCE (Adjacent to Pessac)		
La Mission Haut Brion	*	
La Tour-Haut-Brion	*	
Laville-Haut-Brion		*
VILLENAVE-D'ORNON		
Couhins		*
LEOGNAN		
Haut-Bailly	*	
Carbonnieux	*	*
Domaine de Chevalier	*	*
Fieuzal	*	
Malartic-Lagravière	*	*
Olivier	*	*
MARTILLAC		
Smith-Haut-Lafitte	*	
La Tour-Martillac	*	*
CADAUJAC		
Bouscaut	*	*

PREMIERES COTES DE BORDEAUX

This strip, running abreast Graves and clinging to the opposite (right) bank of the Garonne, is even narrower and makes some pleasant white and red wines. Though there are no châteaux for a visiting list, the minor road D10 runs quite close to the river in places, from Bordeaux to Saint-Macaire. Watching the tugs go by I have lunched very pleasantly in the garden of the Hostellerie, A la Varenne at Cambes, a dozen miles from the city, being disappointed to find when thinking of staying a night, that all the bedrooms were on the other side, close to the road.

Beyond Cadillac the communes are Loupiac and Ste. Croix-du-Mont, where the wines are white and sweet; imitating, but far from emulating, true Sauternes from across the river in Graves.

Labrède
Château de la Brède,
Proprietress: Comtesse de Chabannes,
33650 Labrède.
Tel. (56) 21.80.49.

The village of Labrède is about twenty kilometres from Bordeaux on N113, the road to Toulouse.

The château, birthplace and home of Montesquieu (1689–1755) has no great vineyard, although a little white Graves is made there. Montesquieu (Baron de la Brède) inherited vineyards in Martillac, Olivier and Léognan through his father, and others in Armagnac from his mother. Vignerons the world over will warm to the great philosophical historian who wrote, 'I'm trying to write a book which will sell – I don't mind if it's good, bad or indifferent provided it helps re-establish my vineyards after this disastrous year.'*

La Brède is now a national monument, a turreted château with a moat, set in the English-style park that Montesquieu created. Concerts take place and I recall Alfred Deller, the counter-tenor, and his group of singers performing there at the Bordeaux festival, which takes place annually in May.

Hours: Daily 0900–1200; 1400–1800 (1700 from 1 December to 1 March). Closed Tuesdays, Saturdays and Sundays and in January.
Introduction: None necessary; open to the public 3.50F. for the grounds and château, 2.50F. for the park only.

Léognan
Château Haut-Bailly,
Proprietor: M. Daniel Sanders,
33850 Léognan.
Tel. (56) 21.75.11.

From Bordeaux take the Toulouse road N113 for 11 kilometres, turning right on to D111 for Léognan at le Bouscaut. Look out for Château Haut-Bailly on the right after 4 kilometres and one kilometre before Léognan.

Haut-Bailly makes no white wine but its red is certainly the best of this important commune and one of the best half dozen red Graves. They say its proprietor eighty years ago cleaned out his vats just before the vintage with the finest Cognac and then said he forgot to take it out. The wine was wonderful!

From Histoires Véritables quoted by Anthony Rhodes The Princes of the Grape.

M. Daniel Sanders runs both it and Le Mayne, a Barsac property, where he lives. He speaks French, English and Dutch and when he's away his Chef de Culture, M. Michel Charritte, will deputise as host, given notice. Although we've always understood that it was sacrilege to telephone any Frenchman between 12 and 2 p.m., this it seems is the only time to catch M. Charritte when he comes indoors for lunch! But M. Charritte only speaks French – and with his mouth full, we hope, between 12 and 2.

Hours: Monday to Friday 0900–1200; 1400–1730.
Introduction: By appointment; best arranged through Gilbey S.A., Château Loudenne, Saint-Yzans-de-Médoc, 33340 Lesparre-Médoc. Tel. (56) 41.15.03.

Château La Louvière,
Proprietor: M. André Lurton,
33850 Léognan.
Tel. (56) 21.75.87.

From Bordeaux take the Toulouse road N113, forking right at Pont de la Maye seven kilometres from the centre on to N651, which passes through Léognan in another seven kilometres. Turn left at the crossroads in Léognan on to D111, which winds its way back to N113, passing La Louvière on the left after Haut-Bailly.

This large estate of 62 hectares has 50 of them under vine. The wine is a good dry white Graves but not a classed growth. The origins of the vineyard go back at least to 1550 and just before the Revolution the monks of Chartreuse were looking after it. Chief interest however is the château; a rich, early nineteenth-century Mayor of Bordeaux had it built by Louis, architect of the City's much admired theatre.

Hours: Monday to Friday 0900–1200.
Introduction: Appointments made by telephone to the château with Madame Boyer (daughter of the proprietor, Monsieur André Lurton) are preferred.

Pessac
Château Haut-Brion,
Proprietors: Domaine Clarence Dillon S.A.,
Domaine Clarence Dillon S.A., 33600 Pessac.
Tel. (56) 48.73.24.

Leave Bordeaux by the Route de Pessac (N650 to Arcachon). Cross the big boulevard, George V, into the cours du Maréchal Gallieni, which becomes Avenue Jean-Jaurès two kilometres from the big boulevard where the boundary changes from Talence to Pessac. The château is a hundred metres or so on the right.

Pride of Pessac, south-western suburb of the city, Haut-Brion was the only claret from outside the Médoc to be included in the 1855 classification. 'The only Gentile, or at least trans-Jordanian, to be put alongside the pure Israel of Lafite, Latour and Margaux in the very top rank of the Chosen People,' observed that first-of-the-wine writers, Professor Saintsbury. And when a classification of Graves came at last in 1953, there was no reason to deny Haut-Brion a place head and shoulders above the others, although today La Mission Haut Brion makes a strong challenge.

There have been many owners – as well as many distinguished customers. In 1663, Samuel Pepys met the wine for the first time at the Royal Oak, Lombard Street in the City of London, spelling it 'Ho Bryan'. In 1804, after State confiscation and division during the Revolution, Talleyrand, the Foreign Minister, bought it, selling again at a nice profit four years later. Great years for Haut-Brion came after 1840 when a piece of the vineyard lost in 1770 was bought back.

In 1935 the American banker, Clarence Dillon bought the property, forming the company, 'Domaine Clarence Dillon S.A.' in 1958.

| Hours: | Monday to Friday 0900–1100; 1400–1700. |
| Introduction: | By appointment preferred; please write or telephone at least 24 hours in advance. |

Talence
Château La Mission Haut Brion,
Proprietors: Société Civile des Domaines Woltner, 67 rue Peybouquey, 33400 Talence.
Tel. (56) 48.76.54.

Leave Bordeaux by the rue de Pessac (N650 to Arcachon). Cross the big boulevard George V into the cours du Maréchal Gallieni. Then, one kilometre from the boulevard continuing towards Arcachon, take the rue Peybouquey on the left, after the bus station. The entrance to the château is about 500 metres, on the right, at the corner by the railway bridge.

A suburb, less than five kilometres from the city centre, is an unlikely place in which to find a famous vineyard, yet it is easy to visit and certainly worth while. A statue of St. Vincent in the grounds, his mitred cap awry, reminds us of the Lazarite Mission, the seventeenth-century owners, whose order was founded by him. The patron saint of French vine growers is a little dishevelled because, pretty thirsty according to the legend, he asked for leave of absence from heaven. Alas, he broke his leave, being found hopelessly 'plastered' in the cellars of 'La Mission' and was turned to stone for such unsaintly conduct.

For many years until his death in 1975, Henri Woltner, whose father had bought the property in 1919, was the dedicated Director, increasing the excellence of the wine to the point where it now commands the highest price below that of the first growths. Vitrified steel *cuves* for fermentation at controlled temperatures replaced wooden vats long ago, with marvellous results, but not all proprietors favour them, nor the capital expenditure necessary. The Woltners' second vineyard Laville-Haut-Brion makes a superb dry white Graves and a third, La Tour-Haut-Brion, a good red Graves.

The chapel, a collection of Holy Water stoups and another of Delft dishes left by Dutch merchants are all worth seeing. The wines spend their two or three years in oak casks in the château, but are bottled in other cellars in Bordeaux. Tasting the young wines from the cask is possible at the château for up to 8 people. Otherwise up to 15 can be accepted at a time. English is spoken; no wines are for sale at the château itself. To buy bottles, contact the négociant office, Woltner Frères, 17 cours du Médoc Bordeaux. Tel. 29.28.08.

| Hours: | Monday to Friday 0900–1100; 1400–1700. Closed annually 14 July–20 October. |
| Introduction: | Letter of introduction should be obtained from a reputable wine merchant and appointment arranged by letter or telephone to the château or through agents: Findlater Matta Agencies, Windsor Avenue, Merton Abbey, London SW19 2SN. Tel. 01–935 9264. |

SAUTERNES

Introduction

THE 'NOBLE ROT'

Nobody really knows where and when it was discovered that some rather rotten, mouldy old grapes picked late can make a glorious, sweet dessert wine far more luscious than the wine made from healthy, ripe grapes of the same vineyard picked earlier. It might have been in Hungary, in the Tokay vineyards during the seventeenth century or, in the Rhineland at Schloss Johannisberg, where the labourers did not dare start picking until their ecclesiastical lord and master gave the word and – circa 1716 – he happened to forget for a week or two. It could have been in Sauternes but this seems less likely because there, the significance of the *pourriture noble*, or noble rot, responsible for this phenomenon was not realised much before 1860.

Given a fine, warm but rather humid September, the grapes are attacked by a mould, *botrytis cinerea*, causing them to shrivel, which evaporates the water content and concentrates their sugar. Picking requires individual selection, going through the vineyard half a dozen times or more between September and December. Rain on any scale puts paid to further operations and curtails the crop. Small wonder that these dessert wines are expensive!

In Tokay they use their late-gathered, Aszu wine to sweeten wine already made from normally ripe grapes. In Germany these *beerenauslese* and *trockenbeerenauslese* wines are so rich and rare that prices are huge, even though the wines only achieve 7–8 degrees of alcohol against 10–11 degrees for a typical table wine. The great Sauternes wines however reach 14–17 degrees quite naturally, and the output of the Sauvignon and Sémillon grapes far exceeds anything the Rhine Riesling, much further north, can do.

Sauternes, about thirty miles out of Bordeaux off N113, is only seven miles by four and is part of Graves. Getting there from Bordeaux has been much expedited by the completion of the A61 autoroute to Langon. There are five parishes – Barsac and Preignac on the main road, Sauternes, Bommes and Fargues off it, four to five miles to the south. From Preignac two minor roads, D109 and D8, cross the railway close to Preignac station, running parallel the length of the district, the principal châteaux being a good three or four miles from the point where the railway is crossed.

The proximity of the Garonne and the streams running through Graves probably give rise to the misty Septembers required for the *pourriture noble*, yet a good year for Sauternes occurs at best one in three and a great year one in ten, or even longer. So rich are the wines of the *crus classés* that a glass or two served really cold with dessert – a pear or a peach perhaps – is enough for most people. The newcomer should certainly beware; dashing down by the bowl may well be followed by dashing down the corridor.

SUGGESTED TOUR

(*Michelin* Maps Nos. 75 and 79)
Entre-Deux-Mers – Sauternes – Graves
Leave Bordeaux by N136 and N671 to Créon, then south-west through the pleasant hilly country of Entre-Deux-Mers, 'between the seas' of Garonne and Dordogne, on N671 to Sauveterre-de-Guyenne. In this village turn right on to N672, which leads down to the Garonne at Saint-Macaire. Sauternes lies across the river and Château de Malle, a classified historical monument (p.102) could be visited. The D8 road through Fargues and the Sauternes district reaches Villandraut in under ten miles where the ruined castle is another historical monument. It was built by Clément V, the Bishop of Bordeaux, who became the first Avignon Pope in 1305, giving first his name to Château Pape-Clément and later his title to Châteauneuf-du-Pape, see p.57.

Villandraut is 35 miles from Bordeaux and a rural return can be made on D110 to Balizac and Origne, thence northwards on D116E to Cabanac-et-Villagrains. Total distance about 100 miles (say 160 kilometres).

A shorter tour could be made, cutting out Créon, by taking the right bank river road from Bordeaux to Saint-Macaire.

Further Information

Maison du Vin, Château des Ducs d'Epernon, 33410 Cadillac.
Tel. (56) 62.01.38.

Barsac
Château Coutet,
Proprietor: M. Rolland Guy,
33940 Barsac.
Tel. (56) 62.91.11.

From Bordeaux, take the Toulouse road, N113, for about 37 kilometres to Barsac.* Turn right in the centre of this village, make for the station and cross the railway *south* of it. On the other side, the D114 road runs south along the railway line for about 500 metres and the second turn on the right after the level crossing, leads to the château in another 500 metres.

At the time of the 1855 classification, Yquem, Coutet and Filhot all belonged to the Marquis Bertrand de Lur-Saluces – with de Malle also in the family. No wonder a pleasant air of august, orderly elegance lives on at all these delightful places! Parts of the two châteaux, Yquem and Coutet, go back to the thirteenth century and the wells in their courtyards are identical.

In M. Rolland Guy, owner since 1925, Coutet has a dedicated owner, who is said sometimes to have followed the little wine he has permitted to be exported in cask to make sure that it was being bottled properly by responsible recipients. Coutet and Climens are the two Barsac commune wines among the nine that formed the premier class in 1855. Since Barsac has its own appellation within Sauternes, Château Coutet-Barsac is often the wording of the label. A charming château to visit but no English spoken.

Hours:	Monday to Friday 0900–1200; 1500–1900.
Introduction:	By appointment. Please write or telephone at least 48 hours in advance to the château.

Preignac
Château de Malle,
Proprietor: Comte de Bournazel,
Preignac, 33210 Langon.
Tel. (56) 63.28.67.

From Bordeaux the quickest and easiest way now is to take the autoroute, A61, to Langon (toll 4F.). From the Langon exit retrace steps on D116, following carefully the direction signs marked 'Malle'. The alternative from Bordeaux is to take the Toulouse road, N113, for about 40 kilometres to Preignac. Turn right there on D8E, following similar direction signs for about two kilometres to the château.

Château de Malle's wine is one of the twenty-four Sauternes Grands crus classified as such in 1855. The château itself, built in the seventeenth century by a direct ancestor of the present owner, Comte de Bournazel, is a classified historical monument, to which he devotes much care and attention. Its original furniture and the formal Italian garden with balustrades, statues and a theatre are pleasing features.

With Sauternes so dependent on a fine autumn, the energetic and enterprising Count has planted part of the estate with red grapes, which now produce an A.C. Graves wine, Château de Cardaillan. He also makes a dry white Bordeaux Supérieur, Chevalier de Malle.

Hours:	Easter to 15 October 1500–1900 every day, including Saturdays, Sundays and national holidays but closed Wednesdays. Admission 5F. (3.50F a head for groups).
Introduction:	None necessary.

Château de Suduiraut,
Proprietor: M. Léopold Fonquernie,
33210 Langon.
Tel. (56) 63.27.29.

From Bordeaux, take the Toulouse road, N113, for about 40 kilometres to Preignac.* Turn right on D8E; go to the station and cross the railway. From this point Suduiraut is about 3 kilometres and has a long drive half right after passing the hospital.

Sood-we-row, a difficult pronunciation for Anglo-Saxons, is a charming, imposing place making a Sauternes that comes at least a close second to its august, next-door neighbour Yquem. Indeed, M. Merlet, the Régisseur, points to Yquem declaring his own ground is lower and hotter and that it makes the greater Sauternes.

The vineyard comprises 100 hectares in a parkland estate of about 200 hectares. For visitors the main attraction will probably be the gardens, designed by Le Nôtre of Versailles fame, that lie behind the elegant seventeenth-century house. M. Léopold Fonquernie, an industrialist, bought Suduiraut in a neglected state in 1940 and this 1er cru wine now repays much effort, though hardly the cash.

Hours:	Monday to Friday 1000–1200; 1500–1700.
Introduction:	By appointment; best arranged through Gilbey S.A., Château Loudenne, Saint-Yzans-de-Médoc, 33340 Lesparre-Médoc Gironde. Tel. (56) 41.15.03, who usually buy a large proportion of the output. (M. Merlet requests two or three days' notice and telephone calls to him between 1130 and 1215 are best.)

---

*Quicker and easier now is to follow autoroute A61 to Langon (toll 4F.) *returning* on N113 to Barsac.

*Quicker and easier now is to follow autoroute A61 to Langon (toll 4F.) *returning* on N113 to Preignac.

Sauternes
Château Filhot,
Proprietress: Comtesse Durieu de Lacarelle (née de Lur-Saluces),
33210 Langon.
Tel. (56) 62.61.09.

From Bordeaux take the Toulouse road, N113, for about 40 kilometres to Preignac.* Turn right on to D8E, go to the station and cross the railway south of it. From this point continue on D8E for six kilometres, turning right towards the village of Sauternes at the D125, Mazères-Sauternes, crossroads. In another 1·2 kilometres several roads meet. The second on the left leads in five hundred metres (past a pond) to the château, which is in fact marked on *Michelin* Map No. 79.

There is nothing 2me classe about this parkland estate of 315 hectares, which includes the vineyard of 55 hectares. Even this status, accorded to the wine in 1855, is generally considered to be gross underrating. The huge house with its pair of wings and the lake in front might be an English stately home.

The property was bought by Romain de Filhot in 1709 and the last Filhot had bought Château Coutet as well before being condemned to death in the Revolution. In 1807 his heiress married the Marquis de Lur-Saluces, which explains why later on the wines of Yquem, Filhot and Coutet all came to be sold as 'Monopole des Crus du Marquis de Lur-Saluces'.

In the crisis of 1939 Château Coutet was sold and the Comtesse Durieu de Lacarelle bought Château Filhot, with the right for her and her heirs to use her maiden name, Lur-Saluces, granted by her brother the late Marquis Bertrand de Lur-Saluces, then owner of Yquem. The Countess is now represented at Filhot by her grandson Comte Henry de Vaucelles.

The 1970 vintage was the first to be matured for three years, not in wood but in new glass-lined containers and the wine is thought to be the better for it. It has the reputation of being the least sweet of all Sauternes and is one to go for in good Sauternes years.

Hours: Sunday to Saturday 0900–1200; 1400–1800.

Introduction: By appointment; best arranged either by a recognised wine merchant or through the C.I.V.B., 1 cours de 30 Juillet, Bordeaux, Tel. 44.37.82 or through the Syndicat d'Appellation d'Origine de Sauternes, Tel. Sauternes (56) 62.60.37.

*Quicker and easier now is to take autoroute A61 to Langon (toll 4F.) and then to follow the direction signs to Sauternes.

Château d'Yquem,
Proprietor: Héritiers de Lur-Saluces,
33210 Langon.
Tel. (56) 62.61.05.

From Bordeaux the quickest and easiest way now is to take the autoroute, A61 to Langon (toll 4F.). From the Langon exit, follow the direction signs to Sauternes. These lead to the Preignac-Villandraut road, D109, which runs along the Yquem vineyard and the château can be seen on the right.

The name Lur-Saluces is almost synonymous with great Sauternes. In 1785 when the family was living at Château de Malle, Count Lur-Saluces married the last successor of the Sauvage d'Yquem family and in 1807 his son married one of the (Château) Filhot daughters.

In 1787 Yquem acquired a client of the highest distinction, Thomas Jefferson, who spent five years in France after achieving some notable humane reforms as Governor of Virginia. Two hundred and fifty bottles cost him 340 livres, including packing and delivery.

The châteaux of the Sauternes district were first classified with those of the Médoc in 1855, Yquem being put in a top class by itself, a position which nobody seriously disputes today. Count Alexandre de Lur-Saluces now runs the estate, having taken over from his uncle, who died in 1968. Of the Domaine's 175 hectares, 100 comprise the vineyard, quality being rated of such importance that twenty of these either lie fallow or are not bearing after replanting. Only a proportion of the year's wine will be considered good enough for the château label. The rest will be sold just as 'Sauternes'.

Like Lafite, the turreted château has a quiet orderly yet informal air about it. In summer it is floodlit and can be seen for miles around. Concerts of chamber music are held in the grounds during the Bordeaux festival in May. But it is wine, not music, that has drawn the visitors in their hundreds, because it really is something of a miracle that plain fermented grape juice from some slightly rotten, over-ripe grapes can be turned into such an incredible deep golden nectar.

Spelt d'Yquem but spoken of as Yquem, the Count himself writes that he knows of no rule governing the use of the particle.

Hours: Monday to Friday 0900–1200; 1400–1700. Saturday 0900–1200.

Introduction: Tours of the cellars are largely limited to members of the wine trade and those closely connected with it. The annual production (only 82,000 bottles on average) is sold to the big Bordeaux houses, who act as world-wide distributors. Individuals wanting an introduction to Yquem should approach a good wine merchant, who is advised to seek an appointment through one of the Bordeaux shippers.

SOME CELEBRATED SHIPPERS

Introduction

LES NEGOCIANTS–ELEVEUR

Since the early eighteenth century, Bordeaux has attracted both Frenchmen and other Europeans to set up in business as wine shippers. Their correct title is *négociants-éleveur;* both as buyers and sellers they are négociants and in their cellars they mature, bring-on or 'elevate' the wines they have bought.

Formerly they bought almost all their Bordeaux wines in wood, the link between them and the château proprietors being the brokers or *courtiers.* Prime unit in the Bordeaux market was the *tonneau,* equivalent to four Bordeaux hogsheads of 225 litres each and it is from the tonneau that our words ton and tonnage are said to be derived.

With prosperity in Europe, and claret drinkers increasing across the Atlantic in the U.S.A., indeed across the Pacific to Japan, proprietors began to undertake much more château bottling to ensure authenticity. Simultaneously shippers were becoming far less inclined to bear the expense of maturing large stocks. Instead they tend now to make agreements direct with château proprietors, buying the entire crop, or a large part of it, for a period of years direct from this property and that.

Though château names may have greater appeal, the Bordeaux wine trade could not have been successfully established world-wide without the shippers and this last part of the Bordeaux section gives some of those who welcome visitors. With the exception of Bardinet, Barton et Guestier and Gilbey, they are in the City itself, mostly established by the river that still takes their wares to their more distant customers.

Blanquefort
Bardinet S.A.,

B.P. 513, Domaine de Fleurenne, 33290 Blanquefort.
Tel. (56) 90.92.95.

A change from wine, this old company moved out of Bordeaux in 1975 to new premises on the *Zone Industrielle* in a 14-hectare park at Blanquefort, 15 kms from Bordeaux, on the road to the Médoc.

Having established themselves as liqueur distillers in Limoges in 1857, Bardinet had moved to Bordeaux at the end of the century, the better to import and distribute rum, notably Rhum Negrita (with the creole girl on the label). Their other products include Bardinet brandy, cordial syrups and ready-mixed rum punches, some of which visitors should see and sample.

Hours: Monday to Friday 1430–1630. Closed August.

Introduction: By appointment. Letter arranged through agents (for Rhum Negrita and Bardinet brandy):
United Rum Merchants Ltd., Battlebridge House, 97 Tooley Street, London SE1.
Tel. 01-407 3522.

Barton & Guestier S.A.,
Château de Dehez, 33290 Blanquefort.
Tel. (56) 44.51.44.

Between 1725–1780, Thomas Barton, an Irishman of Anglo-Irish stock established a shipping house, made a fortune and was probably the first 'foreigner' to buy a Médoc property (Le Boscq). Indeed, from a list of a dozen leading shippers in Bordeaux in 1787, Barton is the only name that would recur in a list made now. During the Revolution, Hugh, the grandson who had succeeded Thomas, was imprisoned, escaping to Ireland with his wife, while Daniel Guestier, his Bordelais partner, ran things till it was safe to return.

So the partnership continued from one generation to the next until, only in recent years, the last Guestier was killed in a car crash and Ronald Barton retired to his St. Julien home, Château Langoa*. The firm now belongs to

---

*The Pergamon Press published in 1971 *Fide et Fortitude* by Cyril Ray, telling the story of Langoa and Léoville-Barton, which Hugh Barton had bought in 1821, a hundred and fifty years before.

Seagram, the Canadian Company, which has established these new headquarters, covering over 200,000 square feet at Blanquefort, fifteen minutes by car from Bordeaux, where the *Route des Châteaux* through the Médoc begins.

Visitors see the cellars, bottling plant, etc., and can buy wines. An English hostess usually receives them.

Hours:	Monday to Friday 1400–1700.
Introduction:	Not necessary, casual visitors welcome. Any special arrangements could be made through agents: Cock Russell Vintners Ltd., 17 Dacre Street, London SW1H 0DR. Tel. 01–222 4343.

City of Bordeaux
Calvet,
75 Cours du Médoc, 33003 Bordeaux.
Tel. (56) 29.75.75.

The Cours du Médoc begins on the Quai des Chartrons, running away from the river in the general direction of the Médoc.

Calvet began on the Rhône at Tain l'Hermitage in 1818, Jean-Marie Calvet being the son of a doctor, who practised at Anse, near Lyons. During the next fifty years the firm progressively bought and sold wines from many parts of France, the Bordeaux business being handled by an agent. In 1870, Octave Calvet, who had succeeded his father, built the first Bordeaux headquarters, covering a large area along the Cours du Médoc, more or less midway between the railway station, where Médoc wines arrived, and the quay from which they could be shipped.

In 1966 the offices here were completely destroyed by fire, although the cellars below them were unharmed. Rebuilding, completed by 1969, included new chais behind the new offices.

Calvet are unique among the big Bordeaux houses in that they own no châteaux themselves. They preferred to establish contacts that earned them a great reputation for their own bottlings of châteaux wines, particularly of crus bourgeois and, more recently, V.D.Q.S. wines from various regions.

Jean and Hubert Calvet, great grandsons of the founder, assisted by Bruno a cousin, run Calvet today.

Hours:	Monday to Friday 0900–1230; 1400–1700. Visitors however are strongly recommended to arrive shortly before 1100 or 1500, at which times interesting tours are conducted by multi-lingual hostesses.
Introduction:	None necessary. Any special arrangements could be made through agents: Sherry House Ltd., 92 New Cavendish Street, London W1M 8LP. Tel. 01–580 0301.

Cruse & Fils Frères,
124 Quai des Chartrons, 33000 Bordeaux.
Tel. (56) 29.35.83.

Herman Cruse, youngest son of a Danish family, began here in 1819 and the firm has never moved. A brilliant business man, at the time of the revolution of 1848 he bought Médoc wines of the 1847 vintage prodigiously and cheaply. This probably financed the buying of Laujac (p.90), Pontet-Canet (p.88) and Giscours (p.81) during the next thirty years. Later the family acquired d'Issan (p.82) La Dame Blanche and Le Taillan (p.80) – altogether a fine Médoc 'portfolio', for which the firm acted exclusively as 'stockbroker'.

Christian Cruse, who died in 1974, handled the British account for over fifty years and was a leading figure in Bordeaux with a profound professional knowledge. His son, Edouard Cruse takes an increasing share in the work and there have been many other members of the family in this celebrated company.

During a 45-minute visit, reception room, maturing cellars and bottling plant are shewn. The Cruse family speak English so well themselves that they are likely to insist that an English-speaking guide is available.

Hours:	Monday to Friday 0830–1200; 1330–1700.
Introduction:	Visits may be arranged by telephone – ask for the hostess. Those wishing to take a letter of introduction may obtain one from: Rutherford Osborne Perkin Ltd., Harlequin Avenue, Great West Road, Brentford, Middlesex. Tel. 01–560 8351.

Ed. Kressman & Co.,
72 Quai de Bacalan, 33000 Bordeaux.
Tel. (56) 29.24.30.

Edouard Kressman, a German, set up his own company in 1871 after working with other firms from 1858. The offices are on the left bank, the Quai de Bacalan being a continuation of the Quai des Chartrons, downstream towards the Pont d'Aquitaine.

The company is still a family concern, dealing in the fine wines and brandies of France, particularly those of Bordeaux and Armagnac; their products go out to fifty countries. They own the Graves Château La Tour Martillac, rated a Grand Cru classé for both its red and white wine. The latter is made from some of the oldest vines in Bordeaux, planted in 1884, and Jean Kressman is justifiably proud of this, the yield being small and not very commercial. The firm also have a bottling plant at Moulis in the Médoc.

Visitors see over the cellars and may taste one or two wines.

Hours:	Monday to Friday 0900–1200; 1500–1700. Closed August.
Introduction:	Casual visitors can be received but visits by appointment through a letter of introduction from agents are much preferred. Percy Fox & Co. Ltd., Bearbrook House, Oxford Road, Aylesbury, Buckinghamshire. Tel. Aylesbury (0296) 5966.

Min Marceau,
57 rue Minvielle, 33000 Bordeaux.
Tel. (56) 29.11.31.

This is a comparatively small family company, established for over one hundred years; their office and cellars are fifteen minutes' walk due north from the Place Gambetta, in the centre of Bordeaux. They ship in wood and in bottle to many countries and are pleased to shew people round and give them a wine or two to taste.

Hours:	Monday to Friday 1000–1200; 1500–1750.
Introduction:	By letter from agents: Rawlings Voigt Ltd., Waterloo House, 228/232 Waterloo Station Approach, London SE1 7BE. Tel. 01–928 4851.

Maison Sichel S.a.r.l.,
19 Quai de Bacalan, 33300 Bordeaux.
Tel. (56) 29.52.20 or 29.23.08.

The name Sichel has been synonymous with fine wine for over a century. The firm began in Mainz, starting a Bordeaux branch in 1883 on the Quai de Bacalan, a continuation of the Quai des Chartrons. With France and Germany at war in 1914, Herman Sichel made the Bordeaux house a separate company. After the war, his sons, Herbert and Allan, joined – Allan becoming a notable figure in the London wine trade and author of *The Penguin Book of Wines*, published just after his death in 1965. Peter Sichel, Allan's son, is now head of the firm in Bordeaux.

The Mainz side, known the world over for its Liebfraumilch Blue Nun, ran independently until 1967 when the two houses merged once again.

Bordeaux wines bought by Sichel are aged in cask in the cellars, which extend for almost half a mile under this fine eighteenth-century building on the river. The bottling plant is now out of the city, at Ludon-Médoc. In 1966 the company was the first to set up a Vinification Centre (at St. Maixant), enabling *grapes* to be bought from the growers so as to make wine under the best possible conditions. Château Palmer (p.83) and Château d'Angludet (p.81), where Peter Sichel and his family live, both belong to the company.

An English-speaking person is always to hand at the main office and, given sufficient notice, visitors can be shewn over any of the places mentioned above, as well as other Bordeaux châteaux.

Hours:	Monday to Friday 0900–1130; 1400–1630.
Introduction:	By appointment. Letter of introduction from agents: H. Sichel & Sons Ltd., 4 York Buildings, London WC2N 6JP. Tel. 01–830 9292.

St. Yzans-de-Médoc
Gilbey S.A.,
St. Yzans-de-Médoc, 33340 Lesparre.
Tel. (56) 41.15.03.

The purchase of Château Loudenne by Walter and Alfred Gilbey in 1875 gave them a Bordeaux base, with a landing place far to seaward of the city itself. Having bought the property for £28,000, they proceeded to spend £64,500 on a new *cuvier*, a new cooperage and a new chai, with its own tramway to their quay.

In June 1975, when Edward Heath was the centenary guest of honour there, Cyril Ray wrote for the occasion: 'The vast chais at Loudenne are the warehouses for Gilbey Société Anonyme, which ships all over the world the wines it gathers from every corner of the Bordelais, ranging from the simplest petits châteaux to the noblest of Loudenne's Médoc neighbours. Though the chais are Victorian, and the exterior of the château itself elegantly eighteenth-century, modern houses for the estate's workers form a model garden village: the 120 hectares or so of farmland (30 of them under vines) are scientifically supervised.'

For route directions, hours, introduction, etc., see Château Loudenne, p.89.

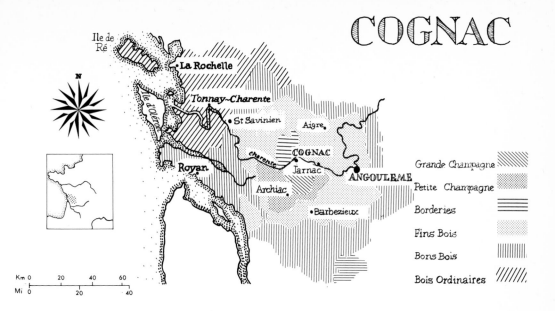

COGNAC

Introduction

GETTING THERE

Cognac, a town with about 25,000 inhabitants, lies on the left bank of the river Charente, 65 miles north of Bordeaux and 290 miles from Paris.

By road: Cognac is about 45 minutes from Angoulême (26 miles) on N10, the Paris-Bordeaux road. The drive from Bordeaux-Mérignac airport takes about two hours.

Royan, the resort at the mouth of the Gironde estuary, from which a large car ferry plies regularly across to the Médoc, is only 40 miles away; La Rochelle and the off-shore islands 60 miles or less.

By rail: There is a connecting service to Angoulême on the Paris-Bordeaux line.

By air: The nearest airport served by regular London and

Paris services is Bordeaux-Mérignac but Cognac does have its own airfield where charter flights, as well as private planes belonging to firms like Martell and Hennessy, land and take off.

By sea: Passages in wine boats from the Port of London to Bordeaux, still possible in the early 60s, have ceased. Much more wine (and Cognac) is now bottled at source, being transported to Britain by road and car ferry.

A WINEFIELD ON CHALK

Like Champagne, Cognac is a winefield on chalk covering much of Charente and Charente Maritime. In 1152, when Eleanor of Aquitaine married England's future Henry II, the two provinces were part of her dowry and for over two hundred years made English history. We have good reason to be grateful to the French Plantagenet queens; homesick in Britain for their native wine and food, they encouraged the wine trade by granting export licences to La Rochelle and other Atlantic ports. The wine was Charentais wine, the river enabling it to reach sheltered bays in the lee of the Ile d'Oléron, where English ships could load the precious cargo from the smaller river craft.

But Charentais wine was low in alcohol and travelled badly; with heavier duties imposed to pay for constant wars, the trade was killed gradually by stronger competitors – the wines of Bordeaux. By degrees the idea came to distil the local wines. With the Huguenots entering the trade increasingly during the seventeenth century, exports increased, until around 1700, the entire production was going abroad. The good news that Cognac brandy, left long enough in casks of finest oak, developed a matchless bouquet and flavour had travelled fast, particularly in Britain and northern Europe.

The vine that makes the rather weak, acid white wine is called the Ugni Blanc or Saint-Emilion des Charentes. The best results come from two small patches of very chalky soil – Grande Champagne and Petite Champagne, south of Cognac town. North of the town, there is the next best patch called the Borderies. After that, the quality declines progressively moving outwards through the less chalky, once-wooded lands of Fins Bois, Bons Bois and Bois Ordinaires.

DISTILLING THE WINE

In the entire region there are said to be 70,000 small farmers making and distilling their wines into cognac. The operation, involving two distillations in a pot still and cleaning the boiler between them, will take sixteen hours. Two hundred gallons of wine should literally boil down to 20 gallons of cognac – raw and colourless, but at 120° proof, strong enough to take the roof off any taster's mouth. The art is handed down from father to son; it needs to be, for distilling is a two-man job. In every distillery – even in the great new ones which the big firms build, Hennessy's at Le Peu for example – there is a small bedroom, like a Captain's sea cabin below the ship's bridge, provided for brief sleep. A farmer will sell his cognac, for a price agreed earlier at vintage time, to a shipper with whom he and his family have dealt, sometimes for generations. Some is likely to be kept as an investment. Thus, distributed over the region, there are always stocks of brandy of all ages, which can be sold by mutual arrangement at any time.

All brandies mature in the cask for three years and some for fifty or more. The raw spirit, slowly taking tannin from the Limousin oak, changes colour and, equally slowly, develops its bouquet and taste. Most brandies are blends of different years and from different districts; having insufficient age to have gained much colour from the wood, their colour is imparted by a little caramel. True cognac is dry, smooth and *without* after taste.

DISTINCTION WITH AGE

Three-Stars indicate the firm's standard quality, a blend of average age four years perhaps. But stars and letters tend to be out of favour because one firm's Three-Star could be better than another's V.S.O.P. Hennessy have gone over to 'Bras Armé' (from the family crest) instead of Three-Star, while Martell prefer 'Médaillon' for their V.S.O.P. quality.

LIQUEUR BRANDY

There is no specific definition but the term generally implies qualities *above* Three Star. V.S.O.P. should indicate a cognac of seven-twelve years' blend age, the initials standing for 'Very Special Old Pale'.

Fine Champagne is a blend from the two great districts, Grande and Petite Champagne, which must contain 50 per cent. from the former. 'Old Landed' brandy is liqueur brandy, which wine merchants in Britain used to ship and mature in bond for up to twenty-five years until they deemed it ready for bottling. Much of it was of a single year, i.e. vintage liqueur brandy, but by French law this is no longer practicable. In cask, brandy evaporates and has to be topped up. The French, with their usual commendable logic, ask, 'How can anybody be sure that a vintage cognac is topped up with the same quality of the same vinage?'

NAPOLEON NONSENSE

'Napoleon' brandy continues to deceive people. As a brand name it is perfectly legitimate but the term in no way indicates great age, quality, or value. A cobweb-covered bottle of wine may have some or all of these qualities, but brandy does not *improve* in the bottle, only in the cask. Brandy of 1815 would have been bottled certainly no later than 1865. It could taste no better than the same quality and style of brandy of 1927 that might be bottled in 1977, also after fifty years in cask. Once bottled, brandy will keep as long as the cork lasts, which should be around twenty-five years, but it will not improve. Though every reputable wine writer points out that 'Napoleon' brandy is nonsense, there has been at least one case in Britain of a disreputable wine merchant pretending otherwise on television. 'Napoleon' and similar cobweb-covered bottles occasionally fetch £25/£50 at wine auctions, thereby demonstrating the old adage that a fool and his money are soon parted.

For the drinker, the preliminaries are very simple. A measure of Three-Star with ice and a slice of lemon, topped up with ginger ale is a 'Horse's Neck' and a most refreshing drink, though it has never cured my hangover. Liqueur brandy is spoilt by dilution, though coffee or plain water drunk separately are very satisfactory with it. A medium-sized balloon glass, allowing the palm and fingers of one hand to be 'cupped' round the base, permits fine cognac to be warmed slowly, allowing the nose to savour the rising bouquet. More drastic methods, like pouring hot water into the glass first and then emptying it, which waiters in Spain seem to fancy, are not recommended.

HOLIDAY PROVINCE

The Charentais province claims the best butter and some of the best melons of France. Overlooked by tourists tending to press on south for more sunshine, this rolling landscape of corn and vine, with grey stone red-roofed villages and little Romanesque churches, is a peaceful place for a holiday. Through it the Charente passes slowly to the sea; daily in summer, a river boat (with meals served) goes from Cognac to Saintes and Saint Savinien and back.

On the Atlantic coast, where there are oyster beds galore, the three islands – Oléron, Aix and Ré – are worth a visit. Oléron, the largest, is now joined to the mainland by a fine road bridge.

The old naval port of La Rochelle is delightful. Few towns can boast a more agreeable walk on a sunny day than that from the Place Verdun to the old port, down through the Parc Charruyer and back under the arcades of the Rue Albert 1er, the main shopping street.

Historically, Cognac is noteworthy as the birthplace of François I, the French monarch who met Henry VIII on the Field of the Cloth of Gold. From his plinth in the town's main square he surveys the scene. Fifty miles to the northward, in 1745 at Fontenoy, there was a battle which marked the end of the age of chivalry. It was the last in which the two commanders began by exchanging ceremonious salutes before each offered the other the advantage of firing first.

No wonder Dr. Johnson was to write, 'He who aspires to be a hero must drink brandy'!

RECOMMENDED DRINKING

Checking the fermentation of fresh grapes by adding brandy has been practised by the Cognaçais for at least three centuries, as it has up the Douro in Portugal. There the result is port; here it is a pleasant rather sweet aperitif, Pineau des Charentes. Locally they drink it 'straight'; Cyril Ray (see Recommended Reading) suggests six of Pineau to one of vodka served very cold.

RECOMMENDED READING

Cognac by Cyril Ray Peter Davies London 1973.

Further Information

Bureau National du Cognac, 3 Rue Georges Briand, 16100 Cognac.

Cognac
Cognac Exshaw S.A.,
127 Boulevard Denfert-Rochereau, 16100 Cognac, Charente.
Tel. (45) 82.40.00.

This is a new address, Exshaw having moved from Bordeaux to Cognac in the early 1970s. Denfert-Rochereau is the main boulevard of the town connecting the bridge to the main Place, François Ier.

John Exshaw – an Irishman like James Hennessy – established his business and cellars in Bordeaux in 1805 and there they remained until 1975, the business having been sold to Otard.

Operating from Bordeaux did not prevent his firm from selecting the best brandies from the Charente and the best oak from Limoges, in which to mature them. John Exshaw's brandies became justly famous among the *cognoscenti*.

In Victorian times the Exshaw bottle bore no label and was quoted as 'gold value' in India. Even before the Suez Canel opened in 1869, the Exshaws had become the leading cognac exporters to the Far East, the previous cargo being carried across the Isthmus by camel.

Visitors are shown the cellars and bottling plants.

Hours:	Monday to Friday	0930–1200; 1400–1700.
Introduction:	By appointment. Letter should be obtained from agents: Rutherford, Osborne & Perkin Ltd., Harlequin Avenue, Great West Road, Brentford, Middlesex. Tel. 01–560 8351.	

Société Jas. Hennessy et Co.,
1 rue de la Richonne, 16100 Cognac.
Tel. (45) 82.14.12.

There should be no difficulty in finding the Hennessy establishment, which is easily seen from the bridge crossing the Charente when coming into Cognac from the north (N141 and 731).

In 1740, Richard Hennessy, third son of Charles Hennessy Squire of Ballymacmoy, County Cork, settled in Cognac, founding in 1765 the first Hennessy company, later re-formed by his son James in 1813. Today, members of the family still play an active part in a huge company which owns five hundred hectares of vineyards, controls over thirty-one distilleries and stocks ninety million bottles of cognac.

Hennessy stocks in wood are believed to be amongst the largest of all the cognac companies; in fact, they claim to hold the largest stocks of aged cognacs in the world.

Distilling being a winter operation, carried out mainly in small distilleries and country farms, the visitors' tour here covers the main reception, blending and bottling departments. And with the precious liquid flowing freely from this vat to that, the scene is impressive. Since 1974 Hennessy have provided a boat to take visitors across the Charente to the ageing warehouse of La Faiencérie, where the tour continues among the thirty thousand casks, in which their cognac slowly matures.

The tour is conducted in English and souvenirs are on sale.

Hours:	Monday to Friday 0830–1100; 1345–1630.
Introduction:	Normally none necessary, but prior notice is appreciated for large parties, or for special arrangements and should be made through: Gilbey Vintners Ltd., Gilbey House, Harlow, Essex CM20 1DX. Tel. (0279) 26801.

Martell & Co.,
16101 Cognac Charente.
Tel. (45) 82.44.44.

Martell and Hennessy are the giants of Cognac and the Martell warehouses are a few hundred metres further downstream. The Martells hail from the Channel Islands and Jean Martell came to Cognac from there in 1715, being 'senior' to the Irishman by fifty years. The company has remained in the hands of the family, a continuity that is rare and seldom equalled in other businesses, in France or elsewhere.

Martell has the largest export market for cognac, 80 per cent. being exported, with Britain still a pretty good customer. China is another, though at one time there was a crisis when new crates without nails were introduced. The

Chinese dealers had made good profits by selling the nails as well as the Cognac. The company quickly included a packet of nails with each new crate and the inscrutable Oriental smile reappeared.

Only about an eighth of Martell's requirements come from the 5000–6000 hectares that they themselves own, about three thousand independent farmers being needed to supply the rest. Recently they took over Noilly, the French vermouth company; with Hennessy and Moët linked, the announcement, late in 1975, that Martell would become a public company was not unexpected.

Visitors are shewn the warehouses, blending gallery, bottling floor and shipping bays. The tour can be conducted in different languages and samples may be bought.

Hours:	Monday to Friday 0900–1100; 1400–1700.
Introduction:	None necessary. Agents (England Wales): Matthew Clark & Sons Ltd., 183/5 Central Street, London EC1V 8DR. Tel. 01–253 7646.

Château Paulet S.A.,
26 route de Segonzac, 16100 Cognac, Charente.
Tel. (45) 82.00.02.

Cognac Château Paulet has been made and sold by four generations of the Lacroux family. In Britain during the 1950s the agency was held by Peter Dominic, then a small private company in which I became Sales Director, so I am well qualified to testify to the excellence of Château Paulet and the hospitality of the family, who speak excellent English.

Visitors see the cellars, distillery and vineyards. Young and old cognacs may be offered to taste. Pineau des Charentes (the aperitif of freshly pressed grape juice on a cognac base) is also likely to be offered, for this is rather a Lacroux speciality as well.

Hours:	Monday to Friday 0930–1130; 1430–1630. Closed 15 July – 15 August.
Introduction:	By appointment. Letter of introduction obtainable from agents: John Barnett & Son, 12 Dyke Road, Brighton BN1 3FE. Tel. (0273) 23444.

Maison Prunier S.A.
3 Avenue General Leclerc, 16101 Cognac, Charente.
Tel. (45) 82.01.36 and 82.18.62.

In 1701, M. Jean Prunier, leading citizen of La Rochelle and a renowned brandy expert, started to ship wines and brandies across the world, especially to Great Britain and Holland. It was only in the early nineteenth century that the family moved to Cognac, establishing the Maison

Prunier as we know it today, still owned by Jean Prunier's descendants. Besides the main buildings close to the station, the company owns ageing cellars in Grande and Petite Champagne.

La Vieille Maison, the fine old timbered house in the middle of Cognac, also belongs to Prunier.

Visitors are shewn round the cellars and can also see distilleries about ten kms. out of the town. English spoken.

Hours:	Monday to Friday 0930–1130; 1430–1630. Closed August.
Introduction:	By appointment. Letter of introduction obtainable from agents: H. Parrot & Co. Ltd., 3 Wapping Pierhead, Wapping High Street, London E1 9PN. Tel. 01–480 6312.

E. Remy Martin & Co.,
20 rue de la Société Vinicole, B.P. No. 37, 16102 Cognac, Charente.
Tel. (45) 82.16.03.

Founded in 1724, Remy Martin specialises in Fine Champagne Cognac V.S.O.P. and qualities superior to this. V.S.O.P. is a quality usually superior to Three-Star and Remy Martin claim that half the V.S.O.P. brandy drunk throughout the world is their marque. Fine Champagne is a blend of the two best qualities of Cognac (Grande Champagne and Petite Champagne) and Remy Martin hold large stocks.

The chais are at the southern end of the town not far from the station and the visit is likely to include the ageing cellars, bottling plant and despatch department. There are English-speaking guides.

Hours:	0930–1200; 1400–1700. Closed three weeks in July and at Easter.
Introduction:	By appointment. Letter should be obtained from agents: Hedges & Butler Ltd., Hedges House, 153 Regent Street, London W1R 8HQ. Tel. 01–734 4444.

L. de Salignac & Co.,
Domaine du Breuil, rue Robert Daugas, 16100 Cognac, Charente.
Tel. (45) 82.14.26.

Antoine de Salignac, gentleman farmer, founded this company in 1809 when there were few distillers in Cognac. By 1855, they were shipping to Sweden, Norway, Finland, Germany and Britain and Salignac cognac is now well-known in most parts of the world.

Welcomed by an English-speaking guide, visitors are shewn the cellars and bottling plant. Tastings and purchases are possible.

| *Hours:* | Monday to Friday 1400–1600. |
| *Introduction:* | By appointment. Letter should be obtained from agents. Holders are then asked to give 48 hours' notice of arrival: Mr. Guillaume de Jarnac, Hiram Walker International Company, 74 Brook Street, London W1Y 2DE. Tel. 01–629 1581. |

Cognac (near)

Cognac Brugerolle,

Distillerie de Mathra S.A., 17160 Mathra, Charente.
Tel. Mathra-près-Cognac 60.

This small, yet old company may not be a byword throughout the English-speaking world, nevertheless its cognac has been winning Grands Prix at European concours very regularly from the earlier years of this century.

It was in 1812 that Jean Brugerolle of the Auvergne moved north-west to settle in Mathra, a village 24 kilometres due north of Cognac, and successive generations of the family have contrived to direct the firm he began.

Visitors are shewn the distillery, vineyards and ageing cellars.

| *Hours:* | Monday to Friday 1400–1600. |
| *Introduction:* | By appointment. Letter should be obtained from agents: Southard & Co. Ltd., 213 Newnham Terrace, Hercules Road, London SE1 7DR. Tel. 01–928 2642. |

Jarnac

Delamain & Co.,

5 and 7 Rue J & R Delamain, 16200 Jarnac, Charente.
Tel. (45) 83.00.13.

About 1815, the firm of Ransom and Delamain split into two – Thomas Hine (see next entry) and Delamain. Both have retained their headquarters in Jarnac, the region's second town and both have consistently sold superb brandies. While Hine is now controlled by The Distillers Company, Delamain still remains a family concern, specialising in the very best crus of Grande and Petite Champagne. Their best known marque is Pale and Dry Grande Champagne, found in restaurants where the best is still appreciated.

Visitors see over the cellars and are offered a tasting of different vintages.

| *Hours:* | Monday to Friday 1000–1130; 1500–1700. Closed August. |
| *Introduction:* | By appointment through agents, who will send instructions, invitation card and route plan on request: Mentzendorff & Co. Ltd., Asphalte House, Palace Street, London SW1E 5HG. Tel. 01–834 9561. |

Cognac Hine S.A.,

16 Quai de l'Orangerie, 16200, Jarnac.
Tel. (45) 83.00.08.

Jarnac, 14 kilometres east of Cognac, is the second town of Cognac brandy, where several distinguished firms have their headquarters. Amongst them is Thomas Hine, known since its acquisition by The Distillers Company of Great Britain, as 'Cognac Hine S.A.'.

Thomas Hine was born in Beaminster, Dorset, in 1775. At the age of eighteen he went to France, finding work in Jarnac with a brandy house founded in 1763. He married Françoise Delamain and in 1796 became a partner in his father-in-law's company, Ransom and Delamain. After suffering some persecution during the Napoleonic wars, having refused to relinquish his English nationality, he gave his name to the business in 1817, five years before his death. His direct descendants manage the company today.

For visits, only individuals or very small parties can be considered. A letter of introduction must be obtained, well in advance of the proposed visit, from the distributors who will send a booklet on request:
Buchanan Booth's Agencies Ltd.,
1 Oxendon Street, London SW1Y 4EG.
Tel. 01–930 8202.

112

GERMANY

Germany

Introduction

GETTING THERE

By road: From Britain the best route to the Rhineland is via Dover to Ostend, taking the motorway(s) that pass south of Brussels and close to Liège, Aachen and Köln. From Calais the same motorway network can be joined by driving up the coast to Dunkirk. From the Belgian coast the distance to Koblenz, where the picturesque roads along the Rhine and Moselle begin, is about 250 miles.

Dutch and Belgian ports can also be reached by regular car ferries from Middlesbrough, Hull, Great Yarmouth, Felixstowe, Harwich and Sheerness.

From Calais and Boulogne, the alternative to motorways is the slow 280-mile journey through Arras, Cambrai and Sedan and then through Luxembourg to Trier, the oldest town in Europe and an excellent start for the Mosel.

From Paris by N3 it is 250 miles to the frontier at Saarbrucken, after which Autobahn E12 leads into the Palatinate in about 50 miles and to Worms in about 80. Bad Kreuznach, wine town of the Nahe tributary, is also about 80 miles, providing the most direct route to the Rheingau, though a ferry is involved.

By rail: The overnight service from Harwich to the Hook of Holland leaves London at 2000, arriving at Koblenz about noon next day and at Mainz an hour later. Various day services from London (Victoria), via Dover-Ostend take about 12 hours.

From Paris (Est) there is a good service to Frankfurt, which reaches Mainz (377 miles) in about eight hours.

In summer a weekly car sleeper service operates between Genoa and Köln, which is sixty miles north of Koblenz.

By air: British Airways and Lufthansa flights between Frankfurt and London and Frankfurt and Manchester are daily, flight times being one-and-a-half to two hours. Frankfurt is only 25 miles from Wiesbaden, making the Rheingau little more than thirty minutes' drive along the autobahn.

By water: The Rhine cruises, operated by modern ships designed for the river, with cabins that become living rooms by day, are widely acclaimed.

The regular service is between Rotterdam and Basle

taking four/five days, but there are many variations, such as train and ship to Basle, Rhine cruise to Rotterdam, and back via Harwich. There are also autumn cruises on the Mosel during the vintage; full particulars from:
K. D. German Rhine Line, 15 Frankenwerft 5 Köln 1. Tel. 20881,
represented in Britain by:
Paul Mundy Ltd., 11 Quadrant Arcade, Regent Street, London W1R 6EJ.
Tel. 01–734 3308.
A 16-page booklet illustrated in colour and a booking form are available from travel agents.

It would be as well to admit at the outset of this chapter that I have only paid one visit to the wine districts of Germany. As I have explained, my visits to the wine districts of Europe have had to be mainly during holidays and from 1950 to 1974 my wife and I never missed a year without such a trip, taking our own car through France to Portugal, Spain or Italy, and occasionally beyond, by sea to Yugoslavia, Greece or Tunis.

Germany was unfortunately low priority. Neither of us spoke German and my wife found the wines too sweet for her very dry taste. German baroque, however, was another matter, a magnet in fact that, en route for Italy in 1970, drew us to Germany.

Having passed through Luxembourg and spent the night in Trier, we set off for a quick look at the wine districts early next day. The morning soon passed; up the Saar, down the Mosel, turning away at Bernkastel-Kues in order to reach the Rhine at Saint Goar for lunch. In the afternoon we took a passing glance at the Rheingau, moved on through Rheinhessen, got lost in Mainz and reached Bad Dürkheim in the Palatinate by early evening. But the 'baroque timetable' demanded Würzburg that same night, exhausted we made it after dark.

In one day we had seen the five principal wine regions of Germany – Mosel, Rheingau, Rheinhessen, Rheinpfalz and Franken – and only the possibility of being certified insane has prevented me from notifying *The Guinness Book of Records.*

Such is the fame of Hock and Mosel that the extent of

GERMANY

Köln
Bonn
Ahr
Koblenz
Rheingau
Wiesbaden
Frankfürt
Rüdesheim
Eltville
Bingen
Mainz
Winkel
Bernkastel
Mosel
Mosel
Nahe
Nierstein
Darmstad
Main
Franken
Würzburg
Trier
Ruwer
Nahe
Bad Kreuznach
Rheinhessen
Wiltingen
Saar
Worms
Mannheim
Neckar
Heidelberg
Bad Durkheim
Wachenheim
Speyer
Deidesheim
Rheinpfalz
Karlsruhe
Württemberg
FRANCE
Stuttgart
Strasbourg
Baden Baden
Neckar
Baden
Danube
Freiberg
Staufen

0   20   40   60   80 Km
0   20   40 Mi

SWITZERLAND

Germany's vineyards and wine production may be thought to be much greater than it is. In fact total production is about one-tenth that of France, which makes the twenty-nine German entries in this Guide proportionally higher than the one hundred and fifty odd French entries.

What the twenty-nine do not lack is variety. They include great properties still in private hands such as Schloss Vollrads, as well as State-owned and Church-owned cellars. There are two Co-operative wineries, two distilleries, three museums and – because beer is the German national drink and Munich such a city of fine buildings and good restaurants, with a Beer Festival of world renown – the Löwenbräu brewery.

GERMAN WINE LAWS

New wine laws took effect with the 1971 vintage – a great year as it turned out to give them a good start. Quality is controlled by recognising three grades: *Tafelwein, Qualitätswein* (Q.b.A.) and *Qualitätswein mit Prädikat*. Geographical considerations establishing eleven regions subdivided them further into districts, villages and sites.

All wines aspiring to Qualitätswein submit to official analysis after bottling; those approved carry a control number on the label.

Each grade (Kabinett, Spätlese, etc.) is now subject to exact minima of 'must' weight (Öchsle) and wines awarded such grades should truly correspond in appearance, smell and taste with a typical wine of identical origin.

By these means the new laws can give a moderately well sited vineyard the chance to reach a higher grade in a good year, given skilful husbandry by the owner. And conversely, even the great can fall to a lower grade through lack of skill, attention, or in a year of very bad weather.

The geographical aim, successfully achieved, was to reduce a bewildering array of place names, though newcomers to the subject, confronted by what still remains a bewildering array, may be forgiven for doubting it.

GRAPE VARIETIES

Germany's great wines are, without exception, white. The Rhine Riesling is the noble grape making almost all Mosel-Saar-Ruwer wines and nearly all the great hocks. But the drawback of quality is insufficient quantity; the Riesling is neither a big cropper nor an early one. To eradicate these weaknesses, much experimenting has been done with new species, bringing forth the Scheurebe and Müller-Thurgau, crosses between Riesling and Silvaner, and Riesling and Riesling. Plantings of these vines continue to increase.

Silvaner (spelt with a 'y' in France), Traminer and Gewürztraminer (nobody knows the difference between these last two that make such marked spicy wines) are planted extensively in the Palatinate. Morio-Muskat is a

Silvaner cross with the burgundy grape, Pinot blanc; Ruländer is known elsewhere as the Pinot gris and as Tokay d'Alsace. Both varieties, which make good white wines, may be met in Germany.

MANNER OF HARVESTING

The following are other important terms found on labels:
Kabinett – An above-average table wine, originally good enough to grace the grower's own cellar or 'Kabinett'.
Spätlese – 'late picked', i.e. the grapes are left on the vine for about two weeks after the vintage begins, thereby increasing the sugar content.
Auslese – 'specially selected', i.e. the grapes are not necessarily 'late picked', but only the ripest bunches are chosen.
Beerenauslese – only the ripest single grapes from the ripest bunches are selected, grapes infected with the 'noble rot' (called *Edelfäule* in Germany and *Pourriture noble* in France – see under Sauternes). These wines are always sweet and very expensive.
Trockenbeerenauslese – grapes are selected singly, as for a Beerenauslese, having been left on the vine even longer, until they are half-dried like raisins. Such wines – virtually concentrated essence of the grape – are very sweet, rich, rare and highly expensive.

Whereas the last two terms definitely indicate a degree of sweetness, the same cannot be said of Spätlese and Auslese. In the majority of cases they will be sweeter than a wine without these descriptions, but this is not always so.
Eiswein – the grapes are picked when frozen, a condition which greatly increases the gravity of the 'must' and raises quality.

MANNER OF BOTTLING

Erzeugerabfüllung – bottled by the producer.
Aus eigenem Lesegut – bottled by the producer from his own vineyard.
Aus dem Lesegut des Winzers – 'From the vintner's own vineyard'; may be used by a bottler who has bought the wine from a grower, given the latter's permission.

SEKT

The new laws also cover the making of sparkling wine, drunk in Germany in vast quantities. The lowest quality is *Schaumwein* (foaming wine) made to sparkle by impregnation in a tank with carbon dioxide gas; the grapes may be, and usually are, imported from other countries where they are more plentiful, such as Italy or Yugoslavia.

Qualitätsschaumwein, better known as 'Sekt' (derived from 'sack'), is of much higher quality. It has to be made by a secondary fermentation (as in champagne) and must achieve 10 per cent. alcohol by volume, which is a reasonable strength for a table wine. It has to be matured for at least nine months before bottling and each bottle must carry a number alloted to it on passing official tests before bottling.

Better still is *Prädikatssekt*, in which 60 per cent. of the grapes must be recognised varieties grown in Germany and even better still is a luxury class called *Prädikatssekt mit Geographischer Bezeichung.*

THE CO-OPERATIVES

As far as wine is concerned the word Co-operative in Germany, as in other European wine countries, denotes a group of small growers, not consumer organisations as in Britain. The movement in Germany, first formed in 1868, had grown by 1900 to 113 wine co-operatives, by 1930 to 334 and by 1968 (the centenary year) to 507.

The advantage to Germany's 58,000 small grower-members is that each is saved the expense of making his own wine. Each co-operative winery, staffed by full-time experts, fixes the date for the vintage and buys the grapes from each member. Offering higher payment for greater quality acts as an incentive; there is healthy rivalry when A learns what degrees *Öschle* his neighbours B, C and D have achieved. General standards have been increased by these means and of course by blending members' wines with one another.

The co-operative wineries are usually modern, with machinery the last word in hygienic automation. In Britain most wine merchants and retail chains offer Co-operative German wines, which can range from humble *Tafelwein* to noble *Beerenauslese.*

Their welcome to visitors is usually, as elsewhere, warm and generous. Entries for the new Mosel-Saar-Ruwer and the Franconian Co-operative will be found in this section of the Guide. Here is a list of the others which the public may visit during working hours. Twenty-four hours' notice by letter or telephone call is appreciated. That at Gau-Bickelheim is the wine centre for some six thousand Rheingau and Rheinhessen growers and is very modern and up-to-date.

Rheingau/Rheinhessen:
Hauptkellerei Rheinischer Winzergenossenschaften eGmbH., 6551 Gau-Bickelheim.
Tel. (06701) 453.
(From Mainz take B42 in the direction of Bad Kreuznach.)
Nahe:
Zentralkellerei der Nahe-Winzer eGmbH., Brückes 2–8, Postfach 535, 6550 Bad Kreuznach.
Tel. (0671) 2361.
Baden:
Zentralkellerei Badischer Winzergenossenschaften eGmbH., 7814 Breisach/Rhein.
Tel. (07667) 139.
(On the road from Freiburg to Colmar.)
Württemberg:
Württembergische Weingärtner-Genossenschaft eGmbH., Postfach 1260, Raiffeisenstr. 2, 7141 Möglingen.
Tel. (07141) 41067.

At one time the Co-operatives had a centralised export organisation in Mainz (Deutsche Interwein), making it easy to obtain information. Devolution now seems to have set in and they all act independently with less satisfactory results where correspondence is concerned.

AHR, BADEN AND WÜRTTEMBERG

These regions are omitted in the pages that follow. Ahr, the small tributary of the Rhine a little north of the Mosel, makes some rather ordinary red wines hardly exported at all. Of much more interest are the Württemberg vineyards, along the Neckar (another Rhine tributary), north of Stuttgart and those of the Ortenau, south of Baden-Baden across the Rhine from Alsace. Sound inexpensive white wines from these districts had some success in Britain before the oil crisis curtailed spending power and the addresses of their co-operatives are included in the list above.

RECOMMENDED READING

German Wines S. F. Hallgarten Faber & Faber 1976.
The latest book on German wines and an exhaustive study. The very high price will unfortunately restrict the readership.
Moselle O. W. Loeb and Terence Prittie, Faber & Faber, London 1972.
Rhineland Wineland S. F. Hallgarten Arlington Books 1965.
German Wines Heinrich Meinhard Oriel Press, Newcastle-on-Tyne 1971.
Germany: West Germany and Berlin, *Michelin Green Guide 1967,* English edition from bookshops.

COURSES AND CONCERTS

Throughout the summer, Kloster Eberbach in the Rheingau becomes a hive of activity. In addition to the arrangements for day-to-day visitors (p. 125) the German Wine Academy, celebrating its fourth year in 1977, holds 6–day seminars each month, finishing the season in October with a post-graduate course for former pupils. Over three hundred people attended these wine courses in 1976, about two hundred Americans, nearly a hundred Japanese, about twenty-five Britons and a sprinkling from many other countries. The price inclusive of hotel accommodation, meals, lectures, visits, excursions and taxes was DM860.00 a head.

Correspondence should be addressed to German Wine Academy, Reisebüro A. Bartholomae, D–6200 Wiesbaden, Wilhelmstr.8.

Orchestral concerts of classical music take place at 1700 every Sunday from May to September inclusive. Works by J. S. Bach, Mozart, Handel and many other composers were played in 1976 by leading German artists, tickets varying from DM8 to DM10.

INTERPRETERS

In Trier and other centres the local tourist office (*Verkehrsamt*) can sometimes provide English-speaking cellar guides.

National holidays in Germany are given below. Please remember when planning your trip that, not only will premises be closed, but that it is not uncommon to remain closed for an extra day (e.g. Friday after a Thursday public holiday) to make a long weekend break.

1 and 6 January, Good Friday and Easter Monday, 1 May, Ascension Day, Whit Monday, 17 June, 1 and 17 November (Repentance Day, date varies), 25 December and Boxing Day.

Further Information:
Deutsche Wein Information
65 Mainz
Fuststrasse 4
Postfach 3860.
Tel. (06131) 23104
issue a useful short guide to German wines and compre-hensive list of wine festivals annually. There is now a branch in London:
Wines from Germany Information Service
(Mr. Geoffrey Godbert)
15 Thayer Street
London W1.
Tel. 01–935 8164.

The German Tourist Information Service also should be able to provide help and information:
German Tourist Information Bureau
61 Conduit Street
London W1R 0EN.
Tel. 01–734 2600.

MOSEL-SAAR -RUWER

Introduction

A left bank tributary, joining the Rhine at Koblenz, there are vines and wines almost every few hundred yards of the 314 winding miles achieved by the Mosel from its French source in the Vosges.

Between Trier and Koblenz it succeeds in turning sixty miles into two hundred, with the vineyards (hacked out of rock and slate) appearing first on this side then on that, as its course twists and turns in a series of loops and bends. Only the warmth thrown off the rock on to the vine makes wine possible in this most northerly of European wine regions. Cultivation of the slopes, often inaccessible to cart or horse, is more suited to monkeys than men. Whether it is the exercise, or the wine, or both, the inhabitants have always been free from gall stones, so that Mosel wines have often been recommended by doctors for kidney and bladder troubles in less energetic mortals.

The charm of the wine, which is traditionally bottled in green bottles, lies in its greenish tinge, bouquet, 'slatey' fresh taste and low alcoholic strength. Before the days of breathalysers and blood tests, you could slake a summer thirst by the bottle without undue anxiety about driving home.

Lacking sunshine compared with the other German regions, all further south, the wines tend to be too acid in poor years, particularly those of the Saar and Ruwer, the large and small tributaries joining the Mosel above and below Trier respectively. But in good years, these achieve a perfect balance, even excelling those of the Mosel itself, which can become too sweet.

Trier, so rich in Roman remains, where Hugh Johnson*

---

*Hugh Johnson *World Atlas of Wine* p.128.

describes the atmosphere in the Bischöfliches cellars as more an act of charity than mere vulgar trade, is a good place to stay. Run up the Saar valley to Saarburg, seeing Kanzem, Wiltingen and Ockfen; take the road up the Ruwer to Kasel and the deviation up to Avelsbach.

The 38-mile drive from Trier to Bernkastel-Kues follows the right bank and is charming, but the stretch below Bernkastel-Kues to Cochem is more beautiful, dramatic and full of famous sites. Immediately below Bernkastel-Kues constitutes 'the magic four miles' down to Zeltingen, a stretch that includes Wehlener Sonnenuhr, most famous site name on the river. The sundial, one of several intended to give passing boatmen the time, was put here in 1842, roughly when the Prüm family, who had been farmers in the valley since 1156, took to wine growing. Prüm or Bergweiler-Prüm (the double-barrelled name results from several inter-family marriages) own a part of Wehlener Sonnenuhr and have become much respected, like Schorlemer, in the higher echelons of Mosel.

By the river road, Bernkastel-Kues to Koblenz is 80 miles, about five hours by car if all the sightseeing stops recommended in the *Michelin Green Guide* are included. It is a worth-while trip; for beauty, the only other wine river region in Europe to emulate the Mosel was that of port wine up the Douro (q.v.), until recent hydro-electric schemes turned the Douro, for better or worse, from a rapid river into a series of lakes.

Boat Services: From June to September inclusive, there is a variety of services between Trier, Koblenz and intervening towns. Enquire locally or from Köln-Düsseldorfer agency, Rheinwerft, Koblenz.

Bernkastel-Kues
Zentralkellerei Mosel-Saar-Ruwer,
5550 Bernkastel-Kues, Gewerbegebeit.
Tel. (06531) 6063–4–5.

This modern co-operative is easily found less than a kilometre out of Bernkastel-Kues on 53, the Wehlen road. It is one of some seven representing the wine co-operative organisation, begun in Germany in 1868 and now having about 60,000 grower-members*. Working a third of the national wine vine acreage, their wines are highly regarded both at home and in the export markets. There is at least one co-operative in each district (visitors always welcome) but this modern one, only recently completed, is the biggest wine-making establishment of the Mosel-Saar-Ruwer region. Four thousand members growers provide the grapes and, the plant can make and store 25 million litres from them.

Visitors taste the wines and see a film with English dialogue.

---

*The addresses of six other co-operatives in Germany, which welcome visitors and a note on the co-operative wine organisation are on p.118.

Hours: Casual visitors can be accepted on Tuesdays and Thursdays at 1500. At other times from Monday to Friday a limited number of visitors can be received if arrangements are made well in advance. Normal working hours are: Monday to Friday 0800–1330; 1400–1700.

Koblenz
Deinhard & Co. KGaA.,
5400 Koblenz/Rhein und Mosel, Deinhardplatz 3.
Tel. (0261) 1041.

The Mainzerstrasse closely following the left bank of the Rhine is conspicuous on the *Michelin* or other town plans of Koblenz. Deinhardplatz is a continuation of it and the firm's headquarters look across the Schloss towards the Rhine. Approach by the red routes (*Guide Michelin* plan), shewn also in dotted red on the town plan, which Deinhard London usually provide with letters of introduction. Visitors can park in the courtyard or in the platz.

It was in 1794 that Johann Friedrich Deinhard, the penultimate of fifteen children, inherited a cellar in Koblenz, later having the good sense to marry the pretty daughter of the town's richest wine merchant after the business had just survived the town's occupation by French soldiers, who drank well but rarely paid.

In the 1820s, young Anton Jordan persuaded London Society that Deinhard's Sekt was much better for them than champagne, while his colleague Julius Wegeler enjoyed a social advantage because his mother had inspired Beethoven's Overture *Leonora No. 3*. Soon the two young men were joined in London by a third, James Hasslacher, and since 1865 when the last Deinhard died, the Hasslachers have run the London side and the Wegelers (with some Hasslachers) the Koblenz.

Twice, the two factions have had to fight for their respective countries on opposing sides, subjugating national prejudice when the wars were ended in order to revive their common interest. Today this business, exporting to over eighty countries, is a by-word for fine Rhine and Mosel wine, making survival through subjugating national prejudice a moral for 'British non-Europeans' in wider fields perhaps!

Visitors see the old cellars where the vintage wines mature and, less than a kilometre and a half away, the most modern Sekt plant in Europe.

---

*In 1975 Deinhard presented to the nation a small Beethoven museum in the house where his mother was born at Ehrenbreitstein, a small village across the Rhine from Koblenz. Besides Beethoven memorabilia, there are some mementoes of Dr. Karl Baedeker, distinguished son of Koblenz and almost the patron saint of travel guide editors. A seven-day Wine and Music Tour by Thomas Cook includes cellars, vineyards and the house.

Hours:	Monday to Friday 0900–1200; 1400–1600.
Introduction:	By appointment. Letters of introduction are obtainable from Deinhard's in London or through stockists of Deinhard wines. Post card or telephone giving 48 hours' notice of intended time of arrival is then appreciated.
U.K. Agents:	Deinhard & Co. Ltd., 29 Addington Street, London SE1 7XT. Tel. 01–261 1111.

Trier

Vereinigte Hospitien,

Güterverwaltung, 5500 Trier/Mosel, Krahnenufer 19.
Tel. (0651) 76051.

The Hospitien lie on the Mosel, right on the river bank, some 300 metres downstream from the old Roman bridge.

In 1804 Napoleon merged the mediaeval hospital foundations, which had been secularised in 1794, into the Vereinigte Hospitien (United Hospital Foundations). The land owned by this, now public, foundation lies in and round Trier and on the middle Mosel and the Saar. The profits of these properties enable the foundation to continue the original charitable work of the institutions from which it was formed. In addition, the Vereinigte Hospitien own and administer seven important vineyards, in Serrig, Wiltingen, Scharzhofberg, Kanzem, Trier, Piesport and Zeltingen. The wine is made in these central cellars, which are among the oldest in Germany.

Mid-September to early December is too busy a time here for tastings, although visits can be accepted. At other times of the year, tastings on payment can be offered to parties of from ten to forty people. Charges are proportional to the number of people and the quality and number of wines on sample. Wines may be bought to take away.

| Hours: | Monday to Friday 0800–1230; 1430–1600. |
| Introduction: | Where parties and tastings are involved several weeks' notice should be given to allow time for a written authority, signed by the Director of the Hospitien Dr. Pilgram, and his cellar master, to be sent back to the applicant. Parties of less than ten not requiring tastings should give 24/48 hours' notice of arrival by letter or telephone. |

Verwaltung der Bischöflichen Weingüter,

1 Gervasiusstrasse, 55 Trier.
Tel. (0651) 72352.

These ancient Bischöfliche cellars are on the corner of Gervasiusstrasse and Rahnenstrasse. Proceed via Südallee into Weberbachstrasse as far as the Stadbibliothek, near the town centre. At least half-a-dozen 'Establishment' bodies own great vineyards around Trier, the State, the Municipality and the Church, to name but three. This, however, is an ecclesiastical entry, leading to the office and cellars from which all the vineyards belonging to the Church (Cathedral, Priests' College and a Catholic School) are administered. In German these are respectively Hohe Domkirche, Priesterseminar and Konvikt.

The joint administration, only set up since 1970, bottles all the wines here, although they continue to be sold under their traditional labels. Their vineyards cover a hundred hectares and cellar capacity is about fifteen million bottles.

| Hours: | Monday to Friday 0930–1200; 1400–1630. |
| Introduction: | Not strictly necessary. 24/48 hours' notice of arrival by postcard or telephone is appreciated. |

Verwaltung der Staatlichen Weindomänen,

5500 Trier, Deworastrasse 1.
Tel. (0651) 75946.

This is the headquarters and cellars of the wine production of the Municipality of Trier. Keep to the double carriageway roads that go round the town – Südallee coming from the Luxembourg direction, Nordallee from the Koblenz direction – which enter the Ostallee at different ends. The *Michelin* town plan marks Sichelstrasse at the northern end of Ostallee. Sichelstrasse leads to Deworastrasse.

The Municipality first entered the trade in 1896, buying vineyards in the Saar at Ockfen, extending them to Serrig further upstream and to Avelsbach, a hill with some famous sites eastward of the town and so near to the Ruwer, the Mosel's other wine tributary, that its wines are classed as such.

Wines may be tasted here at DM5 to DM20 a head and bought too. An English-speaking guide is available, given notice.

| Hours: | Monday to Friday 0730–1645. |
| Introduction: | None necessary; open to all for short visits. Tastings, however, require 24/48 hours' notice by letter or telephone. |

Wiltingen (Saar)
Weingut Egon Müller,

Scharzhof, 5511 Wiltingen Saar.
Tel. (06501) 2432.

The vineyards of the Scharzhof, two kilometres east of Wiltingen and a little nearer to Oberemmel, are the most famous of the Saar. Extending to about a dozen hectares they face south on a steep slope. Monastery-owned from 1030 to the Revolution, Napoleon put them up for auction in 1797 and the buyer (for a song) was Jacques Koch, great-great-grandfather of Egon Müller.

When Jacques Koch died, French laws gave the estate to his six children in equal shares. Then, when the French had gone, the two holy daughters gave theirs to the cathedral so the Church was back in wine business again.

In spite of such sub-divisions, Egon Müller, whose father married a Koch, holds the main part and, say Loeb and Prittie*, has made the best wine since the war. Others would seem to agree; at a Trier auction in 1968, £27 was paid for one bottle of his Trocken-beerenauslese 1959. Tastings are offered free but no wines are offered to the public for sale.

Hours: Monday to Friday 0900–1200; 1400–1600.

Introduction: By letter, which any good wine merchant should be able to supply.

Weinguter Apollinar Joseph Koch,
Scharzhof, 5511 Wiltingen/Saar.
Tel. Klosterbergstr. 126.

As explained under Weingut Egon Müller, when the Government of Napoleon decided to sell, by auction, vineyards confiscated during the Revolution, Jacques Koch, formerly a monk, set up in business here having acquired the Scharzhof for a trifling £645.

Parts later went to his children in equal shares, one direct line passing to Apollinar Koch, who was making the best wines of the Scharzhof in the years between the wars until, in 1936, he committed suicide rather than live in Nazi Germany. The three principal owners of this great vineyard are thus Apollinar Koch, Egon Müller and the Trier Cathedral, which adds the prefix 'Dom' to 'Scharzhofberger' on their labels.

Apollinar Koch also make very fine wines from the parts they own in half a dozen of Wiltingen's best sites. No tastings.

Hours: Monday to Friday 0900–1200; 1400–1600.

Introduction: By letter, which any good wine merchant should be able to supply.

RHEINGAU

Introduction

It is astonishing how small great wine regions can be. Sauternes is seven miles by four and the Rheingau, Germany's greatest, chiefly comprises a 12-mile front along the Rhine, with the vineyards ascending the slopes to a depth that is less than three miles. Additionally within the region, but detached fifteen miles to the eastward on the river Main, are the vineyards of Hochheim (hence the term Hock), but the total area is still no more than 7,590 acres, about a fifth the size of Rheinhessen across the Rhine. No other region can quite match the delicacy, elegance and staying power of the Rheingau wines, where the Rhine Riesling grape reins supreme and the output an acre is less than elsewhere in order to maximise quality.

The Guide's entries include three show places, the

Kloster Eberbach, Schloss Vollrads and Schloss Johannisberg.

Strictly the region begins north of Rüdesheim at Assmannshausen, almost the only village in Germany famed for its *red* wine, due more to an accident with a monkey than to its great merit. In German *Affe* is a monkey and *Ave Thal*, short for *Ave Maria Thal* (Valley of Our Lady), a village, home of the wine. 'V' being pronounced 'F', the wine became *Affenthal*, selling particularly well when the growers had the wit to emboss a monkey on every bottle.

Noisy, boisterous tourist-ridden Rüdesheim comes next, the 'Blackpool' of the Rhineland. Yet in spite of all the wine bars and the bands that are far from pleasing all tastes, everybody seems to agree that the Asbach distillery is a visit not to be missed and the wine museum, close by in the Brömserburg castle, is also worth while.

The villages of Winkel, Mittelheim, Oestrich, Hattenheim and Erbach all have many fine sites, among them

*O. W. Loeb and Terence Prittie – *Moselle*.

Erbacher Markobrunn. That Markobrunn may be better known than the others could be due to water. Midway between Erbach and Hattenheim there is a famous old well with Doric pillars and Roman lettering. With immemorial disputes between the two communities about boundaries, and to which of them the Markobrunnen site belonged, 'Erbach the water and Hattenheim the wine' became a humorous tag. But, truth to tell, it is the wines of Rauenthal, a mile or so above Eltville, that command higher prices than most of the others.

Kiedrich, a good wine village above Eltville, has an English connection in that Sir John Sutton Bt. came to live here from 1857 to 1873, becoming the church benefactor to the extent of 240,000 guilders. It contains Germany's oldest organ (1630) and another with fourteenth-century pipes.

The Hochheim vineyards, on the Main further east still, received a boost in the English-speaking world from Queen Victoria. She visited the site of her favourite wine in 1850; they renamed it Koenigin-Viktoria-Berg and erected a monument to her. A charming gesture, but not a great vineyard.

Eltville
Kloster Eberbach, administered from
Verwaltung der Staatsweingüter im Rheingau, Schwal-bacherstrasse 56–62, 6228 Eltville-am-Rhein.
Tel. (06123) 4155–6.

This former Cistercian abbey, sited at the upper limit of the vineyards, dates from 1135 and is now owned by the Hessian State. From Rüdesheim, a turning left off the B42 wine road at Hattenheim leads directly to it; or, coming from Wiesbaden, turn right at Eltville and approach from the village of Kiedrich. Routes have good direction signs. This is in fact part of the two to three hour tours from Wiesbaden to Rüdesheim given in the *Michelin Green Guide* of Germany (English edition 1967).

Beautifully preserved, Kloster Eberbach houses in what was once the lay brothers' refectory, a collection of ancient wine presses and is the resting place for all the wines of the 32-hectare Steinberg vineyard, which has belonged to the abbey since the monks first planted it in the twelfth century.

From 1 April to 31 October, the monastery can be visited between 1000 and 1730, charge 0.60DM (1974).

Tastings: These are held hourly for visitors on Saturdays, Sundays and public holidays from 1330–1730. Duration 30 minutes, cost 3.50DM for three wines and pro rata. These tastings can be preceded, or followed by, the tour of the monastery.

For Larger Parties: These may be arranged in the summer by writing to the Administrator (address above). Parties of 17–20 taste in the Tasting Room, 21–120 in the refectory and 121–350 people in the lay dormitory. Minimum for these parties is 17 people. Duration one hour. A ten-wine tasting costs 20.00DM and a 12-wine tasting 25.00DM a head.

Introduction: None necessary unless festive wine tastings are required for groups of from 17–350 persons. These cost 20.00–25.00DM a person and are a feature.

Johannisberg
Schloss Johannisberg,
6225 Johannisberg/Rheingau.
Tel. (06722) 80.27.

Schloss Johannisberg, conspicuous on the hill, lies 1600 metres due south of Schloss Vollrads (q.v.). Coming from Rüdesheim on B42 the turning up the slope, marked 'Johannisberg' comes sooner, on the outskirts of Winkel, though almost any turning between Geisenheim and Winkel leads up to the village of Johannisberg. The Schloss is a few hundred metres outside the village; the avenue leading to it has green signposts.

Until 1802, when the monasteries were secularized, the schloss had been a Bénédictine monastery since the twelfth century. Prince William of Orange then had it for five years until Napoleon gave it to Marshal Kellermann. When Napoleon fell, the Congress of Vienna in 1815 gave it to the Emperor of Austria, who bestowed it as a fief on his Premier, Metternich, provided he gave back to the Austrian crown a tenth of the wine yielded annually. This has been done to this day, Otto von Habsburg, who lives near Munich, being the current recipient. Metternich himself lived at the Schloss in 1851 for a short while after his retirement. The present Fürst von Metternich continues the tradition of making one of the Rheingau's greatest wines from the precious 35-hectares of his inheritance.

Visits to the cellar, with or without tasting, are only possible, Monday to Friday, by appointment. The church, terrace and vineyards are open to visitors without prior arrangement. A *Weinstube*, selling Schloss wines by the glass or the bottle, is open daily from Easter to 31 October inclusive; Hours 1230 to 2200.

Hours: Monday to Friday 0900–1200; 1300–1700.
Introduction: By appointment; write or telephone in advance.

Rüdesheim
Asbach & Co. Weinbrennerei,
6220 Rüdesheim-am-Rhein.
Tel. (06722) 121.

Coming from the north (B42), make for the Brömserburg castle (Wine museum), which is on the road about a hundred metres before reaching the traffic lights at the centre of the town. Asbach visitors have parking reserved for them alongside the castle, which is useful when Rüdesheim becomes really congested in summer.

Otherwise, what with the bars and the cable cars passing over the vineyards to look down upon the Rhine from the summit of the Berg, there are plenty of attractions for drinkers. Asbach is certainly one. Some 135,000 people come each year to this distillery, founded in 1892 by Hugo Asbach, who coined the word *Weinbrand*, since used to describe all German brandy made solely from distilled wine, and aged in accordance with German laws.

The firm is well known for its Asbach Uralt brandy and liqueur chocolates. Tours, in English, last about three-quarters of an hour. There are tastings with purchases possible. 'Not to be missed, everything clearly explained and easily understood', wrote a correspondent about this visit.

Hours:	Monday to Thursday 0900–1100; 1330–1545. Friday 0900–1100.
Introduction:	None necessary. Should any special arrangements (e.g. a large party) be required, they should be made with the agents: J. C. McLaughlin Ltd., 124–6 Denmark Hill, London SE5 8RY. Tel. 01–733 1161.

Brömserburg Rheingau-und Wein-Museum,
6220 Rüdesheim-am-Rhein, Rheinstrasse 2.
Tel. (06722) 2348.

Brömserburg, the old ruined castle on the Rhine just downstream from the town centre, is a conspicuous landmark whether arriving by road from the north or by one of those large Rhine steamers, which ply between Switzerland and Holland. The Asbach distillery (q.v.) is close by and has a good-sized car park for visitors.

Originally a moated castle, Brömserburg reached its present cubic form about 1200, after several centuries of alteration and rebuilding. The Metternich family, who were to be given Schloss Johannisberg (q.v.) one hundred and fifty years later, took it over for a while in 1668. Later it became just a private house, being bought eventually by the town in 1940, in order to establish this wine museum.

Exhibits include a unique collection of wine containers, some from ancient Egypt, Asia and Greece, wine casks of Etruscan and Roman origin, a splendid collection of ancient drinking vessels dating back 2,000 years and a fascinating range of ancient cooperage tools.

Entrance Fees: Adults 1.50DM; Groups (10 people and over) 1.00DM; School groups 0.50DM.

Hours:	Daily 0900–1200; 1400–1700.
Introduction:	None required.

Winkel
Schloss Vollrads,
6227 Winkel/Rheingau.
Tel. (06723) 3314.

From Rüdesheim proceed on B42 in the direction of Wiesbaden as far as Winkel. In the centre of this wine village, at the barrels of the wine-tasting stall, take the turning left and follow the sign marked Schloss Vollrads, keeping right at the fork about 500 metres further on after the railway level crossing.

Unlike its illustrious neighbour Schloss Johannisberg, perched majestically on the slope for all to see, Schloss Vollrads conceals itself modestly in a little dip some two kilometres above Winkel and the Rhine. The whole property, one of the largest wine estates still in private hands, has been owned and inhabited by the Matuschka-Greiffenclau family since the fourteenth century and the wines are among the noblest of the Rheingau. Count Erwein, who had been running the estate for some years, succeeded to the property in 1975 on the death of his father, Count Richard, who had been President of the German Winegrowers' Association from 1948–1964 and was a much loved and respected figure.

The courtyard and parts of the garden and vineyards are open to visitors, but not the Schloss because the apartments are occupied.*

Hours:	Monday to Friday 0730–1200; 1300–1600.
Introduction:	By appointment only, for individuals and groups, who are asked to telephone or write well in advance.

*Cyril Ray contributed a monograph, 'Schloss Vollrads' to Christie's *Wine review 1975* describing the history of the property in more detail. The *Review* is published from 8 King Street, St. James's, London SW1.

NAHE

Introduction

Driving down the Rhine from the north on the right bank, the mouth of the Nahe at Bingen is clearly seen over on the other side before reaching Rüdesheim. Its source lies away in the Hunsrück hills and, as if to follow the fashion set by its northern neighbour, the Nahe also goes in for a loop or two, though none quite so formidable as those of the Mosel.

The Nahe vineyards are to be found on the last fifteen miles of its journey, bordering both banks, beginning at Schloss-Böckelheim, a village not a house or castle. The next two miles, as far as Niederhausen, include the Kupfergrube ('copper-mine') and Hermannshöhle sites, about the best known along the valley. Norheim and Bad Munster, on the next two loops, make good wines, second to these.

The Nahe, spanned by a pretty bridge with houses on it, next divides the pleasant spa of Bad Kreuznach in twain. The town's vineyards lie to the north. They too make good Nahe wines, the sites nearest the river, notably Kauzenberg and Brückes, being considered the best.

The wines themselves are light as hocks go and should rarely be kept long – like a fine Rheingau wine for example. Yet they have something of the character of this big brother across the Rhine and something of the fragrance and charm of Mosel wine. The short drive through the vineyards is a détour well worth-while. The Guide provides a chance to sample at the cellars of Anheuser and Fehrs in Bad Kreuznach, centre of the Nahe trade.

Bad Kreuznach
Anheuser & Fehrs,
6550 Bad Kreuznach, Postfach 576.
Tel. Bad Kreuznach 28201.

This company's headquarters, lying back from the road with a garden in front, are on the right in the Brückestrasse, 400 metres before the first road bridge when approaching from Bingen, which is about twenty minutes' drive downstream from Bad Kreuznach.

Founded just over a hundred years ago as a two-family partnership dealing in Rhine wines, Anheuser and Fehrs is now a big concern, respected particularly as Nahe wine specialists. Altogether they own 60 hectares of fine sites in the Nahe valley, the majority at Kreuznach, the remainder divided between Schloss Böckelheim, Niederhausen, Norheim and Winzenheim. Herr Egon Anheuser, the present head, is a big game and deer stalking enthusiast, who has one of the finest collections of heads in Germany.

Visitors – parties of 5 maximum – see over the cellars and can be shewn some of their vineyards.

Hours: Monday to Friday 0930–1130; 1430–1630. Closed most of August.

Introduction: By appointment; letters obtainable from agents:
H. Parrot & Co. Ltd., The Old Customs House, 3 Wapping Pierhead, Wapping High Street, London E1 9PN. Tel. 01–480 6312.

RHEINHESSEN

Introduction

Crossing the Rhine on the bridge south of Wiesbaden, the road for Worms emerges from a complicated pattern of autobahn junctions. This is Rheinhessen, bounded by the Rhine to the north and east, and by the Nahe to the west, a region 20 miles by 30 perhaps, with a hundred and fifty villages and double that number of fair-sized estates. Guntrum, Balbach, Franz Karl Schmitt and the State Domaine are leading owners.

For the tourist, Rheinhessen is flat and uninteresting except along the Nierstein/Oppenheim stretch, close to the river, where there are sites that make wines comparable with the best of the Rheingau. Good wines are also made away to the westward around Bingen, where the term for a 'corkscrew' is a 'Bingen pencil'. The origin is embedded in a meeting of clergy held there once upon a time. The presiding bishop demanded a pencil, whereupon every one of his obsequious listeners each dived a hand into his pocket and came up with a corkscrew.

In Worms, 17 miles south of Nierstein, there is the Church of Our Lady (*Liebfrauenkirche*) surrounded by a few small vineyards, not much bigger than a London square, from which Liebfraumilch originated. In time of course it became any German white wine at all, but the new (1971) laws decree that Liebfraumilch must be made wholly from grapes grown in Rheingau, Rheinhessen, Nahe or the Palatinate and must be a QbA wine.

Mainz
C. A. Kupferberg & Cie.,
Kupferberg Terrasse, 6500 Mainz-am-Rhein.
Tel. (06131) 1051.

Kupferberg Terrasse is a street about 600 metres southwest of the railway station and is clearly shewn on the *Michelin Deutschland Guide* plan of Mainz. Parking is possible but the street is one-way (see plan).

Mainz is the oldest wine town of the Rhine and it was here that young Christian Adalbert Kupferberg, aged twenty-six, began in 1850, extending the cellars downwards until they occupied seven levels underground. Believed to be the deepest in the world, these cellars house millions of bottles of Kupferberg Gold and other sparkling wines at a constant temperature of 12°C.

In 1892, Prince George, later George V, paid them a visit as a result of which Kupferberg Gold became popular with the House of Lords. This is a wine made entirely from Rhineland grapes, being fermented in the bottle by the champagne process in these cellars. The company makes six other lesser sparkling wines, exporting a high proportion of them all.

Hours:	Monday to Thursday 0930–1130; 1400–1530; Friday 0930–1130.
Introduction:	None necessary. Postcard or telephone call giving notice of arrival is appreciated. Other enquiries to agents: Gilbey Vintners Ltd., Gilbey House, Harlow, Essex CM20 1DX. Tel. (0279) 26801.

Nierstein
Weingut Bürgermeister Anton Balbach Erben,
6505 Nierstein, Mainzerstrasse 64.
Tel. (06133) 5585.

Superbly sited on the Rhine, Anton Balbach's office and cellars are on the main road (B9) coming into Nierstein from Mainz and the name is clearly seen in big red letters on the right about 500 metres before reaching the town centre.

The company was founded in 1868 by Anton Balbach who was then the Mayor of Nierstein. Many of the great Nierstein sites known today were still wooded and the Mayor was among those who cleared and planted the famous Pettenthal vineyard, which is by the side of the B9 road midway between Nackenheim and Nierstein.

Owning a total of 18 hectares in such sites as Pettenthal, Hipping, Oelberg and others of the highest class, Anton Balbach (with Franz Karl Schmitt and the State Domain) ships the finest wines of the region, many of them to Britain and the U.S.A. The male line of the family having died out, the Bohn family now run the firm; Manfred, the son, having taken a degree in viticulture in Germany has recently been in Britain to study the Trade and learn the language.

Visitors are likely to find a guide who speaks English and French; wines will be sampled and can be bought to take away.

Hours:	Monday to Friday 1000–1200; 1400–1600.
Introduction:	By appointment. Anton Balbach wines are usually listed by Justerini & Brooks. For letter of introduction write to: Justerini & Brooks Ltd., 1 York Gate, Regent's Park, London NW1 4PU. Tel. 01–935 4446.

Weingut Louis Guntrum,
Rheinallee 62, D–6505 Nierstein, Rheinland-Pfalz.
Tel. (06133) 5101.

The Weingut is easily found on the Rhine side of the main road (B9) between Nierstein and Oppenheim, which are roughly halfway between Mainz and Worms.

Well-known growers and shippers of fine Rhine wines, this family firm started in 1824. Hermann Guntrum, the present head, is fourth in line. They own fine vineyards in Nierstein, Oppenheim, Dienheim and Nackenheim. Guntrum wines are shipped to over sixty countries and Herr Guntrum, speaking excellent English himself, sees that guides are laid on for English-speaking visitors.

Hours:	Monday to Friday 1000–1200; 1400–1600.
Introduction:	Casual callers can be accepted though 24/48 hours' notice by letter or telephone is much preferred. Any special arrangements should be made through agents: Sherry House Ltd., 92 New Cavendish Street, London W1M 8LP. Tel. 01–580 0301.

Introduction

From the general direction of Paris leave the autobahn Saarbrucken-Kaiserslautern-Mannheim at the exit for Grünstadt; Bad Dürkheim is then less than ten miles. From the north the best approach is from Worms, the Weinstrasse begins at Zell and goes on through Grünstadt.

Being well south of Rheingau and Rheinhessen, the Palatinate not only gets more sunshine but has a mild climate, thanks to the protection of the Haardt mountains. The area under vine, 48,000 acres, is a little more than that of Rheinhessen and the annual production of each is about 1½ million hectolitres. As in Rheinhessen, the Silvaner provides a third of the crop, with Müller-Thurgau production varying from 25 per cent. here to 37 per cent. in Rheinhessen.

The term 'Pfalz' for this flat plain is derived from the Latin 'Palatinus', the first of the seven Roman hills upon which Augustus built a palace. Thenceforward, buildings in which Roman Emperors stayed during their travels became (in German) 'palast', 'Pfalz' being a later corruption.

The wine road *(Deutsche Weinstrasse)* runs for fifty-four miles, starting south-west of Worms and continuing through gaily painted villages all the way to the monumental gateway at Schweigen on the Franco-German frontier. Along this sunny plateau, orchard as much as vineyard, wine for many local people is a way of living and of life. For the visitor driving a car, the problem may be how to stop sampling rather than how to start. Most of this wine is suited to local drinking, but either side of Bad Dürkheim – from Dackenheim to Neustadt – constitutes the Middle Haardt where the great wines are made. From north to south, these are the village names to conjure with: Kallstadt, Ungstein, Dürkheim, Wachenheim, Forst, Deidesheim and Ruppertsberg.

Dürkheim, a town of 12,000 people rather proud of their casino, is the centre of the Palatinate wine trade. An even hotter source of revenue than the casino is the Sausage Fair, which takes place annually around the third week of September. It all began in 1442 when the villagers fed and wined a passing pilgrimage. Today, the fair needs an 11-acre fairground to house the sausage stands, the wine tents, the go-karts, the Dodgems and the coconuts. There are wines galore to be tasted and, judging by the quantities of pigs, cattle and chickens said to be eaten, the few thousand visitors, who come from all over Germany, must have eaten fit to burst by the time the final fireworks display tears into the sky.

The Sausage Fair is described as the largest of the wine festivals. There are plenty of these in Germany – all excellent publicity, not forgetting each wine queen, with wine princesses in support, carefully elected for the occasion.

In the Palatinate among the growers' names three stand out: Bürklin-Wolf, Bassermann-Jordan and Buhl. Between them they own a majority of the best sites. The Buhl and Bassermann-Jordan cellars are both in Deidesheim, which is not only considered to be the prettiest place on the *Weinstrasse* but, save Forst, has the best sites – Hohenmorgen, Langenmorgen and Leinhöhle among them. The Forst sites are at least their equals; Kirchenstuck, Ungeheuer and Jesuitengarten spring to mind, Jesuitengarten being the one that everybody remembers.

Bad Dürkheim
Johannes Karst & Söhne,
Weingut und Weinkellerei, 6702 Bad Dürkheim. Tel. (06322) 2103.

Bad Dürkheim, principal town of the Rheinpfalz or Palatinate, has only 12,000 inhabitants, all of whom know Heinz Karst, so, to find his cellars, just stop in the main square and ask.

In this family firm, Karst has succeeded Karst for at least two hundred years and many of the oak barrels in these cellars were made by members of the family, who were master coopers as well as 'master makers' of wine. For twenty years Heinz, the present head, has had the honour of tapping the first cask for the wine queen and mayor to taste, when Bad Dürkheim holds its great Sausage Fair and wine festival from the second to the third weekends each September.

Herr Karst speaks a little English. Visitors taste and can buy his wines after seeing the cellars and bottling plant.

Hours: Monday to Saturday 0900–1200; 1300–1800. Closed 20 July to 10 August (approx.).

Introduction: Herr Karst receives all visitors and a letter or telephone call to him personally is essential to make an appointment. This can be arranged by his principal importers: Gilbey Vintners Ltd., Gilbey House, Harlow, Essex CM20 1DX. Tel. (0279) 26801.

Deidesheim

Dr. v. Bassermann-Jordan'sches Weingut,

6705 Deidesheim, Postfach 20, Rheinpfalz.
Tel. (06326) 206.

The village of Deidesheim is on B271 *(Deutsche Weinstrasse);* from the Market Place take the Kirchgasse, which runs direct to the gate of this estate.

The late Geheimer Rat, Dr. Friedrich von Bassermann-Jordan, author of a number of scholarly books on the history of the vine, was once described as the Dean of German wine producers. An under-rating, one might say; his vineyard sites of Deidesheim, Forst and Ruppertsberg constitute the cream of the Rheinpfalz and could almost be termed a diocese.

Andreas Jordan (1775–1848), grandson of a French wine grower from Savoy, began the business at the end of the eighteenth century and may be regarded as starting quality wines in the Palatinate. Dr. Friedrich von Bassermann-Jordan was his great grandson. The company's sites are planted almost entirely with Riesling and thousands of wine lovers each year visit the cellars and the extensive wine museum, which are in subterranean caves winding under the village of Deidesheim.

Tours of the cellars should be arranged in advance stating whether tastings (for which a charge will be made) are to be included or not. Visitors should report to the office (address above).

Hours: Monday to Friday 0800–1200; 1400–1800.

Introduction: Advance notice by letter through a British wine merchant is preferable. Failing that at least 48 hours' notice could be given direct by telephone.

Weingut Reichsrat von Buhl,

6705 Deidesheim, Postfach 86, Rheinpfalz.
Tel. (06326) 210.

Deidesheim is on B271 *(Deutsche Weinstrasse)* and the offices and cellars of von Buhl are in the centre of the village.

This first-class Palatinate company goes back at least to 1849 when Franz Peter Buhl inherited it. Franz Eberhardt Buhl, who was chief counsellor to the Bavarian royal family and ennobled by the Bavarian King, died in 1921 and his widow carried on until she died in 1952. The present owner is Georg Enoch Reichsfreiherr von und zu Guttenberg.

The company owns a hundred hectares in the best vineyards of Deidesheim, Forst, Ruppertsberg and Königsbach, planted chiefly with the Riesling vine.

Hours: Monday to Friday 1000–1130; 1400–1600.

Introduction: None necessary but visitors must send a post card or telephone beforehand. For parties, tastings with a cold collation can be arranged in the Hotel Reichsrat v. Buhl or the Hotel Haardt, both near the offices.

Speyer

Historisches Museum der Pfalz mit Wein museum,

672 Speyer, Grosse Pfaffengasse 7.
Tel. (06232) 2185.

The Romans found their way north by the easiest routes, up the valleys, Rhône and Rhine for example. Having built the roads, the legions could sometimes be usefully employed planting vineyards on either side. One of these roads went north from Basle, via Speyer and Worms, to Mainz and until Speyer was destroyed in 1689 the city was a wine trade centre. Lying 23 kilometres east of Neustadt, Speyer is a short détour from the Weinstrasse and much the same from the E4 autobahn if heading north or south from or to Basle.

The early association with wine explains why it has a wine museum, with ten rooms of exhibits, said to be the oldest wine museum in the world although only founded in 1910. The major exhibits are almost entirely local, which makes their variety all the more remarkable. Pride of the collection is an original Palatinate wine bottle of about 300 A.D. There is also a perfectly preserved sixteenth-century wine press trough and a cask tapped on the frozen Rhine for the Elector Carl Theodor in 1766.

Adults 1.00DM; groups of 10 or more 0.50DM per person; military personnel 0.50DM; school children 0.20DM; members of the Historisches Verein der Pfalz are admitted free.

There is a printed guide with layout and sixty illustrations, available only in German for 1.00DM and a recorded guide in English or French.

Hours: Daily 0900–1200; 1400–1700. Closed 24 December – 1 January inclusive.

Introduction: None required.

Wachenheim
Weingut Dr. Bürklin-Wolf,
6706 Wachenheim/Weinstrasse.
Tel. Bad Dürkheim (06322) 8956.

Wachenheim is on B271 *(Deutsche Weinstrasse)* between Bad Dürkheim and Deidesheim. The Bürklin-Wolf office and cellars are at the centre of the village.

This company ranks with the best known wine growers and shippers of Germany. Dr. Albert Bürklin-Wolf has been Vice President of the German Wine Growers' Association since its foundation and a passionate advocate of cultivation to make wines of the highest possible quality.

As long ago as 1597 Bernhard Bürklin, Mayor of Wachenheim, is recorded as owning extensive vineyards and in 1875 Albert Bürklin, who later became Vice President of the Reichstag, married a grand-daughter of Johann Ludwig Wolf, who had developed his family estates into model vineyards; the Golden Age for Bürklin-Wolf thus began.

The vineyards cover some ninety hectares of fine sites, 88 per cent. Riesling in Wachenheim, Forst, Deidesheim and Ruppertsberg.

For visitors there are the mediaeval vaulted cellars of the Kolb Hof, modern cellars, the ancient taproom and the park with views of the vineyards, the Rhine valley and the Middle Hardt.

Hours: Monday to Friday 0900–1200; 1300–1700.

Introduction: The main house, cellars, etc., may be visited by prior arrangement.

FRANKEN (FRANCONIA)

Introduction

In mediaeval times Franconia, stretching along the valley of the Main, was a huge province which even included Mainz and Worms, beyond the Rhine in Rheinhessen. Interest in Franken wine, however, has always centred on the capital city, Würzburg, superbly sited on the river and much prized for its Baroque architecture. From Frankfurt or Nürnberg it is about 80 miles, with its own exit from the autobahn that connects them.

Political development in Franconia was unusual in that from the twelfth century until 1802, power was vested in a series of Prince-Bishops. From 1261 to 1720 they lived in the Marienberg fortress standing sentinel over the city, a familiar feature from postcards and posters, above the river on its hill, the Leistenberg.

Below the Marienburg the orderly vine rows of the Leiste vineyard descend steeply to the river. As far back as the sixteenth century the Leiste and the Stein vineyards (a mile or two downstream visible from the fifteenth-century bridge), made sweet Beerenauslese wines, which grew old very prettily in bottle, like Malmsey Madeira, for centuries. These 'Steinwein' came to be associated wrongly with *all* Franconian wines, which are not sweet but dry.

Even now producers still sell Franken wines, labelled 'Steinwein', in the flat flagon-shaped *Bocksbeutel*, although this is illegal. Steinwein by law must come only from the two Würzburg vineyards mentioned.

The Franken vineyards are divided into only three districts – Mainviereck, Maindreieck and Steigerwald. Although the village names are numerous, the area under vine is almost exactly the same as in the Rheingau. The grapes are Silvaner 46 per cent. and Müller-Thurgau 41 per cent.

The result is pleasant dry white wine, better for drinking with food than the delicate sweeter hocks (the Germans themselves often prefer the latter, not at table but after it). Goethe went so far as to take the cure at Carlsbad with a 60-litre cask as part of his luggage; his daily consumption during the cure was reckoned at two litres. Franken wine was his favourite.

The vineyards keep close to the valley of the Main from Schweinfurt, 20 miles north-east of Würzburg, to Aschaffenburg, 40 miles north-west of it, the river covering a good hundred miles, Mosel style, in a series of loops between the two towns. In Würzburg there are two

famous charitable vineyard-owning foundations, the Juliusspital started in 1576 by the Prince-Bishop, Julius Echter and the Bürgerspital (Citizens' Hospital) founded in 1319. Both still function as homes for old people and each has its own *Weinstube* (café-wine bar) in the city. Their wines are excellent but dear.

In 1744 the Prince-Bishops moved their residence from the Marienberg fortress to the new Residenz, a magnificent Baroque palace that Balthazar Neumann, the architect, had taken twenty-four years to complete. The grand staircase, Tiepolo's paintings and the white and gold of the little church *(Hofkirche)* make the Residenz one of the sights of Europe.

The rule of the Prince-Bishops ended in 1802. A dozen years later their substantial wine interests and those of the Dukes of Franconia became the property of the Bavarian State. These subsequently included the State Cellars and the Vine Cultivation Institute in Würzburg, and the College of Wine at Veitschochheim. In 1952 all three were merged in a new body the Bayerische Landesanstalt für Wein, Obst und Gartenbau (The Bavarian Institute for Wine, Fruit and Horticulture). The Director, Dr. Hans Breider, well-known grower and wine book author, retired in 1973 and has been succeeded by Dr. Eichelsbacher.

The State Cellars (see below), architecturally perhaps the finest wine cellars in existence, extend under the Residenz.

Würzburg
Bayerische,
8700 Würzburg, Residenzplatz 3, Postfach 296.
Tel. (0931) 50701.

The offices, sales, distribution departments and main cellars of the Bayerische Landesanstalt für Wein, Obst und Gartenbau (The Bavarian Institute for Wine, Fruit and Horticulture, see above) are all situated in the Rosenbachpalais, by the Residenz.

From the main railway station, go via Kaiserstrasse and Theaterstrasse to the Residenzplatz. Looking towards the Residenz, the Rosenbachpalais is on the left.

The Hofkeller, extending under the Residenz and built circa 1720, can claim architecturally to be the finest wine cellar in the world. Especially impressive are the three huge 1784 vats (50,000 litres and two of 25,000) and the baroque wooden spindle press of 1748. But the Institute moves with the times; the stock of 800,000 *Bocksbeutel* here will have been filled by quite the latest in automatic bottling plants.

Groups large and small are welcome. Wine tastings lasting 2–3 hours can be arranged at a cost from 25DM upwards but at least two weeks' notice is required in writing.

Wines may be bought and English-speaking guides are available.

Hours:	Monday to Friday.
Introduction:	Applications to visit should be received in writing by the Director (Dr. Eichelsbacher) at least two weeks in advance.

Bürgerspital zum heiligen Geist,
8700 Würzburg, Theaterstrasse 19.
Tel. (0931) 50363–4.

Approach from the railway station via Kaiserstrasse to Theaterstrasse. At No. 19 go through the gateway; you can park in the courtyard.

Dedicated to the Holy Ghost *(zum heiligen Geist)*, this Foundation, dating from 1319, was originally a home for the aged. It owns and runs the oldest vineyard in Würzburg, is believed to have been the first to bottle in the *Bocksbeutel* and celebrated its 650th anniversary in 1969.

Nowadays the Infirmary (Director Heinz Zeller) comes under the Würzburg City Council. In the old peoples' home the average age is eighty-two, ascribed to age-old custom of a free glass of wine from the cellars with every meal.

The Gothic infirmary church (1371) should be seen as well as the cellars and, of course, the oldest vineyard. This is the famous Würzburger Stein, which has a wonderful view of the old bishops' town.

The cellars may be visited by small or large groups, with or without tastings. English-speaking guides; small purchases of wine from the Laden Ecke at the corner of Theaterstrasse and Semmelstrasse.

Hours:	Monday to Friday 1000–1100; 1400–1500.
Introduction:	None necessary. Open to all but one to two days' notice is required, by letter or telephone.

Würzburg (near)
Gebietswinzergenossenschaft Franken E.G.,
8711 Repperndorf.
Tel. (093.21) 5163.

Repperndorf is a village 16 kilometres east of Würzburg, on the main road (No. 8) to Nürnberg, and 4 kilometres short of Kitzingen, a town of 18,000 inhabitants.

The Wine Co-operative of Franconia, started in 1959, has a membership of 2,200 growers and a total of 900 hectares under vine. This is their wine-making centre and headquarters. One hectare being about 2·5 acres, the *average* holding comes almost exactly to one acre per grower – far too small for profitable operation. By providing the centre and a small staff of experts to buy the grapes, vinify and then blend, excellent wines are made.

In Britain their Fränkischer Weissen, white dry and in the traditional *Bocksbeutel*, is worth trying, particularly by anybody finding white burgundy too costly.

Juliusspital Weingüt,
8700 Würzburg, Juliuspromenade 19.
Tel. (0931) 51610 and 50067.

From the main railway station take the Kaiserstrasse, then turn right into Juliuspromenade and go as far as the east wing of the Juliusspital to Klinikstrasse. Here on the ground floor are those popular Würzburg institutions the Juliusspital wine bars; offices and cellars are at No. 5 Klinikstrasse.

Founded in 1576 by Bishop Julius Echter von Mespelbrunn, this charitable foundation supports a hospital with five hundred beds and an old people's home with two hundred and twenty beds, income from the sales of its wines making a substantial contribution.

The Fürstenbau, the Hospital Church, the rococo apothecary and the old wine cellar, 250 metres long lined with fine carvings and by rows of barrels, all combine to make a visit of unusual interest.

With a fortnight's notice wine tastings for up to 60 people can be given in one of two tasting rooms, or in the cellars by candlelight but only German is spoken.

Hours: Monday to Friday 1000–1100; 1400–1500.

Introduction: Tours and tastings can only be conducted (in German) for groups, by prior arrangement made at least 14 days in advance.

Mainfränkisches Museum,
Festung Marienberg, 8700 Würzburg.
Tel. (0931) 43016.

The superbly sited castle and fortress of Marienberg, with the Schlossberg and Leiste vineyards sloping down to the river Main and the city of Würzburg on the other bank, appears on many a post card. It was originally built as a residence for the Prince-Bishops, serving this purpose until 1719 when they moved to the Residenz in the centre of the city.

In 1950 the wine press hall was opened in the old arsenal, built from 1708 to 1712. The museum contains a large collection of wine glasses, goblets, etc., dating from the Middle Ages to the present, besides numerous large wine presses and other antiques connected with wine growing. Each year the Franconian *Weinprämierung* is held in this wine press hall.

Fee 1.50DM. Free the first Sunday of each month.

A printed guide is available in German and English. Photography is permitted but tripods are prohibited.

Hours: Daily, including Sundays and Public Holidays: April to October 1000–1700; November to March 1000–1600.

Introduction: None necessary.

GERMANY MISCELLANEOUS

Munich
Löwenbräu Munchen,
Nymphenburgerstrasse, Munich.
Tel. (0811) 5200–1.

Munich, capital city of the German state of Bavaria, is served by air, rail and road from all parts of Europe. Tramways 1, 4, 21 and 27 pass close to the Löwenbräu Brewery and passengers should alight at the Stiglmaierplatz.

The brewery was founded in 1383 and is one of the oldest of seven in Munich. One third of the Löwenbräu production is sold abroad, this brewery exporting more beer than any other in Bavaria.

Visitors are shewn the brewhouse, Lagerkeller and the bottling plant. There is also a short film, which can be shewn.

Hours: Monday to Thursday 0900–1200; 1400–1600. Friday 0900–1200.

Introduction: By appointment, which can be arranged direct by telephone if in the locality or, before departure, by letter through the agents:
J. C. McLaughlin Ltd., 124–6 Denmark Hill, London SE5 8RY.
Tel. 01–733 1161.

Staufen im Breisgau
Alfred Schladerer GmbH,
713 Staufen im Breisgau.
Tel. (07633) 6032.

The Black Forest, which forms the heel of West Germany, is only divided from Alsace by the Rhine and both regions are famous for their 'fruit' brandies, a term used to describe distillates of fruits other than the grape.

Staufen, where Alfred Schladerer began this business in 1844, is a small town about half-an-hour's drive south-west of Freiburg and not much further from Colmar. The firm got off to a good start and has grown to be the leading producer and exporter of white fruit brandies in Germany. Kirsch, from the local cherries, is the best-known of these but there are also Williamsbirne (Poire William), Brombeergeist (Blackberries), Mirabelle (Golden Plums), Zwetschgenwasser (Quetsch) and Himbeergeist (Raspberries).

Schladerer distil them all in a modern distillery situated in a charming old house behind the offices and are pleased to shew visitors over it.

Hours:	Monday to Friday 0930–1130; 1430–1700.
Introduction:	By appointment. Letter of introduction obtainable from agents: Walter S. Siegel Ltd., 43/44 Albemarle Street, London W1X 3FE. Tel. 01–499 3872.

Steinhagen (near Bielefeld)
H. W. Schlichte,
4803 Steinhägen/Westphalia, Kirchplatz 26.
Tel. (05204) 131.

Steinhagen, half-way between Düsseldorf and Hannover, is Germany's 'capital' for Steinhager (Schnaps), a Westphalia speciality. The motorway Düsseldorf-Hannover is the best approach. Leave the motorway by the Brackwede/Sennestadt exit (which is west of the exit to Bielefeld) and follow road No. 68 towards Osnabrück; this will lead to the village of Steinhägen in about 15 kilometres.

The Schlichte distillery is the oldest of eleven Steinhäger distilleries in Steinhägen. The Schlichte family, still in control, have lived in the village for nearly four hundred years distilling Steinhäger from grain and juniper berries, which makes it like Genever, the Dutch gin.

Visitors are shown the fermentation room, still house, bottling plant and warehouses.

Hours:	Monday to Thursday 0900–1200; 1400–1600. Friday 0900–1200.
Introduction:	By appointment, which can be arranged direct by telephone (05204–13245–6) if in the locality, or earlier by letter through the agents: J. C. McLaughlin Ltd., 124–6 Denmark Hill, London SE5 8RY. Tel. 01–733 1161.

ITALY

Italy

(Wine regions arranged clockwise: Piedmont – Lombardy – Trentino – Alto Adige – Veneto; thence down the map)

Matta Patter
or
The Bold Bad Barons of Ruddi-Ricasoli

Our eyes were fully open to our awful situation
Ten days before our party planned for wine and cheese
  collation
We needed Barbaresco, Est! Est! Est! and Bardolino
And flasks and *fiaschetti* from Firenze or Torino
To taste the wines of Italy a hundred guests were bidden
But where we were to find them was a secret darkly hidden
Until we wrote to Dominic – a firm we'd like to flatter –
They sent a dozen cases and the Shipper's name was
  Matta
    The Shipper's name was Matta
    The Shipper's name was Matta
The Shipper's name was Matta, Matta, Matta, Matta,
  Matta.

Introduction

Italian wines – like the Italians themselves – have always been good fun. And, with the possible exception of the Soviets, they make more wine than any other nation. The fact that Italy is such an enormous vineyard is not apparent to the visitor because few expanses are given over wholly to viticulture as in France. The Italian grows vines up trellises in any spare corner, or mixes them up with his cereal crop; the result, by the time it reaches your glass out there, is sometimes *beyond* a joke.

In Britain, the story of Italian wines was, until a decade ago, synonymous with the name Matta. Frederigo Secondo Matta had come to London in Edwardian times from his native Piedmont village of Tomango d'Asti. The Mattas were poor and he worked as a sommelier in the Café Royal in those raffish days, learning a great deal about wines from a list that contained no less than one thousand of them.

In 1919 he set up shop in Westminster Bridge Road, so close to Waterloo Station that later on, a London version of 'When Brolio growls, all Siena trembles' became 'When Beeching* breathes, all Bertolli's flasks tremble'. By the late fifties, possibly five out of every ten bottles of Italian wines sold in Britain must have been shipped by F. S. Matta. In Peter Dominic, then an independent company based at Horsham in Sussex, with about fifteen shops in the country towns of southern England, we thought it was time to justify our slogan 'All the World's Wines' and give the Italian sales a little light-hearted stimulation. (Words like 'marketing' and 'sales promotion' were happily still unheard of.) The only reader of 'Matta Patter' who had not heard of *Ruddigore* was probably F. S. Matta himself, but he was possibly also the one Italian who had not heard of *Aida* either. The Bardolinos and the Barberas were only 9/6d a bottle in those days, with 1955 classed growth clarets from 10/6d.

During the sixties, a decade of mergers, Matta joined

---

*Chairman, British Railways Board 1963–1965.

ITALY

SWITZERLAND

FRANCE

Aosta

Lombardy

MILAN

TURIN

Piedmont

Asti

Alba

Barolo

Genoa

Sondrio

Trentino Alto-Adige

Trento

Lake Garda

Verona Veneto

Padua

Venice

Plave

Adige

Po

AUSTRIA

YUGOSLAVIA

Bologna

Lucca

Arno

Florence

Greve

Radda

Castellina

Siena

Rimini

Urbino

Ancona

Tuscany

The Marches

Perugia

Assisi

Orvieto

Umbria

Tiber

Lazio

ROME

Frascati

Foggia

Campania

Bari

Puglia

Naples

Atripaldi

Brindisi

Taranto

Ischia

Mt Vesuvius

Salerno

Capri

Calabria

CORSICA

SARDINIA

Trapani

Partenico

Marsala

Menfi

Palermo

Messina

Mt Etna

Catania

Reggio di Calabria

SICILY

Km 0   100   200

Mi 0   50   100   150

140

the Beecham group, the remarkable old man having retired home to Italy where he died in 1974.

With wine on the up and up, the Italian Government began to recognise that there might be a much greater future for their wines in the export markets. To match fine clarets and burgundies was not within their compass, but they certainly could compete with the second rank, wines of the Rhône, Provence or the Loire, for example.

NEW WINE LAWS

There was much to be done. France and Germany had laws governing nomenclature of wines, boundaries of the regions, the grape species permitted, the quantity each should be allowed to bear and so on. The Italians had practically none. So, in 1963, the Government passed legislation putting Italian wines into three classes.

1. *Denominazione di Origine Semplice* (D.O.S.)
Ordinary wine of the region named – no other requirements.

2. *Denominazione di Origine Controllata* (D.O.C.)
A Consorzio of growers in each region was required to draw up and submit boundaries of the region, grape species, maximum yield permissible and other lesser related matters to a Government Committee in Rome. If proposals were approved, D.O.C. status, indicated in writing or by symbol on each bottle, would be awarded.

3. *Denominazione di Origine Controllata e Garantita* (D.O.C.G.)
Establishing D.O.C. (2) above for regions and districts all over Italy has taken time. This superior class, initially of lower priority, is now receiving more attention. It will be the equivalent of château or domaine bottling, each bottle being guaranteed authentic by the producer and carrying a Government seal.

These new Italian laws, summarised by the initials, D.O.C., cover much else besides districts, vine species and output. In the Barolo country, for example, vine plants must be at least one metre, and rows 2½ metres apart. Wooden posts harbour insects and are to be replaced by concrete. As to growing vines up trees with rows of cabbages here and there, that is banished as completely as the use of chemicals or sugar in the must.

Of course, when travelling, the old conglomerations of trees, vines and cabbages will still be seen, but not in any vineyards granted D.O.C.

D.O.C. wines are allowed to be exported in bulk and bottled outside Italy. From 1966 to 1975 inclusive, nearly one hundred and seventy regions were granted D.O.C. by Government decree, many of their names being quite unknown outside Italy. D.O.C.G. wines, on the other hand, must be estate-bottled and it is to be hoped that this top status will be limited to a small distinguished minority.

With real control asserting itself through these laws, new companies have thought it worth while in Italy to enter the trade, setting up modern premises with modern vinification equipment. It is a great improvement. *Viva il buon vino, sostegna e gloria d'umanita.* Don Giovanni would have approved.

GETTING THERE
By road: Before those magnificent feats of engineering, the Mont Blanc and Great Saint Bernard road tunnels were built, the easiest way from France into Italy's north-west province was over the Mont Cenis Pass. From Chambéry, the N6, climbing gently up the valleys of the Isère and the Arc, is fast motoring and the pass presents no difficulties. On the Italian side, lacking a motorway, the forty-mile journey to Turin is slow.

Using the Mont Blanc tunnel, the approach via Geneva or Annecy is slow and the tunnel toll £5 or more for a medium-sized car. In Italy it is fast going with the motorway beginning at Aosta. From Switzerland, the approach is via the Saint Bernard tunnel and Aosta.

By rail: The Rome Express route is from London (Victoria) and from Paris (Gare de Lyon) via Turin. The equally famous Simplon Express goes from the same stations to Milan through Switzerland.

From early June to late September there are car sleeper trains several days a week, Boulogne-Milan and Paris-Milan.

By air: Milan, under eighty miles by motorway from Turin, is northern Italy's major international airport with connections to all parts of the world. The usual fly/drive service is available.

ARRANGEMENTS FOR VISITORS
The value of visitors for public relations purposes has long been recognised by the large vermouth firms near Turin, and the bigger wine firms elsewhere are following suit. There are, however, difficulties in making arrangements. Few Italians, in my experience, reply to letters. Glyndebourne Opera, I'm told, share this view. Jani Strasser, their chorus master for very many years, declared that this is the reason why Italian trains are so full. People are forced to go to see Auntie, because Auntie has not replied to their letter.

Another difficulty is that some of the establishments are controlled from a main office miles away, in Milan or Rome for example. Yet another is language. Only the volume of visitors to the vermouth companies warrants the employment of multi-lingual guides. Elsewhere, it is often a case of an English-speaking member of the clerical staff doing his or her best, and in remote parts there may not be one.

WINEFOOD GROUP
A number of the entries that follow relate to members of the Winefood group of companies. Arrangements for visits are made through the group headquarters: Winefood, via G. di Vittorio 32-20094 Corsico (Milano). Tel. (02) 4483. Telex 34375. Teleprinter communication recommended.

RECOMMENDED READING
Italian Wines by Philip Dallas, Faber & Faber Ltd., London 1974.
The Wines of Italy by Cyril Ray, McGraw-Hill 1966. Penguin Books 1971.
Viva Vino DOC Wines of Italy by Bruno Roncarati, Wine and Spirit Publications 1976.

NATIONAL HOLIDAYS
National holidays in Italy are given below. Please remember when planning your trip that, not only will premises be closed, but that it is not uncommon to remain closed for an extra day (e.g. Friday after a Thursday public holiday) to make a long weekend break.

1 and 6 January, 19 March, Easter Monday, 1 May, Ascension Day, 2 June, Corpus Christi, 29 June, 15 August, 1 and 4 November, 8 and 25 December and Boxing Day.

Further Information
The Italian Institute for Foreign Trade (I.C.E.)
Heathcote House
Savile Row
London W1.
Tel. 01–734 2411.

PIEDMONT

Introduction

Turin has many hotels as befits a fine city with nearly two million inhabitants. Advance booking is, however, almost essential in summer; Italian cities are popular choices for fairs and business conferences. Arriving after a long day's motoring, frustrated by one-way streets, trying to find one hotel after another in search of rooms, usually ends in desperation in the Gran luxe hotel at twice the price one intended to pay.

Vineyards of quality abound in Piedmont. That Italian aristocrat of vines, Nebbiolo, is first sighted when coming down through the Aosta valley. From Gattinara, a place on the way to the Lakes from Turin, as well as from the Mediterranean coast, Piedmont wines are sound, but her fame rests among the hills south of Turin.

There lies the land of Barolo, Barbaresco and Barbera. Barolo and Barbaresco are both villages, around which the Nebbiolo vine excels, giving red wines that need bottle age, particularly Barolo. Confusingly, Barbera is not a village but a grape, widely grown north and south of Asti, resulting in a light red wine – as popular a tipple in Turin as Beaujolais in Lyons.

If the wines are all Bs, the towns are all As. Alba, Asti, Alessandria and Acqui – join these on a map and you have enclosed the Moscato d'Asti region, in which the Muscat grapes for the sweet, aromatic, sparkling Asti Spumante must be grown. 'Sweet and lovely' would be an appropriate signature tune for this great wine. A little sugar has to be added to produce the effervescence, but no noble rot and repeated pickings are involved as in Sauternes and the Rhineland; the Moscato grape and the climate do it all quite naturally, and a tank is better than a bottle for the secondary fermentation.

Asti and Alba are both on the Tanaro, a sizeable tributary of the Po. Alba, between the river and the hills, is smaller and a more pleasant place to stay. Being the home of the white truffle, the gourmets concentrate here in October for the annual truffle fair. The Savona hotel serves an impressive hors d'oeuvre and there are one or two good restaurants in the little region of Barolo close by, notably the Belvedere, sited amid the vines at La Morra. Vines are trained high in northern Italy, cooler air giving more acidity and better balance to the wines, the requirement not being more heat by pruning close to the ground as in France. The landscape is variegated green, relieved

142

by mellow red-tiled roofs and the grey stone of castles and churches.

The sparkling wine of Piedmont is the famous sweet Asti Spumante from Muscat grapes grown in the Moscato d'Asti region already mentioned. It is a dessert wine, delicious with a peach (grown incidentally to perfection in these parts) fermented quite naturally in tanks and, if well made, fresh and fruity to a marked degree. Moscato d'Asti, otherwise similar, can be half a degree weaker.

Piedmont is also the cradle and home of Italian vermouth, begun by Carpano in 1786 in his bar by the doors of the Turin Stock Exchange. Punt é Mes (Point and a half) was to become not only a technical term, but the market's tipple as the tips and stock exchange tales passed from *Orsi* to *Tori* and on to *Cervi** (which should mean Bear, Bull and Stag to an Italian).

Vermouth of course is a blend of wines specially treated, flavoured with herbs and lightly fortified with local brandy, being sold finally at about the same strength as sherry or port (19° absolute alcohol or 31° British proof). There are no defined areas, the wines in Italy coming mainly from the south. Carpano, Cinzano, G. & L. Cora, Gancia, Martini e Rossi, Riccadonna and Contratto are now said to be the big seven. Four of them will be found welcoming visitors in this guide. All have something to shew but, I fancy, it is the fabulous Martini Museum that draws the crowds.

---

*In Italian, *ribassista* and *rialzista* respectively mean the Stock Exchange variety of bear and bull. I can find no special word for stag.

Luigi Calissano & Figli SpA,
Corso Langhe 5, 12051–Alba (Cuneo).
Tel. (0173) 2420 and 43998.

Calissano is a member of the Winefood group and its premises will be found about five hundred metres from Alba's central square in a south-westerly direction.

The cellars go down four storeys below ground level and the company deals in all the wines of Piedmont. Guides, speaking English they say, are co-opted from the local tourist office when necessary. Wines may be bought.

Hours:	Mondas to Friday 0930–1100; 1500–1730. Closed August.
Introduction:	By appointment. If in Alba, just telephone. For prior appointment, see note on Winefood group p.141.

Canelli (near Asti)
L. Bosca & Figli,
Canelli 14053 (Asti).
Tel. (0141) 81.161.

Canelli is a small town 20 kilometres south of Asti and about 15 kilometres east of Alba. From Turin take the Alessandria motorway, leaving it at the Asti exit. By rail, Canelli is 48 kilometres from Alessandria on the branch line to Bra.

La Bosca, founded here in 1831, is believed to be the first company to produce Asti Spumante. Today they continue the good work, marketing Vermouth and the wines of Piedmont as well.

The premises now consist of three establishments, including the original cellars of 1831. Visitors see Asti Spumante made in the traditional way and the fine library of viticulture and oenology, where some books date from 1600.

Hours:	Wednesdays, Thursdays and Fridays 0930–1200; 1430–1630.
Introduction:	By appointment. Letter obtainable from agents: Capital Wine & Travers Ltd., Central House, 32 High Street, London E15. Tel. 01–534 7536.

Gancia & C. S.p.A.,
14053 Canelli (Asti).
Tel. (0141) 81.121.

Canelli is a small town 30 kilometres south of Asti and about 30 kilometres from Alba. From Turin, take the Alessandria motorway, leaving it at the Asti exit; from Genoa take the Milan motorway leaving at the Serravalle-Alessandria exit; from Milan take the Genoa motorway leaving it at the Tortona-Alessandria exit. By rail, Canelli is 48 kilometres from Alessandria on the main Turin-Genoa line.

Carlo Gancia began this great firm in 1850 and for generations it has been building a great reputation for its Vermouth and its sweet and dry sparkling wines. Their Americano red aperitif is a recent success and Gancia products are now drunk in many parts of the world.

Visitors are shewn round the cellars and winery by English-speaking guides, with free tastings of vermouths, aperitifs and sparkling wines.

Hours:	Monday to Friday 0830–1130; 1430–1730. Closed August.
Introduction:	By letter from agents, after which a postcard giving 48 hours' notice of time of arrival is appreciated. Hedges & Butler Ltd., Hedges House, 153 Regent Street, London W1R 8HQ. Tel. 01–734 4444.

Fontanafredda (near Alba)
Fontanafredda,
Tenimenti di Barolo e di Fontanafredda, 12051 Alba (Cuneo).
Tel. Serralunga d'Alba (0173) 53.00 and 53.92.

Nine kilometres south-west of Alba (take the road towards Barolo), this superbly sited group of buildings set among trees and vines makes a fine colour picture on the Piedmont page of Hugh Johnson's now world-renowned *World Atlas of Wine*. Formerly the Royal Hunting Lodge, still furnished much as it was in 1861 when King Victor Emmanuel II united Italy, the peacocks strutting about outside give Fontanafredda a mildly regal air.

Visitors can view the royal apartments before descending to the cool cellars – floor after floor built later out of the rock – to taste Barbera, Barolo and Fontana, the company's rather special Asti Spumante.

Fontanafredda belongs to a national banking concern, whose directive is Piedmont wines only – and the very best.

Hours:	Monday to Friday 0930–1100; 1500–1730. Closed August – first three weeks.
Introduction:	None necessary provided post card or telephone call giving 24 hours' notice of arrival is sent and received. For large parties (say, over 10 people) make prior arrangements through agents: Gilbey Vintners Ltd., Gilbey House, Harlow, Essex CM20 1DX. Tel. (0279) 26801.

La Morra (near Alba)
Kiola Cantine Batasiolo S.p.A.,
12064 La Morra (Cuneo).
Tel. (0173) 60131.

Kiola is another company based in the Barolo region, which is roughly a rectangle, 10 kilometres by 7, with 800 hectares under vine. Its headquarters are near La Morra, 12 kilometres from Alba. Take the Barolo road from there, following the sign posts to Kiola.

Founded as recently as 1969, Kiola owns a hundred hectares of very good vineyards, all close together in the Barolo region. They are divided into six estates – Batasiolo, Cerequio, Bricco, Zonchetta, Bofani and Boscareto – each of which has its ancient farm house.

Visitors see over the cellars at Batasiolo. The company makes and markets all the usual Piedmont wines, which can be tasted and bought. English and French are spoken.

Hours:	Monday to Saturday inclusive 0900–1200; 1400–1800.
Introduction:	By appointment, which agents will arrange with the Milan office – Via S. Valeria, 5. Tel. 876.764 or 876.854. Write to: Hedges & Butler Ltd., Hedges House, 153 Regent Street, London W1R 8HQ. Tel. 01-734 4444.

Nizza Monferrato (near Asti)
Bersano S.p.A.,
Piazza Dante 21, I-14049 Nizza Monferrato.
Tel. (0141) 71.273.

Nizza Monferrato is a small town 20 kilometres south-east of Asti. Leave the autoroute, A19, at the Asti or Alessandria exit.

Bersano began in Milan in 1893 as a small chain of retail wine shops, changing in due course to viniculture and the company now produces and markets the whole range of Piedmont wines. Signor Arturo Bersano is often present to welcome visitors, not only to his cellars but to his outstanding museum here.

Early wine-making equipment, including some grotesque old wine presses, are spread about gardens and cellars. *Objets d'art* and pictures connected with wine occupy several rooms. Tastings are offered in the cellars, wines can be bought and for lunch there's a pleasant trattoria in the town called 'Da Italo'.

Hours:	Monday to Friday 0900–1130; 1500–1700. Closed August.
Introduction:	By appointment. Letter obtainable from agents: Cock, Russell Vintners, Seagram Distillers House, 17 Dacre Street, London SW1H 0DR. Tel. 01-222 4343.

Pessione
Martini e Rossi S.p.A.,
Casella Postale 475, 10100 Turin.
Tel. (011) 531242 Ext. 47.

Pessione, 24 kilometres from Turin and 4 kilometres off the road to Asti (fork right after Chieri) has always been the home of this great international house of drinks. It began in 1840, in a small way, as Martini e Sola, great advances being made later by Luigi Rossi and his four sons – Teofilo, Cesare, Ernesto and Enrico. Today, surprisingly, it is still a private company, the Rossis di Montelaras being the sole owners.

Of exceptional interest at Pessione is the Martini 'History of Wine Making' museum. This is a unique exhibition of drinking vessels of the ancient Mediterranean – Greek, Etruscan, Roman – their methods of bottling and preserving and their drinking vessels. There are superb examples of ancient Greek and Roman wine vessels, of which the Etruscan Cup 7th–5th century B.C., must be the oldest; there are old-time wine presses and Piedmontese wine wagons and, surprisingly, a display of English glass decanters and labels. One exhibit is a bottle dating from 200 B.C., containing the solidified residue of aromatic wine sealed with a cork from Sardinia – remarkable indeed recalling that the cork was unknown to Dom Perignon in Champagne in the early eighteenth century. The whole collection is insured for five hundred million lire.

On the modern side, the tour begins in Martini's giant export bottling department, proceeding to the 'factories' of liqueurs, of Vermouth and of sparkling wines in turn. Admission (to both) is free. Tastings are offered in the reception room but purchases are not possible.

Hours:	Monday to Thursday 0900–1200; 1400–1700; Friday 0900–1200. Closed August. On Friday afternoons, Saturdays, Sundays and public holidays *the museum only* is open but without guides and no tastings are possible.
Introduction:	None necessary. Telephone at least the day before to the Martini e Rossi Turin office (P.R. dept. 531242 Ext. 47).

Santa Vittoria d'Alba
Cinzano e Cie S.p.A.,
Stabilimento di Santa Vittoria, Santa Vittoria d'Alba (Cuneo).
Tel. (010) 39.172.47041.

Santa Vittoria d'Alba, which *is* virtually Cinzano, is between Bra and Alba, about three kilometres from the latter. From Turin, take the Moncalieri, Carmagnola, Bra road. The Savona motorway is an alternative but takes much the same time. Distance either way is about 50 kilometres.

Cinzano is a vast concern with over thirty companies around the world from Argentina to Australia. A document dated 1707, still in the firm's possession, licensed Giovanni Cinzano to distil brandy and cordials for sale in Pecetto and Turin. Another, given by the Confectioners and Distillers of Turin, records the investiture of the Cinzano brothers as Master Distillers.

A strong, but far from unpleasant aromatic scent, coming from the infusion of herbs being distilled for making Vermouth, may greet the visitor. The old copper stills and the huge wooden troughs full of herbs are pleasing and they will shew you corridors and cellars full of impressive maturing vats containing wines for Vermouth and others for Asti Spumante. As to bottles, storage capacity exceeds a million.

The original cellars once belonged to the House of Savoy, the Royal House from 1056 to the day Victor Emmanuel III abdicated in 1946.

English-speaking guides can be provided if arranged well in advance.

After the tour, there are wines to be tasted in the visitors' bar and doubtless Cinzano, having bought a number of restaurants locally in recent years, will suggest luncheon at one of them, L'Muscatel in Santa Vittoria.

Hours: Monday   to   Friday   0930–1130; 1430–1630. Closed Mid-July to Mid-August.

Introduction: By letter of introduction from: Cinzano U.K. Ltd., 20 Buckingham Gate, London SW1E 6LR. Tel. 01–828 4343.

LOMBARDY

Introduction

East of Piedmont and west of the Veneto the flat monotonous driving along the autostrada south-east from Milan to Bologna gives the impression that the whole province of Lombardy must be as flat as the proverbial pancake. And so it is in the plain of the Po, which comprises about three-quarters of Lombardy. But north from Milan to Lecco and along the eastern shore of Lake Como, the road is climbing slowly towards the Swiss frontier and the Alps. East (N38) from the top of the Lake, along the valley of the Adda, lies the Valtellina, no mean red wine region, where the grape is the Nebbiolo once more, disguised under the name Chiavennasca.

The wine towns are Sóndrio and Chiuro, but being nearly ninety slow miles from Milan, there is no great rush of touring oenophiles to get here. It could even be that one of the wines is called Inferno because it's such hell getting here, though the dark character of the deep valley that does not get much sun is, maybe, a more likely reason. The leading producer in the Valtellina is Signor Nino Negri.

Chiuro (near Sóndrio)
Casa Vinicola Nino Negri S.p.A.,
23030 Chiuro (Sóndrio).
Tel. (0342) 54207.

From Milan take the road northwards to Lecco, which

then passes along the eastern shore of Lake Como before turning eastwards to Sóndrio (138 kilometres from Milan). Chiuro, where there are direction signs to Nino Negri, is 10 kilometres beyond.

This company was founded in 1897 by Signor Nino Negri and his son, Carlo, is now in charge. The Romans, undeterred by the steep sides of the valley, in which there are some wonderful views, first grew wines here.

Today, wines labelled Valtellina must be made 70 per cent. from the Chiavennasca grape, which is a Nebbiolo. The better class is Valtellina Superiore, made from at least 95 per cent. Chiavennasca and this class embraces Sassella, Grumello, Inferno, Valgella and Fracia, all sub-districts or sites.

All these wines are very similar, spending at least two years in wood and one in bottle. Signor Negri also makes a Castel Chiuro, aged for much longer and perhaps the best of them all.

Visitors are very welcome; some of the cellars are very old.

Hours: Monday   to   Friday   0930–1130; 1430–1630. Closed August.

Introduction: By appointment. If in the district just telephone them. For prior appointment see note on Winefood group, p. 141.

TRENTINO-ALTO ADIGE

Introduction

Before considering the wines of Italy's northern province, sandwiched between the Veneto and Austria, it is as well to clarify the hyphenated title. The whole wine district lies in the valley of the Adige, the river which almost encompasses Verona and ends in the Adriatic south of Venice. It begins in the north at Merano, extending downstream to Trento some forty miles away, with Bolzano, the capital, about midway between the two towns.

From Bolzano, Venice and Milan are about 140 and 180 miles respectively, but with unbroken motorway travel from Innsbruck to the tip of Italy, it is all fast going nowadays.

The northern part, which includes Bolzano or Bozen, is the German-speaking Alto Adige, which was part of the Austrian Hapsburg empire until ceded to Italy in 1918. The southern part, the Trentino, around the town of Trento, is Italian-speaking. The province is a joint autonomous region with an Italian-speaking majority.

Thus it happens that, as in the past, the wines of the Alto Adige largely go north over the Brenner Pass, where winter-sporting holidaymakers in the mountainous western half of Austria enjoy them and probably think they come from the Austrian vineyards of the Danube basin, near Vienna. The white wines are mainly sold under their grape names – Sylvaner, Riesling, Traminer. Terlaner, indicative of Terlano, a village above Bolzano which has a good co-operative, is a good label place name.

The best red wine district is Caldaro, around the lake south of Bolzano. Label names vary, from Caldaro (with variations) to the German name for the lake, Kalterer (with variations). Another red wine, fuller and some say better than Caldaro, is Santa Maddalena. These vineyards are east of Bolzano on the Isarco, tributary of the Adige. Good, if not greatly distinguished, is a fair verdict on all these wines; but after a day in the fresh air, well away from the motorway, in this pretty valley or among the Dolomites close at hand, they can be very satisfactory.

Trento

Càvit Cantina Viticoltori Soc. Coop. a.r.l.,
Trento – Loc., Ravina Casella Postale 165.
Tel. (0461) 80155.

Càvit (Headquarters of the Consortium of the Cantine Sociali del Trentino) was founded in 1956. The yearly production of its fourteen associated vineyards, situated in various parts of the Trentino region, is about 60,000 hectolitres, which represents about 60 per cent. of the total production of the Province of Trento. Càvit's production includes all the range of wines of the Trentino and a fine spumante, Gran Càvit. Lago di Caldara or Caldaro, the ruby red table wine is the best wine of this region, with its own D.O.C.

Visitors can see the main cellars, the fourteen associated cellars and the vineyards; tastings can also be arranged on request.

Hours: Monday to Friday 0900–1200; 1500–1800.

Introduction: Write at least three weeks in advance proposing date and time of visit and number in party to agents:
Calvert Wine & Spirit Co. Ltd., Seagram Distillers House, 17 Dacre Street, London SW1H 0DR.
Tel. 01–222 4343.

VENETO

Introduction

The perfect approach to Venice is by sea, through the lagoon on a fine summer morning. Otherwise, the roads to Rome, Florence and Venice are so well trod that I can safely leave 'Getting There' to the reader.

The hinterland of Venice and Verona is an agricultural region that makes over eight hundred million litres of wine annually – double that of Tuscany. Only about an eighth of this output is D.O.C. wine, virtually from three regions north of Verona – Bardolino, Valpolicella and Soave.

Bardolino, village and small district on the south-eastern side of Lake Garda some fifteen miles west of Verona, contributes 14 per cent. of the D.O.C. total, its wine being light, fresh and almost rosé. About 40 per cent. comes from Soave, a village and district a dozen miles east of Verona. Soave, pale and dry, is reckoned Italy's best white wine, though it needs to come from the small Classico area, within the whole, to prove its true worth.

Between Soave and Bardolino are the wooded hills of Valpolicella, another village giving its name to a district, so large this time that this soft red wine, the nearest in style that Italy gets to claret, contributes about 54 per cent. of Veneto D.O.C. wine. The vineyards are mainly scattered and owned by small farmers, who delight in regaling the visitor with home-made salami and Valpolicella in its youngest, roughest state. 'Pure unmixed pig', said my host on one occasion as we ate the proffered ham. I nearly vomited; at home in Sussex, that was the phrase the local farmer had used, as he dumped the 'fruitiest' load of pig manure on my doorstep only a week earlier.

Verona is the place to stay for visiting the three regions. The Torre dei Lamberti from which our entrant at Lazise takes the name, stands majestically in the entrancing Piazza delle Erbe (Square of Grasses), as much a symbol of Verona as the equestrian statue of Can Grande, one of the Scaligere family – 'Princes of the Scala' – who ruled Verona from 1260 to 1387. They were all 'dogs' these despots; there was Mastino II (Mastiff II), Cansignorio (Lord-dog) and Cangrande,(Big dog). 'Cangrande della Scala' (Big Dog of the Staircase) I call him, because after a good dinner here with plenty of Soave and Valpolicella, I can see him guarding the fifteenth-century staircase – a gem of Verona – in the courtyard of the Old Market Place. He sits at the top, as frightening as the dog with eyes as big as saucers in Hans Andersen's *Tinderbox,* and nobody dares pass.

Lazise (near Verona)
Lamberti S.p.A.,
34017 Lazise (Verona).
Tel. (045) 678058.

The name Lamberti is that of a noble tenth-century family of Verona, where their tower, the Torre dei Lamberti still stands. Lazise is a gay, colourful resort on the south-eastern shore of Lake Garda, 23 kilometres from Verona. Leave Verona on N11 to Brescia, branching right to Bussolengo, Pastrengo and Lazise. From the Brenner, leave Autostrada A22 at Affi; from Venice or Milan on Autostrada A4, leave at Peschiera.

Lamberti's impressive cellars at Lazise and their ultra-modern bottling plant at Pastrengo are right in the Bardolino classico district, where they own 100 hectares of vineyards. The company are just as involved with Soave and Valpolicella, where they have various wineries handling grapes, partly from their own vineyards and partly from individual growers. To ensure high quality, it is important to buy the grapes and have control of vinification rather than to buy wines made by a variety of individuals.

Hours:	Monday to Friday 1000–1200; 1500–1700. Closed August.
Introduction:	A member of the Winefood group (see note on p.141) Lamberti wines are sold by Justerini and Brooks, Peter Dominic and Westminster Wine. For letter of introduction write: Justerini and Brooks Ltd., 1 York Gate, Regent's Park, London NW1 4PU. Tel. 01-935 4446. (The Export Dept at Lamberti's group head office in Milan (Tel. (02) 4483) likes to arrange an English-speaking guide ·vhenever possible and requests good warning for this reason.)

Verona
Fratelli Bolla s.a.s.,
Piazzetta Scala 8 – 37100 Verona.
Tel. (045) 23860/23877.

Founded in 1883 by Alberto and Luigi Albano Bolla, this company has become one of the leading shippers of the three classical Veronese wines, supplying hotels, shipping and air lines in many parts of the world. In Britain the firm is well-known, F. S. Matta having been the principal importer for many years.

Germany is now the firm's best customer but Bruno Bolla, the present head, visiting Britain on business recently, was optimistic for the long-term future of Italian wines here.

Visitors are shewn the cellars, bottling plants, etc.

Hours:	Monday to Friday 0930–1130; 1430–1630. Closed August.
Introduction:	By appointment. Letter obtainable from agents: Hedges & Butler Ltd., Hedges House, 153 Regent Street, London W1R 8HQ. Tel. 01–734 4444.

THE MARCHES

Introduction

Having crossed the Lombardy plain from Milan to Bologna, the A14 autoroute parts company from the A1 – the Sole for Florence and Rome – passing through the pleasant province of Emilia-Romagna, known for its sparkling red Lambrusco di Sorbara (near Modena), though there are also some good Trebbiano white wines.

The autoroute reaches the Adriatic near Rimini, serving in the course of the next sixty miles, Pesaro, birthplace of Rossini, Urbino inland, perched on its two hills looking down upon the countryside of The Marches, and the port of Ancona.

Good locally is a dry straw-coloured wine called Bianchello and a ruby red called Conero, but the wine The Marches exports is the white, dry Verdicchio dei Castelli di Jesi, which is in the same class as Soave, its northern neighbour. The district is small – 30 miles long and 15 wide – and it lies 15 to 25 miles inland from Ancona, not too far to be sure that the Adriatic fish – the perfect accompaniment to Verdicchio – are still fresh. Curiously Jesi,* the little town that gives the wine its name is just outside its own district. The trendy shaped bottle that curves inwards at the waist is slowly being given up in favour of a more conventional waistline.

Having paid homage at Urbino to Piero della Francesca and the Ducal Palace of Montefeltro, his enlightened patron, the art lovers can continue the 'Piero pilgrimage' at Borgo San Sepolcro and Arezzo or drive southwards through the heart of Umbria seeing the smaller towns like Gubbio, Todi and Spoleto. Given good weather this drive through the Appennines is delightful and it has the merit of leading past Torgiano, where Dr. Lungarotti's cellars (p.154) should not be missed.

Cupra Montana
Cantina Sociale di Cupramontana,
P.O. Box 6, 60034 Cupra Montana.
Tel. (0731) 78273.

Cupra Montana, marked on larger scale maps, is the centre of the Verdicchio Classico district. Ancona, main town of The Marches is 48 kms, the A14 autoroute (Ancona Nord) is 36 kms and the small town of Jesi 15. A small town with 5,000 inhabitants, a few kilometres to the south of the main road between Jesi and Fabriano, Cupra Montana has this well-run wine co-operative, which has supplied its fresh dry white Verdicchio Classico to Justerini & Brooks for some years now.

Visitors can see the winery, the vineyards and the novel concrete towers, where the grapes of a hundred small farmers can be stored. Verdicchio and other local wines can be sampled and drunk at Fonte della Romita, the local restaurant. The Park Hotel (included in the Italian *Guide Michelin*) is a possible overnight stop. For the return in the direction of Jesi, there is a well marked 'scenic' road worth taking.

Hours:	Monday to Saturday inclusive 0900–1130; 1500–1700.
Introduction:	If in the district, just telephone them. For prior appointment write to principal shippers: Justerini & Brooks Ltd., 1 York Gate, Regent's Park, London NW1 4PU. Tel. 01–935 4446.

---

*Jesi and Iesi are equally correct.

Osima-Scala (near Ancona)
Azienda Vinicola Umani Ronchi s.r.l.,
Osimo SS16 (Ancona).
Tel. (071) 7719 and 7750.

This company is based in Osimo Scala, 18 kilometres due south of Ancona (Strada-Statale 16). It is near Monte Conero, in pretty country close to the Adriatic and backed by the ancient city of Osimo. The best approach is from the Ancona Sud exit of the A14 autoroute.

The firm was founded in 1955 by a group of local vineyard owners. It has its own establishments in Osimo and in Castelbellino, which is in the heart of the Verdicchio Classico district. Wines made from its own vineyards are among those worth trying in its own restaurant in

Osimo, established partly to promote the wines of this region.

Visitors see over the cellars and bottling rooms. Meals are available at modest prices and, time permitting, a look at the vineyards can be included.

Hours: Monday to Friday 0900–1130; 1500–1700.

Introduction: Though none is strictly necessary, more time is likely to be given if the visitor has given advance notice of arrival directly or through agents:
Hedges & Butler Ltd., Hedges House, 153 Regent Street, London W1R 8HQ. Tel. 01–734 4444.

TUSCANY

Introduction

Introduction

Cyril Ray, in his standard work, *The Wines of Italy,* advances three reasons why Chianti became the best known Italian wine abroad. First, the zone is large, making one wine of the same kind; secondly, ownership has always been largely vested in the old and noble Florentine families, rich enough to make good wine and market it; thirdly, the picturesque wicker-covered flask was quickly regarded by romantic foreigners as symbolic of sunny Italy.

Nowadays two types of Chianti are recognised. The first is the young fresh wine, which should be drunk young like Beaujolais and is eminently suited to spending its short life in the traditional wicker-covered *fiasco*. The second is the greater Chianti, made by maturing for up to six years in cask, which needs bottles in order to be laid down to mature for at least another two years. These are the Chiantis worth keeping like clarets or burgundies; they compete with Barolo and Torgiano as Italy's best red wines.

Chianti may be made in seven regions between Florence and Siena, 'Classico' being the core of the whole and the longest established. There is surely no more glorious landscape than this harmonious combination of vine, cypress and olive – a perfect wine region for the visitor, save

perhaps that there is no river – no Mosel or Douro to twist and wind between terraced slopes.

Visiting individual estates is difficult. Few of their aristocratic owners live on these estates and those that do hardly wish to spend the summer answering the doorbell. Cellar men and vineyard workers have their own work to do and only speak Italian. Besides, in a country which grows wine grapes almost anywhere except on top of the Appennines, our northern curiosity may seem as surprising to Italians as an invasion of Britain by earnest Italian students of blackberries would be to us.

Exporting wine, a novelty for most Italian growers ten years ago, has made such progress that sales abroad now exceed £1,000 million a year. Perhaps there is little need to strengthen the marketing organisation as yet. Nevertheless, with forty producing estates in Chianti Classico and a Consorzio, to which most leading proprietors belong, it would be a good plan in summer if one or two properties in turn could be open to the public each week, with transport provided to and from Florence. Such a scheme, on the lines of Britain's 'Gardens Open to the Public' would give pleasure to many and raise funds for charity or other useful purpose.

Those with transport wishing to spend a day in the

Classico region could leave Florence on N222 and head for Greve- and Castellina-in-Chianti. Just a village with three thousand inhabitants, Castellina is the heart of the region, with two small charming hotels, the Villa Casalecchi and the Tenuta di Ricavo, open in summer.

In the middle of the vintage (the date was Sunday 12 October in 1975) the *Lega del Chianti* assemble here in traditional robes for a busy day, first attending Mass and then tasting till evening to select the best ordinary Chianti Classico and the best Chianti Reserva from each of the nine communes that make up the Classico region.

The panel of judges are members of the Italian Association of Oenologists and Master Tasters (O.M.A.V.) and in 1975 they invited David Peppercorn, the well-known English Master of Wine, to join them, enrolling him as a member of the *Lega*.

This Classico tour can continue to Siena directly on N222 or, with more time to spare, via Radda-in-Chianti.

For the return journey, the fast Siena-Florence 'superstrada' skirts Poggibonsi and San Casciano, but if there is no hurry, keep to the old road through Tavarnelle or make a bigger détour to San Gimignano, which has a good dry white D.O.C. wine, Vernaccia di San Gimignano, apart from its rather more celebrated *quattrocento* attractions. Another attractive town, where there is an autumn wine festival is Impruneta, on the outer fringe of the Classico region, only 8 miles south of Florence.

Though none the worse for the change, the traditional white wine of Tuscany is no longer permitted to be called White Chianti. Others likely to be met in the region are Brunello di Montalcino, a big strong chianti from Brunello grapes grown south of Siena, and Vino Santo, the big strong dessert wine, which has a slightly bitter after-taste.

Further Information
Consorzio Vino Chianti Classico,
Via Valfonda, 9,
50123 Firenze.
Tel. (055) 24861.

Brolio (near Siena)
Casa Vinicola Barone Ricasoli S.p.A.,
Castello di Brolio, Brolio-in-Chianti, Siena

Along the wine road, N222, from Florence to Greve and Siena, the gaunt towers of Castle Brolio can be seen from Radda-in-Chianti onwards. The next turning left after Meleto leads to it. From Florence, Brolio is 65 kilometres. Take your time through one of the most beautiful landscapes in the world, lunching perhaps after the visit in Siena's piazza, 15 kilometres further on.

Wine has been made at Castle Brolio since 1000 A.D., with many interruptions prior to 1500 during the constant warring between Guelph and Ghibelline. Five hundred metres above sea level the Castle, commanding the sur-

rounding countryside, was a formidable fortress and observation post. 'When Brolio growls, all Siena trembles' but the Sienese must have cheered rather than trembled when at last they destroyed the place in 1498 and thereafter Siena and Florence joined forces under Medici rule.

No family has done more for Chianti than the Ricasoli. Baron Bettino (1809–1890), who became Premier of the newly United Italy in 1861, spent much of his life at Brolio, keeping an eye on his young wife and experimenting with vines and wines. His method of fermenting in two successive waves led to Chianti as we know it today. Brolio Chianti Classico and Brolio Riserva remain among the finest chiantis obtainable.

An English-speaking guide shows visitors the cellars and they can walk round the grounds.

Hours:	Monday to Friday 0900–1200; 1400–1700. Closed August.
Introduction:	By appointment. Letter obtainable from agents: Cock Russell Vintners Ltd., Seagram Distillers House, 17 Dacre Street, London SW1H 0DR. Tel. 01–222 4343. In Florence the Ricasoli administrative office is: Casa Vinicola Barone Ricasoli, Piazza Vittorio Veneto No. 1, 50123 Florence. Tel. 28.38.55–6–7.

Florence
Marchesi L. & P. Antinori,
Piazza degli Antinori, 50123 Firenze.
Tel. (055) 282.203 and 298.298.

The first Marquis Antinori entered the Florentine Wine Guild in 1385 when Giovanni di Bicci, first of the Medici dynasty, was twenty-five. Thenceforward this old Florentine family has been associated with the production of the best wines of Tuscany.

Gradually they have gone further afield, extending ownership to vineyards beyond Tuscany, Antinori wines becoming well known in many parts of the world.

The address above, in Florence, is that of the head office and cellars where visiting arrangements are made. Their family Chianti Classico estate, difficult for visitors to find, is about twelve kilometres out near San Casciano in the north-west corner of the Chianti Classico district. This and lack of English-speaking people on the estate necessitate the head office contact.

Hours:	Monday to Friday 0930–1200. Closed August.
Introduction:	By appointment. Letter obtainable from agents: G. Belloni & Co. Ltd., Belloni House, 128–132 Albert Street, London NW1. Tel. 01–267 1121.

Pontassieve (near Florence)
Chianti Melini S.p.A.,
50065 Pontassieve, Firenze.
Tel. (055) 83.23.48.

Pontassieve is a small town on the right bank of the Sieve, where that tributary joins the Arno, 18 kilometres east of Florence (Firenze). A1, Autostrada exits are Valdarno coming from the south and Firenze-Sud coming from the north. From Florence itself, take road N67.

The House of Melini dates from 1705. One of a number of wine companies in Italy belonging to Winefood (see page 141), Melini have cellars at San Gimignano, Gaggiano and Castellina-in-Chianti and they own 200 hectares in the Classico district.

Hours: Monday to Friday 0930–1230; 1430–1700. Closed August.
Introduction: If in the district, telephone to make an appointment, otherwise prior letter of introduction obtainable from agents: Rutherford, Osborne & Perkin Ltd., Harlequin Avenue, Great West Road, Brentford, Middlesex. Tel. 01–560 8351.

Chianti Ruffino S.p.A.,
50065 Pontassieve, Firenze.
Tel. (055) 83.23.07.

Founded in 1877 by two cousins, Ilario and Leopoldo Ruffino, members of this old Florentine family, the business has prospered. Riserva Ducale is the company's best Chianti Classico but they now export other Italian wines besides Chianti.

Visitors see the cellars, winery and, if requested in advance, the vineyards.

Hours: Monday to Friday 0930–1230; 1430–1700. Closed one month – about 10 August to 9 September.
Introduction: If in the region, telephone 83.23.07 and ask for Signor Marino Marcheselli, the P.R. manager. Otherwise by appointment, letter obtainable from agents: Hedges & Butler Ltd., Hedges House, 153 Regent Street, London W1R 8HQ. Tel. 01–734 4444.

Siena
Enoteca Italica Permanente,
Fortezza Medicea, 53100 Siena.
Tel. (0577) 28.84.97.

The Enoteca *(Winotheque)* or Wine Library is at the top of the old city, a hundred metres from the football stadium in a tree-lined square off La Lizza.

This is the national wine 'library', where sample bottles of every quality wine made in Italy are kept. Arranged in a wing of the sixteenth-century Medici fortress, many sample bottles are displayed in a beautiful barrel-vaulted hall. On the ground floor there is an elegant tasting bar where every wine in the library can be tasted, though notice may be needed for some of them. The sampling charge is 250 lire or more; bottles may be bought.

Hours: Daily 1100–2400.
Introduction: None necessary.

UMBRIA

Introduction

Shaped a little like Ireland, Umbria is slap in the middle of Central Italy. Driving south down the A1 autoroute, it is lying away on the left for some 65 miles, roughly from Arezzo to Orte. A bulge in the boundary on the western side towards the lake of Bolsena brings that delightful hill town, Orvieto, within Umbria. Good hotels and restaurants make it a good overnight stop. Torgiano, in the centre, between Assisi and Perugia, the capital, is the other Umbrian wine name that has come to the fore since the D.O.C. legislation.

Once a resort of the popes and lying directly between Florence and Rome, Orvieto had advantages in making its golden-yellow, aromatic wines known. But the name became debased and in the mid-sixties I recall a friend, trying to buy more authentic wines directly from growers, discovering that hardly any of the best known shippers of Orvieto owned any vineyards there at all.

Prevention of irregularities was long overdue and it is not without significance that D.O.C. was not awarded to Orvieto Classico and Orvieto Tipico until as late as August 1971. The slightly sweet Abboccato is, I always think, more natural to Orvieto than the dry Secco but it is a matter of personal choice. As so often in Italy, the grapes are mainly Trebbiano and Sangiovese and normal bottles are replacing the dumpy-shaped wickered flasks.

Fresh, lively, round, fruity . . . all the best wine epithets are used in praise of Rubesco Torgiano; the white Torre di Giano is well liked too.

While Dottore Lungarotti and his nephew, Baldo Lucaroni, take care of the Torgiano wine, Signora Lungarotti, the doctor's wife, has brought her own erudition and enthusiam to the creation of a splendid wine museum, visited by 5,000 people in 1974, its first year. The history of wine making in Umbria is illustrated with local maps, prints, amphorae, examples of Majolica earthenware and a very comprehensive collection of wine presses and tools. Material from the Lungarotti and the Severini family archives contribute to a good museum library.

Alas, Torgiano's production, red and white, is estimated at a mere 300,000 litres. Vineyards around Lake Trasimento are being extended, with some hope of making wines of quality. The project is being undertaken by Cantina Sociale di Castiglione at Castiglione Sul Lago, Perugia.

Castiglione-in-Teverina (near Orvieto)
Casa Vinicola Conte Vaselli S.p.A.,
Fattoria di Orvieto-Baschi, Montecchio, Castiglione-in-Teverina, CAP 01024.
Tel. 0761/41305.

Castiglione-in-Teverina, lying between the boundaries of Umbria and Lazio, is ten kilometres south of Orvieto and about one hundred from Rome. Approaching via the Autostrada del Sole from Florence or Rome, the exit is Casello di Orvieto. Proceed thence towards Baschi, turning at crossroads where indicated for Castiglione. For an overnight stop, Orvieto, with good hotels and a first-class restaurant, is strongly recommended.

A hill town – almost a miniature Orvieto – dominating the Tiber valley, the origins of Castiglione-in-Teverina are ancient, certainly Etruscan. The Vaselli company owns extensive vineyards in the communes of Orvieto, Baschi, Montecchio and Castiglione-in-Teverina.

Most of the cellars, excavated out of the tufa (volcanic rock), are four storeys deep. Visitors taste the Orvieto wines; some in oak casks, others in modern cement tanks.

Parties, maximum 20, are welcome but only Italian is likely to be spoken.

Hours:	Monday to Friday, 0930–1100; 1500–1730. April to September inclusive.
Introduction:	Visits are arranged by the Vaselli office in Rome: Vinicola Conte Vaselli S.p.A., CAP-00186, Piazza Parlemento 16, Roma. Tel. (69720) 41.42.43.44, or this can be done through Justerini & Brooks, who list their Orvieto: Justerini & Brooks Ltd., 1 York Gate, Regent's Park, London NW1 4PU. Tel. 01-935 4446.

Torgiano (near Perugia)
Cantine Dr. Giorgio Lungarotti & Co.,
06089 Torgiano (Perugia).
Tel. (075) 82.22.76/7.

About two hours' drive both from Florence and from Rome, via the autostrada A1, Torgiano lies eleven kilometres south-east of Perugia just off the E7 road to Todi. From Rome, leave the autostrada at Orte; from Florence, leave it at Val di Chiana.

Dr. Lungarotti and his father before him have steadily created vineyards of repute here since 1920. Outside Italy, nobody seems to have been aware of this, so that in 1971, when Rubesco Torgiano and the white Torre di Giano were first shipped to Britain, no reference to the place was to be found in English wine books and the quality, particularly of the red (Rubesco), came as a delightful surprise. Three years before, the first D.O.C. awards in Umbria had been given to these two wines.

Both are estate-bottled from Lungarotti-owned vineyards, the red spending two to three years maturing in oak casks and one or two more in bottle, before being shipped.

This is a great place for visitors – beautiful country, ancient cellars, enthusiastic English-speaking hosts, plenty to taste and the wine museum. The museum is their latest enterprise, which drew 5,000 visitors in 1974, its first year.

Hours:	Museum: Daily 0900–1200; 1500–1800. Cellars: Monday to Friday 0800–1300; 1500–1800. Saturday 0800–1300.
Introduction:	All welcome; none necessary.

LAZIO

Introduction

By far the most celebrated wine of Lazio, which includes Rome and stretches south to Campania and north to Tuscany, is Est! Est! Est! of Montefiascone. The story of the German Bishop Defuk has been told a million times and it must have sold ten million bottles. So much so that since D.O.C. required Italian wines to come from their recognised districts Est! Est! Est! seems quietly to have vanished. But in any case the Bishop's demise was due to a sweet Moscato wine, whereas for years now, Montefiascone has been making dry Trebbiano wines.

The other district of Lazio is Castelli Romani, covering about fifty square miles of the Alban Hills south-east of Rome. Full-bodied, honey-coloured with a flavour derived from volcanic soil, the best wines come from around Albano, drunk by popes at their summer residence at Castel Gandolfo, from Genzano south of the Lake and from Frascati to the north, close to the A2 autoroute for Naples. Imbibers seem to agree that they all taste superb in their original cellars but rather different in Rome, where professional oenologists have made them fit for human consumption!

Nevertheless I shall always be grateful for Frascati in Rome. It was after my own first visit that I wrote in *Wine Mine*, ('A Mine of Wine Information' published from 1959 to 1975, usually twice yearly, for the edification of Peter Dominic's customers in Britain):

'When your feet are being blistered to make a Roman holiday and you sink into the nearest café chair with Keat's grave and The Vatican still to be "done", call for a glass of cold Frascati! You may yet make it.'

Frascati
Vini di Fontana Candida S.p.A.,
Casella Postale 52, 00044 Frascati (Roma).
Tel. (06) 944131.

Fontana Candida is in the commune of Monte Porzio Catone, 24 kilometres from Rome and 10 kilometres from the Monte Porzio Cantone exit of the Autostrada del Sole (A2) to Naples.

The company, established thirty years ago, has quickly become known as a specialist in the wines of Frascati, both in Italy and abroad. These normally are for early drinking but the Fontana Candida superiore is made to last longer and is as fine a Frascati as can be made.

Visitors see the cellars, vineyards and may buy wines.

Hours:	Monday to Friday 1000–1230; 1500–1600. Closed August.
Introduction:	By appointment. If in the district just telephone. For prior appointment see note Winefood group, p.141.

Cantina Produttori Frascati, 'San Matteo',
Via di Vermicino 16, 00044 Frascati (Roma).
Tel. (06) 940448.

Dry and *amabile* (sweet) styles of Frascati, are all made at this Co-operative, which ships the dry style to Britain in considerable and growing quantities.

Hours:	Monday to Friday 1000–1230; 1500–1600. Closed August.
Introduction:	San Matteo Frascati is sold by Justerini & Brooks and Peter Dominic. For letter of introduction write: Justerini & Brooks Ltd., 1 York Gate, Regent's Park, London NW1 4PU. Tel. 01-935 4446.

Vini Valle Vermiglia Via Vanvitelli,
20–00044 Frascati (Roma).
Tel. (06) 940.066/940.413.

Valle Vermiglia's cellars are 17.8 kilometres from Rome, along the Via Tuscolana leading to Frascati, and where the road starts climbing to Castelli Romani.

The cellars are modern but the history of the Tuscolo wine, being a typical Frascati, can be traced back to ancient Rome.

Visitors are shewn the cellars and taste local wines.

Hours:	Monday to Friday 0900–1200; 1400–1600. Closed August.
Introduction:	None necessary. Agents are: Hedges & Butler Ltd., Hedges House, 153 Regent Street, London W1R 8HQ. Tel. 01-734 4444.

CAMPANIA

Introduction

Campania, province of Naples, makes 300 million litres of wine a year, but only half a million is D.O.C. wine. Ischia has a D.O.C. for its dry white wine, which is required to come entirely from the island's vineyards. Capri, on the other hand, covers wines made on the Sorrento peninsula – at Ravello for example, high above Amalfi. I have often drunk the rosés, the white wines and occasionally the big red Gran Caruso from Ravello; with pleasure on holidays in Positano. Like the wines of Provence they belong to bikinis, beaches and *bouillabaisse;* bottles taken home so enthusiastically in the car boot never taste quite the same at home after a hard day's work in Wigan on a wet Tuesday. One day the boundaries will be sorted out and Capri given its D.O.C.

Lacrima Christi is a wine of Campania, the grapes being grown on the slopes of Vesuvius. The vineyards – about 2,500 acres – are in parts of the following communes and a glance at a wine map shews how a tour could be invented. This would be the order, taking the road from Naples that goes clockwise round the base of the volcano:

San Sebastiano, Cércola, Póllena, S. Anastasia, Somma-Vesuviana, Ottaviano, S. Giuseppe-Vesuviano, Terzigno, Boscoreale, Boscotrecase (and on the bay itself returning to Naples) Torre del Greco, Ercolano and Pórtici.

The many sweet, sticky concoctions that have disgraced this charming name should draw tears from hardened sinners, let alone Christ. Though the grape can be dried to make a sweet *liquoroso*, the true wine is dry and delicate, with a little sweetness and a flowery fragrance. Mastroberardino, the firm which gave me the commune names above, make it well at their Atripaldi headquarters, about twenty-five miles inland from Salerno. Red and rosé wines, which ought to be sold as 'Vesuvio rosso' and 'Vesuvio rosato', and presumably will be under D.O.C., may be found masquerading as Lacrima.

In 1892, at a Vienna Wine Congress, Lacrima Christi was adjudged the world's best wine. From then until 1948, the old volcano frequently spat sulphuric acid on the vineyards. Dormant since then, some say Vesuvio is dead; perhaps there will be a great Lacrima Christi revival.

Atripaldi
Michele Mastroberardino,
Casa Vinicola, 83042 Atripalda (Avellino).
Tel. (0825) 62.61.23.

Coming from the north on the A1 autostrada, take the autostrada (A16) to Bari before reaching Naples. Leave A16 by the Avellino Est exit. Casa Mastroberardino will be found 2 kilometres beyond.

The cellars date from the early nineteenth century. The family name 'Master Bernard' is associated with *Mastranze,* a sort of mediaeval institute of local 'masters of wine'.

Visitors see the cellars and vineyards and will taste such wines as Taurasi, 'the strong dark monarch of the Aglianico family*', Greco di Tufo, (both D.O.C.), Fiano di Avellino, Hirpinia del Vesuvio and, of course, Lacrima Christi.

Hours:	Monday to Friday 1000–1230; 1600–1800. Closed August.
Introduction:	By appointment. If in the district, just telephone but prior warning by letter is preferred, stating expected date and time of arrival.

Naples
D'Ambra Vini d'Ischia,
via Mergellina, 2–80122 Naples.
Tel. (081) 681732.

Coming from Rome the autostrada exit is south-east of Naples whereas the Via Mergellina is on the north-west side. The Mergellina railway station and the Via Mergellina are both close to the quay from which the ferries leave for Ischia and Capri (see Naples plan, Italian *Guide Michelin*, square AZ).

Philip Dallas (see Recommended Books p.141), describes this House founded in 1888 as, 'for decades the Standard bearer of Ischia wines'.

In addition to Ischia, which has been granted D.O.C. status, Casa d'Ambra produce some other dry wines under their grapes names – Biancolella, Forastera and

---

*Hugh Johnson *World Atlas of Wine.*

Per' 'e Palummo. Visitors see the cellars and, if going to Ischia (ship 90 minutes, hydrofoil 40 minutes) can be directed to their vineyards.

Hours: Monday to Friday Summer 0900–1230;

1630–1830; Winter 0930–1230.

Introduction: By appointment. If in the district just telephone. For prior appointment, see note Winefood Group p.141.

SICILY

Introduction

Second largest wine-producing province of Italy*, the Mediterranean's largest island is good for ten million hectolitres a year. Table and dessert wines from the Moscato grape are made in many parts, the best of the table wines coming from the volcanic slopes of Mount Etna, Etna being a D.O.C. for red, white and rosé.

Marsala, Sicily's internationally known dessert wine, is shipped from the port of Marsala on the island's short stretch of west coast. Trapani, the province which claims Marsala, makes over half of Sicily's wine.

There are international airports at Palermo, the capital, and at Catania; fly/drive hire is recommended. The sightseer's circuit of the island can be made comfortably in five days to a week, the chain of Jolly hotels serving adequately for overnight stops. That in the charming hill town of Erice has spectacular views over Trapani town, on the flat ground below, and is convenient for Marsala.

Though it cannot compare as a wine with port or sherry, Marsala has always enjoyed a fair measure of popularity. It is made from two Sicilian white grapes fortified with grape brandy, being finally flavoured and coloured with *vino cotto* – unfermented grape juice reduced in volume by heating. The different components are kept separately in casks, then blended and then rested again. The wine will be at least four years old before it is drunk.

The inventor – as all English wine books recall – was John Woodhouse of Liverpool, who sent home an experimental consignment in 1773. Ingham and Whittaker set up a rival company later and in 1831 they were joined by

Vicenzo Florio, a rich Calabrian. Today there is one company, Florio, Ingham and Woodhouse, owned by Cinzano.

About a quarter of Sicilians attempt to live by wine, and the co-operatives, producing about half the output of the island, bring new hope to a people who are among the poorest in Europe. One object to progress in impoverished western Sicily was apathy and delay in building a dam on the river Iato, near Partinico. It was here in 1955 that an unknown architect, Danilo Dolci, whose father had been the stationmaster, had settled to live in abject poverty as the peasants lived, in a town where drains, schools and piped water were non-existent. Thenceforward, from 1955 to 1971, he led the people in every form of pacific protest from fasting to marching until that dam was built.

Marsala
S.A.V.I. Florio & Co.,
Stabilimenti di Marsala, 91025 Trapani.
Tel. (0923) 51122.

The headquarters of Florio, Ingham and Woodhouse occupy almost a square kilometre close to the waterfront at the southern end of the town. There is a Motel Agip but the Jolly Hotel at Erice, in a prettier setting some 30 kilometres to the north, is a pleasanter place to stay.

The Florio farms, where Grillo Inzolia and Cataratto grapes are grown, lie in the coastal belt from Trapani in the north to Castel Vetrano to the south. In the hilly regions around Partinico, Alcamo and Balestrate, they grow Grecane and Damaschina, which give the sweet and mellow fragrance to the wine.

---

*The largest is Puglia, the heel of Italy.

Included in the tour of the establishment is a small museum of bottles, the labels of which tell the history of the companies. Among curiosities there is one label 'Hospital Size', which denoted 'a tonic' sold in the States during Prohibition. Written orders from Nelson are still preserved in this town, which chose another Englishman, Thomas Becket, as its patron saint.

Hours: Monday to Friday 0930–1130; 1430–1630. Closed mid-July – mid-August.

Introduction: By appointment. Letters obtainable from Florio agents throughout the world, who can be contacted through your local wine merchant.

U.K. Agents: Cinzano U.K. Ltd., 20 Buckingham Gate, London SW1E 6LR. Tel. 01–828 4343.

SPAIN

Spain

(Wine regions arranged clockwise from Bilbao)

Introduction

Now Funtarabia marks our goal
And Bidassoa shows,
At issue with each whispering shoal
In violet, pearl and rose,
Ere crimson over ocean's edge
The sunset banners die . . .
Yes – Twenty takes to Bourg-Madame
But Ten is for Hendaye.   — Kipling.

Spain, as everybody knows, is the home of sherry, the world's favourite fortified wine. Indeed, in 1967, a British High Court ruled that sherry must come not only from Spain, but from delimited district around the town of Jerez de la Frontera in Andalusia. Wines of similar style made elsewhere must be described as South African sherry, Cyprus sherry, etc.

Far less known is that Spain has as much as one tenth of her agricultural land growing vines for wines, bringing forth 25 million hectolitres of wine a year. Though this is only a third of what France and Italy can achieve and less too than Argentina, it is enough to put Spain fourth in any world order of wine-making countries. Only 6 per cent. of her total is sherry and although Jerez, where many of the shipping firms are British, is the region of greatest interest, I have left it till last. So I have started in the north from Bilbao, where the Southampton carferry deposits so many British travellers, and I have followed a clockwise circuit as I have done previously with other countries.

GETTING THERE

By sea and road: With petrol so expensive, motoring 2000 miles from French channel ports to southern Spain and back looks to be a dwindling pastime. Increasingly motorists are likely to use the Southampton car ferries – to Bilbao (Swedish Lloyd, 37 hours), and to Santander (Spanish Aznar Line, 32 hours). From Bilbao, Spain's best table wine region Rioja is under a hundred miles, and for those heading south, the fast main road from Vitoria to Madrid is close to the region.

By rail: For slow journeys without going to the trouble of buying a horse and cart, rail used to be the perfect transport in Spain. Much, however, has changed. Miranda del Ebro, only 10 miles from Haro and 43 from Logroño, the Rioja region's principal towns, is now only 1 to 1½ hours

from Logroño. Miranda del Ebro is the junction of the Barcelona-Bilbao line (used for exporting via Bilbao) and the main line from the Hendaye-Irun frontier to Madrid. From Irun the fastest train now makes Miranda (110 miles with a climb from the coast) in 2¼ hours.

By air: There are direct flights daily from London and Paris to Madrid, to Bilbao and to Barcelona. To Seville, international airport for Jerez, daily flights from London take six hours via Madrid (two hour stop). Other direct flights, on most days weekly, are to Valencia, Almería and Málaga. Fly/drive car hire can be arranged in advance with British travel agents. For Jerez, fly/drive from London via Gibraltar is no longer possible because the Spanish authorities have closed the frontier.

NATIONAL HOLIDAYS

National holidays in Spain are given below. Please remember when planning your trip that, not only will premises be closed, but that it is not uncommon to remain closed for an extra day (e.g. Friday after a Thursday holiday) to make a long weekend break.

1 and 6 January, Maundy Thursday, Good Friday, 1 May, Ascension Day, Corpus Christi, 29 June, 18 and 25 July, 15 August, 24 September (Jerez only),* 12 October, 1 November, 8 and 25 December.

Further Information:
Spanish National Tourist Office
70 Jermyn Street
London SW1.
Tel. 01–930 8578.

*In additon to these national holidays, many towns in Spain have local holidays not celebrated elsewhere.

SPAIN

FRANCE

La Coruña · Santander · Bilbao · San Sebastian

· Vitoria
· Pamplona

Burgos · Cenicero · Rioja · Logroño
· Fuenmayor · Ebro

PORTUGAL

Douro

MADRID

Tagus

La Mancha

Guadiana · Valdepenas

Córdoba · Montilla
Guadalquivir

Ayamonte · Huelva · SEVILLE
Sanlúcar de Barrameda
JEREZ DE LA FRONTERA · MALAGA
Puerto de Sta. María · Genil · Almeria
· Cádiz
Algeciras
GIBRALTAR

Valencia

ALICANTE
· Elche

Cartagena

CATALONIA

San Sadurni
de Noya

Villafranca
del Panedes
· Reus · BARCELONA
Tarragona
Panedes

0 · 100 · 200 · 300 Km
0 · 100 · 200 Mi

RIOJA

Introduction

Much of the best table wine of Spain – red, white and rosé – comes from Rioja, a contraction of Rio Oja, a small tributary joining the Ebro at Haro. The pronunciation is Ree-oc-a. The Ebro flows south-east across Castille, reaching the Mediterranean eventually south of Tarragona. The best wines comes from the Ebro valley between Haro and Logroño, the region continuing for another forty miles east of the latter, less rainfall reducing the quality.

In spring, it is all a pretty patchwork quilt of green vines, golden corn and brown earth against the distant blue-grey mountain background of the Sierra de Cantabria. Motoring north, I have often meant to spend more time in the region but the call of France has always proved too strong; from Logroño it needs an early afternoon start to cross the Roncesvalles Pass and reach that delightful Hotel du Trinquet, on the Nive at Saint-Etienne-de-Baigorry, in good time for dinner.

In Rioja, the first wine growers' society was formed in 1560 and by 1635 they were taking their wines sufficiently seriously for the City Fathers of Logroño to forbid carriage and carts certain streets, lest their rumblings should disturb the ageing wines below. A century ago, French growers settled here, crossing the Pyrenees to start again after the phylloxera had destroyed their own vineyards. They greatly improved the general standard. Thirty years later, when the remedy of grafting vines on to American root stocks (which were too tough for the insect to destroy) had been proved effective, they went home.

Today the wines of Rioja are to be found in hotels and restaurants throughout Spain. The red Reserva wines are matured in wood for five years or more and the white for nearly as long. Lesser, but still very good red wines spend their first six months in concrete vats followed by one to two years in cask. Rioja wines are blends of different ages, any year appearing on the label (e.g. Cosecha 1962) indicates that of the oldest wine in the blend, not a wine wholly of the year stated.

Names of wines – like Marques de Romeral or Villa Zaco – are invariably brand names, however much they may sound like aristocratic growers or glorious domaines. Reserva or Gran Reserva indicates the best. These practices, confusing to visitors unaware of them, enable the Rioja growers to achieve consistency and very good their older wines can be, particularly when price is considered.

Although letters were written when preparing this book to nine of the leading exporters, only two replies were received, which explains why only two Rioja entries follow.

Cenicero
Bodegas Riojanos SA,
Cenicero (Logroño)*
Tel. Cenicero 8

About half-way between Haro and Logroño, Cenicero is a small town close to the Ebro. A place of sacrifice of nomadic flocks to pagan gods, the Romans named it Cinissarium ('the ashpit'). Founded in 1890, Bodegas Riojanos still retain some of their original buildings for ageing wines, although their whole installation is completely modernised.

George Rainbird, who visited most of the Rioja Bodegas prior to writing *The Wines of Spain* in André Simon's *Wines of the World*, lavished praise on this company, particularly on Medieval, an extra dry white and on Viña Albina and Monte Real, both white and red. These wines are exported widely and are on most restaurant lists in Spain but for some reason business with Britain has never been developed.

The company's gay-coloured brochure welcomes all to visit their castle and cellars tasting what they will.

Hours: Monday to Friday 0900–1200.
Introduction: None necessary.

Fuenmayor
A.G.E. Bodegas Unidas SA,
Fuenmayor (Logroño)
Tel. Logroño 45.02.00.

About 160 kilometres and a comfortable morning's drive from Bilbao, the Rioja region is the Ebro valley from Haro to Logroño and beyond. Fuenmayor, a small town between these two principal wine towns, is 14 kilometres from Logroño. Bodegas Unidas are close to the railway station.

This big company, formed by various mergers, declare they are responsible for about one-third of the Rioja wines

---

*The bracketed word in these Spanish addresses indicates the province.

163

now exported. They have many cellars about the region but visitors are shewn the new bodegas, where the old wines in 20,000 oak casks contrast with the automation that can fill 23,000 bottles in an hour.

During the last dozen years the company has quadrupled the volume of wines exported, which go to over fifty-five countries. In Britain their brands include Siglo (red, white and rosé), Castillo Dorado, Marques del Romeral and Crux Garcia Real Sangría, a summer mixture of red wine and citrus fruit juices. They are usually obtainable from Peter Dominic shops.

Hours: Monday to Friday 0900–1200.
Introduction: None necessary but 24 hours' notice by letter or telephone is requested.

CATALONIA

Introduction

Catalonia is more of a wine region than is realised, though where the wines go has puzzled me ever since, years ago, I sat one warm summer evening in a large café in Barcelona full of local people and every one of them was drinking fruit juice. Up the coast from Barcelona, the red and white wines of the little district around the town of Alella are held in some regard by those who know them. The Co-operative in the town will accept visitors and their large wickered jars of white Marfil, bottled and sealed there, are worth buying. Marfil Blanco, Seco and Tinto, from different firms, are also recommended.

Panadés is a bigger district, running twenty miles inland from Sitges, and centred on Villafranca del Penedès thirty miles south of Barcelona. Jan Read (see Further Reading) describes its wine museum (p.165) as one of the best in the world. Though there is Tarragona-type dessert wine in plenty and some sound local dry white wine, the unexpected is to be found at San Sadurni de Noya in the form of possibly the world's largest sparkling wine installation. The firm, Codorníu (p.165) have been making Spanish sparkling wines by the champagne method since 1872. As to 'the largest' – Moët et Chandon in Epernay are also said to have the largest cellars and their subsidiary Mercier has something like ten miles of cellars and an underground railway through them.

The wine district of Tarragona is a continuation of Panadés, extending down the coast for sixty miles with an average width of about twenty. A small part of it is Priorato, up in the hills, where the wines are so strong that they go mostly to the blending vats of Tarragona and Reus, from which emerges Tarragona, a dessert wine that is not to be despised on account of the one-time sobriquet, 'Poor man's port'. The best of Tarragona is in fact the best Priorato and it does not need to be fortified with local brandy.

The Guide's entry to the Compañia de Carretero at Reus is a good one. The Amigo brothers speak English and many English people know their wines already from the La Vista range of Spanish table wines widely available in Britain since 1969.

Though communications are quicker now that a double-track road runs from Gerona almost to Tarragona, people on holiday on the Costa Brava have no need to go so far to make a wine visit. Perelada, near Figueras, is only about twenty miles from the Franco-Spanish frontier at Port Bou. The estate makes a good sparkling wine and sound table wines. It has a museum, a library and a festival.

Rheus
Compañia de Carretero S.A.,
Calle Gaudí 28, Reus (Tarragona).
Tel. (977) 30.13.22 or (977) 30.55.96.

Reus, in the province of Tarragona, is 14 kilometres inland from the port of Tarragona. Drive through Reus to the Plaza Martires. The Calle Gaudi is just off the Plaza near the Reus Deportivo football stadium. Founded by the Amigo family over a hundred years ago, the present Managing Director is Juan Amigo Domingo. (In the usual Spanish fashion, the mother's name follows the surname.) A brother, Wenceslas; a son, Francisco; and a nephew, Joseph are, together with Juan, the four members of the Amigo family in the business.

Being close to the Madrid-Barcelona main line, the old ancestral home was destroyed by bombs, aimed at the railway, in the Civil War (1936–1939), after which the whole business had to be restarted from scratch. Now they buy, blend, mature, bottle and export most of the Catalonian wines, including La Vista medium white and La Vista full red, brands popular in Britain.

Don Juan speaks very good English and likes to meet British visitors. There are tastings from the wood and visitors usually come away with a bottle or two, when he is about.

Hours:	Monday to Friday 1000–1200; 1600–1800. Afternoons preferred.
Introduction:	Not essential but 24 hours' notice of arrival, by postcard or telephone call, is appreciated. Any special arrangements (e.g. parties over 10) should be made through the agents: Morgan Furze Ltd., City Cellars, Micawber Street, London N1. Tel. 01–253 5263.

Codorníu S.A.,
San Sadurni de Noya (Barcelona).
Tel. San Sadurni de Noya 39.

From Barcelona go south on the Tarragona motorway, leaving it for San Sadurni de Noya by Exit 7. Coming north from Tarragona, leave the motorway by Exit 10.

Casa Codorníu is pleasantly sited on a small river only seven kilometres from Villafranca and 46 kilometres from Barcelona. Here Spain's best sparkling wines have been made since 1872, Codorníu exporting to some seventy countries, in greater quantity, they say, than that of the three leading champagne shippers put together.

The company certainly goes out of its way to encourage visitors; a leaflet with this object emphasises that by arriving before 1030 or before 1630, the visit can be as pleasant and complete as possible. Wines are made by the champagne process, which is likely to be shewn.

Hours:	Monday to Friday. Arrive shortly before 1030 or shortly before 1630.
Introduction:	None necessary. Any special arrangements could be made through the U.K. office: Codorníu (United Kingdom) Ltd., Burlington Buildings, Orford Place, Norwich NR1 3RU. Tel. (0603) 61.86.15.

Bodegas Torres,
C. Comercio 22, Villafranca del Penedès (Barcelona).
Tel. (93) 892.01.00.

Forty-eight kilometres from Barcelona on the road to Tarragona, Villafranca del Penedès is a small town of 20,000 people with a wine museum and numerous bodegas. Among these, Bodegas Torres, established in 1870, is a well-known exporter particularly to the United States. Vina Sol (white dry), San Valentin (white semi-sweet), De Casta (rosé), Coronas (red dry) and Sangredetoro (red dry) are their regular wines; Gran Vina Sol, Gran Coronas, Gran Sangredetoro, Gran Vina Sol 'Green Label' and Gran Coronas 'Black Label' are the Gran Reserva wines.

Hours:	Monday to Friday 0900–1200; 1500–1800.
Introduction:	None necessary; enquiry at the Wine Museum is recommended.

Museo del Vinas y Vino,
Plaza de San Jaime, Villafranca del Penedès (Barcelona)
Tel*. (93) 892.03.58 or (93) 892.11.62.

The museum is to be found in the mediaeval palace of the Kings of Aragon, opposite the cathedral.

The ground floor is set out attractively to illustrate the art of vine growing in Spain, without wholly excluding other foreign wine makers of yore, such as Egypt. Besides scenarios and paintings, there is a superb collection of huge wooden wine presses, dating from the fifteenth century onwards. There are also collections of wine glasses, corks and cork-making equipment, barrels, vats and bottles. One large model gives a comprehensive view of sparkling wine making at nearby San Sadurni de Noya.

On other floors there are paintings by regional artists, records of wine festivals through the years and a collection of labels, fascinating but so extensive that my correspondent, recollecting that his 15 pesetas' entrance fee entitled him to a free dégustation of local wine, went off to the bar in search of it.

Hours:	Daily (except Monday) 1000–1400; 1600–1900. Closed Mondays.

---

*The telephone numbers are those of the Town Hall, which controls the museum, where there is no telephone.

MALAGA

The hinterland of this port and resort, said to be the warmest place in Europe in winter, is a large grape-growing region. Málaga is a sweet dessert wine made from the Pedro Ximenez grape (with some others). It is slightly fortified and blended with boiled-down unfermented grape juice *(arrope)*.

This was the wine known as 'Mountain', much drunk in England two centuries ago; one comes across the name on those attractive old black and white bin labels. The big company is Scholtz Hermanos of Málaga, with whom I have corresponded at times on the subject of visitors. Local enquiries could be made but my impression is that, in a holiday district as crowded as the Costa del Sol, the company does not want to risk being overwhelmed.

As in Provence, it is best to look around; there are always small proprietors of vineyards and of wine shops eager to shew their wares.

VALDEPENAS

All along the Mediterranean coast, south to Gibraltar and up the other side, is wine country of one kind or another. Even Galicia, Spain's north-western province, where it always seems to be raining, can make a million and a half hectolitres a year. The biggest bulk producers are Valencia and the regions of Valdepeñas and La Mancha inland. With plenty of sun and therefore sugar in the grapes, the wines are strong in alcohol, making them suitable to blend with more acid ordinary wine made further north, in France or Germany, for example. Others are suited to Vermouth.

Where there is wine there must be cellars. Valdepeñas, which conveniently is on the E25 main road connecting the whole Seville-Córdoba-Granada-Málaga region to Madrid, has been described as one vast bodega; Bodegas Bilbainas, the largest of the Rioja companies, is among them. 'Ask and keep on asking!' is the best advice I can give to readers stopping in search of a wine visit.

RECOMMENDED READING

Particularly good on table wines is *The Wines of Spain and Portugal* by Jan Read Faber & Faber London 1973.

MONTILLA

Introduction

Twenty-five miles south of Córdoba the little town of Montilla is the centre of a patch of the chalky *albariza* soil, similar to that of Jerez and, being inland, it becomes even hotter in summer. Circumscribe a circle of twelve miles' radius, with centre Montilla, push the top down a bit and you have roughly delimited Montilla-Moriles, the wine region that inspired the word 'amontillado', adopted by the shippers of Jerez long ago. Montilla, however, is not entitled to be called 'sherry'.

From the Seville-Córdoba road, branch roads through the region reach this unspoilt little hill town in some twenty miles. From Jerez, Montilla is 121 miles and from Seville 82 miles. Montilla is also on the main road from Córdoba to Málaga, a route which many people prefer to take from Madrid to Málaga, so that more people than might be supposed stop in the town to stretch their legs around a bodega, or to wet their whistles at a bar or restaurant.

Though there are of course prior claims on the visitor in this part of Andalusia – the Alhambra and the Generalife at Granada, the Great Mosque at Córdoba to name but three – it is pleasant, having killed oneself with sight-seeing, to revive the corpse with a cool, dry Montilla, a wine that attains 16 per cent. alcohol quite naturally without any fortification with brandy. In Córdoba particularly, they seem to recognise this by providing bars at every corner, with much décor of the bullfight on the walls, sawdust on the floor and excellent cool Montilla.

There are two main differences in the making of sherry and Montilla. First, the Pedro Ximenez grape, used for sweetening in Jerez, is pruned short in Montilla to provide the basic wine. Secondly, what was good enough for the ancients is good enough for Montilla growers; they continue to ferment the wines in huge *tinajas**, giant earthenware, or even cement, pots open at the top. For two to three weeks fermentation is furious, dying down slowly until, in the spring, there is classification and transfer to a *criadera*, the nursery where all young wines spend three to five years before being moved again to appropriate soleras.

Montilla makes half as much wine as the Jerez region. Popular today are the Medium, Cream and Golden styles, blended to lesser strengths than the natural dry Montilla and therefore cheaper here because they attract less duty. In wine, if we most admire incomparable excellence achieved with man's minimum interference, then the first prize in Jerez and in Montilla, goes to the dry wines. At the Torre del Oro restaurant in Seville they serve hot prawns in garlic butter; alongside them I carry a memory of two cool copitas – Montilla in one, Manzanilla in the other. Comparisons would indeed be odious.

Montilla
Bodegas Alvear S.A.,
Montilla (Córdoba).
Tel. (957) 650100, 650104, 650108.

Approaching Montilla from the west (Seville) or north (Córdoba), pass the Las Camachas restaurant on the right, and then turn left at the first crossroads into the town of Montilla. Continue up the main street, past Bodegas Perez Barquero on the right, until a little park of orange trees is seen on the right. Alvear lies just behind this park.

This old Montilla house was started in 1729 by Don Diego de Alvear y Escalera. His brigadier grandson, Don Diego de Alvear y Ponce de Léon, gave a great fillip to the business after he returned from the Argentine, where he

had been Chief of the Border Commission between Spain and Portugal for twenty years.

Since 1945, there has been great expansion, including a move to these new headquarters, under the direction of Don Francisco de Alvear y Ward, Count of Cortina.

Visitors learn all about Montilla wines, with complimentary tastings included.

Hours: 1 November – 15 April, Monday to Friday 0830–1330; 1500–1830. 16 April – 31 October, Monday to Friday 0830–1330 only.

Introduction: None necessary. All welcome.

J. Cobos S.A.,
P.O. Box No. 25, Montilla (Córdoba).
Tel. (957) 650182.

This company, founded in 1906, is one of a number in Montilla now grouped as the Montilla and Moriles Wine company, which has a registered office in Bristol and a board with three Spanish Directors and three English. Coming to Montilla from the west (Seville) or from the north (Córdoba), the first bodega seen on the right, next door to the restaurant Las Camachas, is the new Bodega Montialbero. Visitors should report to the Cobos offices within, where an English-speaking guide will quickly be found.

The firm specialises in the Jerez styles of Montilla wines – fino, amontillado, cream, etc. – and Cobos wines will be found in hotels and restaurants in most parts of Spain. The export trade is growing rapidly.

A tour of the bodegas lasts about an hour. Visiting the vineyards, sometimes possible, takes a further hour. Wines are tasted from the cask during the tour and can be bought to take away in the firm's restaurant, Las Camachas, next door.

Hours: Monday to Friday 1000–1300; 1600–1800.

Introduction: None necessary. Visitors are just invited to call. Arrangements for the unusual, a large party for example, should however be made in advance through:
Montilla and Moriles Wine Co. (U.K.) Ltd., Royal Oak House, Prince Street, Bristol BS1 4QN.
Tel. (0272) 26830.

---

*Tinajas is pronounced Tin-ack-erd and the English joke in Montilla is that by the time you've toured both the bodegas and the vineyards you'll feel properly 'tin-ack-erd' yourself.

JEREZ DE LA FRONTERA

Introduction

In an article, published in *Wine Mine* in 1968 a few years before he died, Charles Williams, head of Williams and Humbert in London for many years and the eldest grandson of the founder, described his first visit to Jerez de la Frontera. It was November 1910 and he was eight. A small packet boat, still in service then between Gibraltar and Algeciras, took his parents, his nurse, one aunt, two male guests, himself and all their mountainous luggage from the P. and O. steamer, anchored in the Bay, across to the Reina Cristina hotel at Algeciras. The (now) 86–mile journey to Jerez being extremely tedious, whether by rail or public *diligencia*, Papa had hired a 'great old creaking coach' in which they set off, aunt, baggage and all, at 2 a.m. There were no windows and the clouds of white dust must have made them as white as millers. It took twelve hours, including stops – average 7 m.p.h. approx. A motor bus service began the following year.

Sixty-six years later getting to Jerez still presents problems, partly due to Arabs – men this time not steeds – and changes that may follow the end of the Franco dictatorship.

GETTING THERE:

By sea-road:

To Jerez from:	Km	Miles	Remarks
Algeciras	138	86	Car ferry ports of call – Southampton – Lisbon – Algeciras – Tangier. S.S. *Eagle:* service withdrawn; P. & O. state no plans to resume.
Bilbao	998	622	Swedish Lloyd car ferry 37 hours from Southampton.

To Jerez from:	Km	Miles	Remarks
Cadiz	31	20	Occasional cargo ships only – passengers not cars. Bus or train from Cadiz and a good boat service across to Puerto de Santa María.
Faro	293	182	Airport of Algarve, Portugal 2½ hours from London. Fly/drive hire. Frontier is Rio Guadiano with car ferry across to Ayamonte, where Parador is excellent.
Lisbon	508	317	Good air services from all parts. London 2½ hours. Fly/drive hire.
Madrid	597	371	Air services from all parts. Fly/drive hire.
Málaga	268	167	Good air services from London and Paris. Fly/drive hire.
Santander	995	620	Spanish Aznar car ferry 32 hours from Southampton.
Seville	91	57	International airport. London 3 hours via Valencia. Fly/drive hire. Autoroute Seville-Jerez once you're there.

With petrol expensive and time usually short, I leave the

motorist to plan his journey with the aid of the table above. In summer the passage through the Bay of Biscay is calm much more often than not.

Driving to Seville and Jerez via the Algarve is a pleasant possibility and Faro, the Algarve airport, has been included in the table from a fly/drive aspect. The river Guadiana forms the southern frontier between Portugal and Spain, cars crossing it on a small ferry. Above Ayamonte, on the Spanish side, there is a comfortable modern Parador. Early February, when the Algarve is a mass of white almond blossom, would be the ideal time for this trip.

By rail: Jerez is 421 miles from Madrid on the Madrid–Cadiz line. The best train of the day takes seven to eight hours.

Barcelona to Seville (730 miles), via Valencia and Córdoba, takes twenty-two hours but there are sleepers in all classes.

WINE TOURS

Heritage Travel, 22 Hans Place, London SW1X 0EP, Tel. 01–584 5201, include a seven-day tour, flying London to Seville, among their 'Wines of Europe' tours.

Other travel companies, such as Pegasus Holidays (London) Ltd., 2 Lower Grosvenor Place, London, SW1W 0EG, Tel. 01–828 7554, run long weekends in Seville, flying direct from UK, which include visits to Jerez bodegas.

ORIGINS OF SHERRY

In Andalusia the religious wars between Christians and Moors are reckoned to have lasted from 718 until 1492, when the Moors were finally expelled from Granada. Since 1264, or thereabouts, when the Christians had taken Jerez, the frontier between the warring factions had moved about and the suffix 'de la Frontera' has remained, not only with Jerez but with many smaller towns of the region, bastions against the infidel in their time.

It was a pity the Moors went; these cultivated Moslems certainly did not discourage wine and an admiring glance at their fountains, in the cool courts of the Alhambra or the gardens of the Generalife, shews they had the right ideas about water – a greater joy to the eye than the palate.

Long before this, enterprising Englishmen with a taste for sea air, and a nose for liquor, had settled in this sunny corner of Spain. Chaucer had mentioned 'the wines of Lepe' before 1400. By 1585, production of sherry was sufficient for 'El Draque' (a bogeyman in Andalusian nurseries to this day) to remove no less than 2,900 pipes on one of his Cadiz raids. But if Drake remains the devil incarnate, Shakespeare is a Jerez hero. As recently as 1956, the town erected a memorial to him in the park, inscribing it with John Falstaff's words: 'If I had a thousand sons, the first human principle I would teach them should be, to forswear thin potatoes and to addict themselves to sack.'

That *Sherris sack* was firmly established in London during the Stuarts is more certain from Pepys' diaries than what sort of wine it was. A growing English colony was undoubtedly engaged in the sherry trade from Jerez, but names do not seem to be known until the eighteenth century, presumably because records were lost or, more probably, not kept.

Jerez itself began as Scheres, the Spanish corrupting it to Jerez and the English to sherry. Among wine regions it has as much as any and more than most to offer the visitor. In summer, admittedly, it can become uncomfortably hot; yet, with the Atlantic only nine miles away, the heat is usually tempered by the breeze. There is rain on about seventy-five days in the year, most of it from November to January, with the occasional heavy thunderstorm in summer.

SEASONS FOR VISITING
In summer, the sea remaining pleasantly warm into October, the 'Sherry Coast' or 'Costa de la Luz' is good for a bathing holiday. Chipiona is a pleasant little resort, surprisingly unspoilt by the proximity of the enormous Spanish-American base at Rota, while south of Cadiz the beaches are worth exploring from La Barrosa to Cape Trafalgar. This is the point where Mediterranean meets Atlantic, with hot and cold water laid on to choice. On the way, the hill village of Vejer de la Frontera is worth the climb to see it.

In the town, opposite the park, the new Jerez Hotel, completed early in the 1970s, is luxuriously modern with a swimming pool and a bar long enough to accommodate the wares of all the shippers. More famous and very comfortable is Los Cisnes (The Swans). Here the swans do the swimming in the pool of its sub-tropical garden, on which many rooms look out.

Jerez's two great fiestas are the Spring Feria early in May, with horse sales, bullfights and flamenco and the Fiesta de la Vendimia, the Fête of the Vintage. Elsewhere in Spain this begins on 21 September; in Jerez they make it the weekend nearest 8 September, heralding the first pickings with a Grand Procession and fireworks at night.

SANLUCAR DE BARRAMEDA
Jerez is not the only town of sherry. It was from Sanlúcar de Barrameda that Columbus set sail when he discovered the new world, though today the aspect of sand dunes and shallow water hardly merits the name *port* for this town of Manzanilla, 13 miles from Jerez at the mouth of the Guadalquivir. Not that this matters; its fame is securely founded on its hinterland of *albariza* soil and the salt air, not water, that gives the unique dry delicacy to the wines that mature in its bodegas. Take Manzanilla away from Sanlúcar, even to Jerez or Puerto de Santa María, and within a few months it turns into rather ordinary *fino*. There is too some loss of character when exported and, in common with other sherries shipped to Britain, it is slightly fortified.

PUERTO DE SANTA MARIA
Only nine miles to the south-west of Jerez, Puerto de Santa María is the port of sherry. Though it looks across five miles of the Atlantic to Cadiz, the journey by road (south and then north again) to the end of the isthmus on which Cadiz is built used to be 25 miles* – quite a barrier to land communication in the sixteenth century when the export trade was appreciable. One of the earliest shipments recorded from Puerto de Santa María was to Plemma, believed to be Plymouth, in 1485.

During the nineteenth century the port grew around the Duff Gordons and the Osbornes closely integrated companies that had established their main bodegas there instead of in Jerez. With their fine proportions and white-washed simplicity, sherry bodegas are handsome buildings and in that of San José, built in 1837 with no central columns, Osbornes have one of unusual architectural merit.

THE VERSATILE APERITIF
The most popular of fortified wines is perhaps the most versatile. Sherry can be drunk chilled, 'on the rocks', with tonic, with soda, in hot soup or Bovril and one type or another can be found to suit almost every type of food. In Spain it is delicious with the tempting *tapas* of fish and hors d'oeuvre offered before meals. A smoker's wine too; by nicotine its flavour is hardly diminished at all.

For some three weeks, from early September each year,

---

*A road bridge across the bay now cuts it to about ten miles.

---

they harvest the Palomino grapes from the chalky hillsides and the lower, sandy clay slopes around the three towns. Modern presses in the bodegas now take over from the old presses in the vineyards. Tumultuously at first, the juice ferments until December when, after racking into clean casks, the curious transformation to sherry begins.

'El vino no se hace en la viña, sino en la bota.' 'Wine is not made in the vineyard but in the cask', says the Spanish proverb. Early in the new year, by an accident of nature, in some of these casks a peculiar white scum, *mycoderma vini*, or 'flor', will appear on the wine's surface. Its appearance is unpredictable; wines from one vineyard may grow the 'flor' for a few successive vintages and then cease doing so. Fino wines can only be made from these casks with the 'flor'. The others – those without the 'flor' – will only make oloroso wine.

These – fino and oloroso – are the two basic styles of sherry and, *in their natural state*, both are dry. Finos remain dry. If sweeter wines are required, some sweet wine from Pedro Ximenez grapes, which have been left in the sun for about a fortnight, will be blended with oloroso sherries.

There is of course no reason why the dry, natural olorosos should not be sold as such; sometimes they have been. They are pleasant enough but lack the bite of the finos.

All new wine must be classified according to the class into which it is developing. The *venencia* draws the samples from the cask, pouring the wine into a tasting glass from a height, the contact with the air enhancing the bouquet.

Development of sherry takes place on the open plan. The bodegas are above ground and open to the air. The casks are open too in that they are only part filled. Such treatment would be death to a claret or burgundy, yet sherry not only improves, but usually develops more alcoholic strength, body and colour.

During the year after the vintage, young fino wines are usually moved to a fino *criadera* or nursery, and after a year there into a fino solera. With oloroso wine these periods may be longer.

SOLERA SYSTEM

Vineyards, which may provide fino wines one year and oloroso the next, make life rather unpredictable, so a method of maturing and blending to give consistency, known as the solera system, grew up with the years. Imagine about forty butts in a row on the floor (hence the term solera) of a bodega. These butts are called the first scale and they contain the final blend all ready to be sold.

Imagine above them another row of forty butts, the second scale. These contain a slightly younger blend, though otherwise similar. Imagine further scales, up to a total of four, one above another, each containing similar wine about a year younger. Thus, in the bottom scale is the wine ready for shipment; in the fourth, or top scale, is wine about three years younger.

When wine is required for shipment, it is drawn off from the bottom scale. Wine from the second scale is then drawn off to replace it; wine from the third scale replaces the second and so on, the fourth scale on top being replenished by young wine from the *criadera* or nursery. The *criaderas* form a separate junior school being topped up by the *añadas*, the youngest wines of all.

In practice, a year's requirements can be met by doing these withdrawals two or three times a year. Only a very small proportion of the wine in each butt is then moved but the movement is sufficient to aerate all the wine and benefit the whole solera. An intricate system, expensive in labour yet well suited to sherry for it achieves its purpose.

Every shipper has a number of different soleras in his bodegas from which he makes up his final blends for the market. Sometimes new soleras are laid down, often to commemorate an occasion such as the firm's centenary.

These are some of the salient points about the versatile aperitif. A shipper in Jerez, hearing I was writing this guide, said, 'Please tell them that sherry tastes best before lunch.' I suspect he means he likes his visitors then. In Jerez – probably the last refuge of a gentleman – the siesta should still be sacred.

RECOMMENDED READING

Sherry by Julian Jeffs, new edition 1970 Faber and Faber London.
Sherry The Noble Wine by Manuel M. González Gordon, translation and new material 1972 Cassell London.

Jerez de la Frontera
Croft Jerez S.A.,
Rancho Croft, Box 414, Jerez de la Frontera.
Tel. (3456) 34.66.00.

If approaching by toll road, leave it at exit Jerez-N (North); this will lead you, after crossing the railway line, to traffic lights on the main avenue. Hotel Jerez can be seen opposite, half right. Continue straight on; take second turning on the right (Via Lebrija) which leads to the ring road, immediately opposite Rancho Croft.

If approaching on the old road from Seville, or from Cadiz, take the ring road. Rancho Croft will be found near the north-east end.

With the completion of the last two bodegas in 1975, Croft achieved their whole Jerez company on one site after years of expansion involving the tedious transporting of butts through the busy streets from one bodega here to another there.

The new bodegas, 180 metres long, are built in modern materials, without departing from the traditional style and elegance associated with Jerez. Best known for port, Croft have been blending fortified wine since 1678 and this new Jerez headquarters should be the *Rancho* of (the word means 'home of') Croft Original and all other Croft sherries for quite a time to come.

Tasters and purchasers are more than welcome at this new well-equipped 'comprehensive', as much part of the contemporary scene as the august antiquity of the older Jerez academies.

Hours:	Monday to Friday 1000–1300.
Introduction:	By appointment. If in the district just telephone, but preferably write earlier for letter and route plan to agents: Gilbey Vinters Ltd., Gilbey House, Harlow, Essex, CM20 1DX. Tel. (0279) 26801.

Pedro Domecq S.A.,

PO Box 80, Jerez de la Frontera.
Tel. (3456) 33.18.00.

The bodegas of Pedro Domecq, the largest of the sherry and Spanish brandy companies, are close to the Plaza de Domecq at the south-western corner of the town (see *Michelin Guide* Jerez town plan).

The Domecqs were a very aristocratic French family, still far from Jerez in 1730 when an Irishman, one Patrick Murphy, started in business and was soon joined by a Frenchman, a local general merchant called Juan Haurie. Haurie died in 1794 and then came disaster. In the Peninsular War, Juan Carlos Haurie, a nephew who had succeeded him, became the Quisling of his day. Waxing fat on French army contracts, with the civilian population starving, he was the most hated man in Jerez and when it was all over both he and the business were ruined.

Then it was that Pedro Domecq Lembeye, whose grandmother was a Haurie, began to take charge. Learning the trade with the firm's agents in London, he and John James Ruskin, fellow clerk and cashier, found a backer, forming in 1815, Ruskin, Telford and Domecq to act as the new agents. Born five years later, John Ruskin, writer and social reformer, was destined to steal the limelight, but his father was undoubtedly a very able wine merchant. The firm of Domecq prospered and, with brands like La Ina, Double Century and Celebration Cream, has continued to do so.

In Jerez they run an 'open house' policy, with bilingual guides explaining the process of making sherry and the firm's interesting history. Tours take about an hour. There are wines to taste and souvenirs to buy.

Hours:	1 November–15 April: Monday to Friday 0900–1200; 1500–1730. 16 April–31 October: Monday to Friday 0900–1330.
Introduction:	Although open to all as stated, the company like to give a little more attention to their best customers and welcome those who care to obtain a letter of introduction from their agents: Luis Gordon & Sons Ltd., 9 Upper Belgrave Street, London SW1X 8BD. Tel. 01–235 5191.

González, Byass & Co. Ltd.,

Manuel Maria González 12, Jerez de la Frontera.
Tel. (3456) 34.00.00 to 34.01.00.

The González Byass bodegas (marked on the *Michelin Guide* Jerez town plan) are at the south-west corner of the town, where the road from Cadiz leads into it.

This remarkable firm, producers and shippers of Tio Pepe ('Uncle Joe'), Rosa and other less familiar brands, was started in 1835 by Don Manuel González Angel, the delicate child and weakling in a family of seven, five of them boys. Having succeeded in marrying the daughter of the richest man in Jerez, in the teeth of opposition from his future father-in-law, Don Manuel set about building a solera with no capital. Surmounting all difficulties, by 1855 he needed his English agent, Robert Byass, as a partner; and seven years later the Queen of Spain herself came all the way by coach from Madrid to pay a formal visit.

In spite of the doctors, who were certain of his early demise, Don Manuel lived to seventy-five, dying in 1887 to begin what appears to be almost a tradition of longevity in this firm. Pedro N. González Gordon, 2nd Marquis de Torresoto, died in 1967 in his ninetieth year. His father (1849–1946) made ninety-seven. Robert William Byass (1860–1958) made ninety-eight.

The present bodegas cover a large area; the shell-shaped La Concha, designed by Eiffel for the Queen's visit in 1862, is of architectural interest.

Hours:	Monday to Saturday 0930–1300; additionally October to May inc. 1600–1700. Closed mid-July to mid-August.
Introduction:	Prior notice not essential but preferable. Forty-eight hours' notice by telephone is acceptable, but the company recommend applying (state intended date and time of visit) for a letter of introduction from agents: González, Byass (U.K.) Ltd., 91 Park Street, London, W1Y 4AX. Tel. 01–629 9814/7.

John Harvey & Sons (España) Ltd.,

Alvar Nuñez 53, Jerez de la Frontera.
Tel. (3456) 34.60.00.8.

Alvar Nuñez is a one-way street going east out of Jerez to Arcos de la Frontera and Ronda. The Harvey bodegas are conspicuous on the left.

Bristol Milk is a style of dessert sherry going back to Stuart times, which any wine merchant can market; Bristol Cream, on the other hand, is Harvey's own brand, and they have made it about the most famous sherry in the world. Wine shippers and merchants of Bristol since 1796, Harveys now concentrate chiefly on their world-wide sherry commitments, shipping in bulk from these picturesque nineteenth-century bodegas to Bristol, where final blending and bottling take place prior to world-wide delivery by land and sea. Sherries are also bottled and despatched from the Jerez bodegas to Spanish customers,

172

the E.E.C. countries and others easily accessible from Jerez. Mackenzie and Co., established in Jerez in 1842, now belong to Harveys, who in turn are part of Allied Breweries.

English-speaking members of the staff shew visitors round the bodegas; a tasting of the various styles of sherry is included. Sherry may be bought by the case or by the bottle.

Hours: Monday to Friday 0900–1400.

Introduction: By appointment. Letter obtainable from:
John Harvey & Sons Ltd., Harvey House, Whitchurch Lane, Bristol BS99 7JE.
Tel. (0272) 836161.

M. Antonio de la Riva, S.A.,
Alvar Nuñez 44, Apartado 493, Jerez de la Frontera.
Tel. (3456) 34.18.77.

Alvar Nuñez, a one-way street leads eastwards out of Jerez to Arcos de la Frontera and Ronda. The La Riva bodegas are about three hundred metres on the right.

This company has been involved with sherry for so long that not even the de la Rivas themselves can say precisely when they started. Documents suggest 1776. Another distinction is that son has succeeded father directly from that day to this; yet another is Tres Palmas, a dry fino of the highest class. Doing well in the seventies is their Hispano range of amontillado, Manzanilla and cream sherries with labels depicting Philip II of Spain. La Riva own vineyards in the Macharnudo district on notable *albariza* soil to the north-west of the town.

Visitors are shewn the bodegas, taste the La Riva range of sherries (and Spanish brandies) and may visit the vineyards.

Hours: Monday to Friday 1130–1400.

Introduction: By appointment. Letter of introduction from agents:
Cock Russell Vintners Ltd., Seagram Distillers, 17 Dacre Street, London, SW1H 0DR.
Tel. 01-222 4343.

Sandeman Hermanos y Cia.,
Pizarro 10, Jerez de la Frontera.
Tel. (3456) 33.11.00.

Pizarro is clearly shewn on the *Michelin Guide* Jerez town plan at the nothern end of the town and right at the centre of the plan. Garvey, Sandeman and Valdespino form a little group of bodegas there.

The House of Sandeman was founded in London in 1790 by George Sandeman, one of the many Scots who have left their native land to seek their fortune. By 1792 he had been both to Jerez and Oporto and was doing well, James Duff being one of his sherry suppliers.

From 1823 to 1879 Sandeman's London House represented the Jerez firm of Julián Pemartín, which came to grief when Julián Pemartín, a clear case of *folie de grandeur,* threw one magnificent ball too many and went broke. Sandeman acquired his wines and bodegas and in due course became one of the first shippers to own vineyards, which now amount to 350 hectares in the best areas of *albariza* soil. Their Fino Apitiv and Armada Cream are particularly highly regarded in Spain.

There are English-speaking guides for the tour and tasting, which takes about fifty minutes. Sherries and Sandeman's Capa Negra brandy may be bought.

Hours: 1 October–31 May: Monday to Friday 0930–1230; 1500–1700. 1 June–30 September: Monday to Friday 0900–1300. 1 January–31 December: Saturdays 0930–1230.

Introduction: 48 hours' notice by letter or telephone is requested. Letters of introduction obtainable from agents:
George G. Sandeman, Sons & Co. Ltd., 37 Albert Embankment, London SE1 7UA.
Tel. 01-735 7971/5.

Williams and Humbert Ltd.,
Apartado 23, Jerez de la Frontera.
Tel. (3456) 33.13.00

Coming in from Seville, these bodegas are easily found on either side of the Santo Domingo and close to the Bull Ring.

Alexander Williams learned the trade for a number of years in Jerez with Wisdom and Warter and then married Amy Humbert, whose father was a friend of Mr. Warter. This however did not lead to much of a rise and Mr. Warter was quite adamant, young Williams would never be taken on as a partner. Disgruntled but determined, Williams set up on his own, with £1,000 from his father-in-law and a partner called Engelbach, in the year 1877.

It was not a good time; sherry sales, rising steadily in Britain since the Regency, were now declining. Many other firms were wiped out by depression and the phylloxera. Aided by Gladstone's Act of 1861, which had brought about 'off-licences' and licensed grocers –pastures new and very welcome to wine shippers –Williams and Humbert survived. Dry Sack dates from 1905, As You Like It from 1929, but they say A Winter's Tale (1933) was named in honour of Mr. Winterbottom, a director, not Shakespeare.

Visitors are shewn various bodegas, the bottling plant and a film giving a general picture of the bodegas and of Jerez. They are invited to taste various types of sherry.

Hours: Daily, except Sunday and National holidays 0900–1300.

None necessary, but any special arrangements can be made through agents:
Sherry House Ltd., 92 New Cavendish Street, London W1M 8LP.
Tel. 01–580 0301.

Puerto de Santa María
Duff Gordon & Co., S.A.,
Fernán Caballero No. 2, Puerto de Santa María, Near Cadiz.
Tel. (3456) 86.34.40.

This port of sherry lies a dozen kilometres from Jerez on the toll road to Cadiz. Duff Gordon bodegas are scattered all over the town, but visitors should go to the headquarters above.

The company's early years, from 1768, are striking for the contrast between the founder, Sir James Duff (1734–1815), the British Consul in Cadiz, who was a paragon of virtue, and its most important customer 'Prinny', who was quite the reverse.

It was fortunate that during the Napoleonic wars Cadiz was the only important city on the Peninsula to remain free of French rule so that sherry shipments to London increased satisfactorily, aided by Duff Gordon's diligent agents, Gordon and Murphy, where John James Ruskin (John's father) was clerk until he set up his own firm later, (see Domecq, p. 172).

After Sir James, Sir William Duff Gordon and his son, Sir Cosmo followed in quick succession, the latter being joined by a partner, Thomas Osborne, who had done well on his own, shipping to the U.S.A. The company has remained in the hands of the Osborne family, Osborne being as well known for its Vetérano Spanish brandy as Duff Gordon for El Cid and Santa María cream.

Today, 40,000 butts are spread about twenty-eight cellars, the 1837 San José Bodega being of architectural interest having no central columns, and holding 3,000 butts of El Cid.

Tours take about an hour, with expert guides explaining the processes. Duff Gordon sherries are tasted.

Hours: Monday to Saturday 0900–1200. Closed mid-July – mid-August.
Introduction: Open to all, but the company like to give special attention to their best customers and welcome those who care to obtain a letter of introduction from agents:
Jenks Bros., Castle House, Desborough Road, High Wycombe, Buckinghamshire.
Tel. (0494) 33456.

Sanlúcar de Barrameda
Manual Garcia Monge, S.A.,
Plaza de la Victoria, 2/4 Sanlúcar de Barrameda.
Tel. (3456) 36.04.16.

Sanlúcar de Barrameda, the Manzanilla town and port at the mouth of the Guadalquivir, is 22 kilometres from Jerez across those pleasing chalky slopes thick with the Palomino vine. The Plaza de la Victoria is just off the Plaza del Cabildo, the Town Hall square. The bodegas of Manuel Garcia Monge should be easy to recognise because the building was a convent inhabited by the friars Hospitalarios de San Juan.

The convent was sold by the Spanish Government in 1871, being bought later by Don Manuel Garcia, who has built up the present large export business from scratch. 'Our whole set up is rather old fashioned', writes Señor Nuñex of the company, in excellent English. 'Yet it is unusual perhaps for visitors to see a chapel, refectory and chapter room, which have been turned into a bodega, the original pious paintings remaining intact on the walls and ceiling. And, of course, there are glorious sherries to taste, Manzanilla in particular, at its source.'

Hours: Monday to Friday 1000–1300.
Introduction: By appointment, which could be made locally by telephone, or earlier by letter from elsewhere through agents:
John McGeary & Co. Ltd., 32–34 Borough High Street, London Bridge, London SE1.
Tel. 01–407 8176.

PORTUGAL

Portugal

> 'Now give us lands where the olives grow',
> Cried the North to the South,
> 'Where the sun with a golden mouth can blow
> Blue bubbles of grapes down a vineyard-row!'
> Cried the North to the South.
> *Elizabeth Barrett Browning*

Introduction

GETTING THERE

By road: From the French Channel ports, Lisbon is about a thousand miles and Oporto not much less. The usual route is via Bordeaux and Burgos; but to save petrol, car ferry to Bilbao or Santander (p.168) is recommended. For Oporto it is then straight on to Salamanca and the Fuentes de Onoro/Vilar Formoso frontier, this being a good route for Lisbon too if the wooded hills and valleys of a maritime country are preferred to the open, bare and arid landscape of the Spanish plateau.

Another way to Lisbon is south from Salamanca, entering Portugal at Caia between Badajoz and Elvas, where the plums come from. This is the Madrid-Lisbon route. Badajoz can also be reached from Burgos, via Segovia and Avila. Apart from the architectural attractions of these two ancient towns, the former has one of Spain's best restaurants, the Méson de Cándido.

Coming from the north, after visiting Santiago de Compostella for instance, the river Minho forms the frontier – an enchanting wide green valley, with greyish green hills standing back from it. From the comfortable Pousada (Government inn) at Valença do Minho, the new arrival looks back across the Minho to Tuy and the hills of Galicia, a first glass of Vinho Verde in his hand perhaps.

I have crossed too from Verin to Chaves when making for Pinhão, the very heart of port wine up the Douro. It was March and after going to Leon, we had taken minor roads to La Bañeza and Pueblo de Sanabria, chilling our bottle of picnic wine in the melting snow that still remained in sunless crevices along the hilly northern border.

Yet another route and, I think, the shortest to the Douro valley, is via Zamora and Bragança. After Burgos and Valladolid, a lesser but fast road leads off from Tordesillas, where there is a first glimpse of the Douro flowing close by its pleasant Parador. From here it is 110 miles,

through deserted agricultural country to the frontier bridge over the river Sabor, running south to become 'a tributary of port wine'. A climb out of the wooded valley into the sunshine, with Bragança a few miles ahead, opens into the enchanting country of Tras-os-Montes.

Bragança, a little white town whose ruling family once gave England a queen (Catherine married Charles II) and Portugal a dynasty, has a cathedral and an interesting museum with worked stones and carvings of Celt-Iberian times. The Pousada de São Bartolomeu is a suitable overnight stop; there are still 160 miles of pleasant, but slow driving to Oporto, through a sunny version of the Scottish Highlands, the next Pousada being a hundred miles further on, sited, with spectacular views, at the top of the Marão pass beyond Vila Real.

To see the port vineyards, turn south off the main road beyond Murça to Alijó where there is a Pousada. Bragança to Alijó is about seventy miles.

Road maps of Portugal, on which wineries are clearly marked, are obtainable at frontier posts or by writing to National Tourist offices such as Casa de Portugal, in London, (see below for address).

By sea: For car ferries, see Spain p.168.

By rail: Paris to Lisbon is 1,174 miles via Irun, Salamanca, Vilar Formoso and Coimbra. The fastest train, the Sud-Express, takes about 26 hours.

For Oporto, change at Salamanca or Fuente de San Esteban. The line crosses the Spanish-Portuguese frontier at the town of Barca d'Alva whence it follows the twisting course of the Douro wandering its way across Portugal. The train stops at every little station amid the great hillsides, covered with terraced vines as far as the eye can see.

The late G. R. Delaforce, giving a *Saintsbury Memorial Oration* in London in 1971, described this journey which

PORTUGAL

George Saintsbury* never made. Before air travel, it was one that English boys of Oporto families had to do several times a year, between home and school, unless they took a ship from Liverpool, cutting holidays shorter and in winter probably being seasick. 'Wog' Delaforce – as he was known throughout the Trade – insisted on reciting the names of the stations. And what names! Pocinho, Freixo, Vesúvio, Vargellas, Ferradosa, Alegria, Tua, Cotas and Pinhão. Would that he were still alive – this amiable last eccentric of the British wine trade – to give a lesson in pronouncing them, in an accent as English as Winston Churchill's French.

From Pinhão, Oporto is another eighty miles 'down the line', a line that continues to cling to the river for some miles beyond Régua. Visitors and tours often go up from Oporto on the 0750, arriving at Pinhão about 1045, and returning by the late afternoon train. For longevity the engines beat port wine hollow – they're German, vintages 1880–89. Up the Douro, walking along the line is often the quickest route from A to B. Should you meet a train in a tunnel, which will be single track, just lie down! They say it soon passes over your head.

By air: Lisbon is well served by numerous air lines from the principal cities of Europe and from the U.S.A. Flight time from London is about 2½ hours.

To Oporto from London, British Airways run at least a weekly service with reduced prices in summer. There is a daily service via Lisbon throughout the year.

Between Lisbon and Oporto there are frequent flights daily; flying time is about thirty minutes.

NATIONAL HOLIDAYS
National holidays in Portugal are given below. Please remember when planning your trip that not only will premises be closed, but that it is not uncommon to remain closed for an extra day (e.g. Friday after a Thursday public holiday) to make a long weekend break.

1 January, Corpus Christi, 15 August, 5 October, 1 November, 1, 8 and 25 December.

RECOMMENDED READING
The Wines of Spain and Portugal Jan Read Faber & Faber 1973.
The Englishman's Wine Sarah Bradford Macmillan 1969.
Portuguese Wine Raymond Postgate J. M. Dent 1969.

Further Information
Casa de Portugal,
Portuguese Tourist and State Information Office,
20 Lower Regent Street,
London SW1.
Tel. 01–930 2455.

---

*George Saintsbury (1845–1933) author of *Notes on a Cellar Book*, was the first Englishman to write a book of this kind.

DOURO (Province of port)

Introduction

The advantage of coming to Portugal by Verin or Bragança is that it leads into the remote Tras-os-Montes, so remote in this north-eastern corner that it is otherwise likely to be missed.

From its source 7,389 feet up in the Sierra de Urbion (a Spanish mountain range seen in the distance from another wine region, Rioja), the Douro flows about 485 miles to the Atlantic at Oporto. After crossing Castille, it forms the frontier for seventy miles before becoming wholly Portuguese for the final 134 miles from the frontier at Barca d'Alva.

In these upper reaches its course is spectacular through great gorges with formidable rapids. Roads being of the 'Birmingham by way of Beachy Head' variety, the flat-bottomed *barcos rebelos* used to sail fruit and port wine down to Oporto, even after the railway was built. Only since the war, with the growing reliance on the road tanker, have these picturesque vessels become rare.

Though sailing vessels can hardly stage a comeback, there seems little doubt the Douro will become a considerable waterway. Upstream from Oporto and in Spain, the river has been dammed for hydro-electric power. When three more dams (out of five) are finished, the Douro will provide half the country's electric power. The new reservoirs, irrigating much arid farmland, will also be the outlet for the tin and wolfram to be mined in the north-east corner of Portugal.

THE PORT COUNTRY

The demarcated zone for port covers some fifty miles from the Spanish frontier to a point below Régua. The best quintas begin at Vesúvio – Senhora da Ribeira, do Vesúvio, Taylor's de Vargellas and Quinta da Ferradosa. It was in the rapids of the Cachas that Baron Forrester, who mapped the Douro and did great things for the port trade, was drowned in 1862. He was wearing a belt full of golden sovereigns, with which to pay the wine farmers, who distrusted the escudo. The boat capsized. His companions, Baroness Fladgate and Dona Antonia Ferreira,

were saved; supported by their crinolines, the wind floated them to safety; the Baron, weighted with gold, was drowned.

Alegria, Cockburn's do Tua and Graham's de Malvedos come further downstream and at Pinhão Croft's Roeda, da Silva's Noval, Eira Velha, Bom Fin and Quinta La Rosa lie at the centre of the port wine region.

When the dams and the reservoirs are all finished, no doubt passenger vessels will take tourists through this valley of rugged grandeur, where the man-made stone walls retaining the soil on these steep terraced vineyard slopes have to be seen to be believed. Meanwhile, with a car, make for the Baron Forrester Pousada at Alijó, above Pinhão or the Parque Hotel at Lamego, these being the only overnight stopping places I would recommend.

From Alijó the road makes a spectacular ten-mile descent. As the vineyards are beginning, the Douro comes in sight far below at Pinhão. At this point, you may observe an old English shippers' custom of stopping your car and drinking to 'the river of gold' in a glass of port. It should be observed – so I have been told – whenever the Douro is sighted, a pretty bibulous task on some journeys here where it pops up round every other mountain; but at least the breathalyser is still unknown in Tras-os-Montes.

Pinhão, a village of no great credit to the works of man, is hardly likely to detain the visitor – except at the level crossing gates where the pleasure of 'playing trains' has yet to pall on local railway men. But lifting the eyes to the hills is another matter; the terraced vineyards are everywhere, not only beautiful, particularly in autumn when the leaves turn to red, but a remarkable feat of human labour as examination of the great stones in the soil will confirm.

The greatest of these vineyards are nearly all on the right bank, which gets the morning sun. Barely a mile upstream is Croft's Quinta da Roeda, approached across a highly dangerous level crossing with no gates, the cause of a few 'near misses' for guests at vintage time in the past. The Pousada management should be able to arrange a walk round but do not expect any English-speaking guides. Except at vintage time in September, when the shippers come up from Oporto to take up residence and entertain their trade guests and friends for a week or two, they are rarely there.

To reach Oporto, continue along the road which follows the left bank of the Douro for fifteen miles to Régua.

Soon after Régua, as the road climbs over the Quintella Pass from Mesão Frio to Amarante, it is goodbye to the Douro until Oporto. If you are still following the precept of a glass of port whenever you sight the river, there is a last opportunity crossing the great iron bridge on the way back to Oporto from the Lodges at Vila Nova de Gaia. It was built in 1881, five years after Eiffel's rail bridge further up river. The traffic is formidable, the pathway narrow and the 200 foot drop over the railings looks immense. Walking across, we thought that 'One for the Bridge' was definitely one to be missed – particularly after calling on the Lodges.

THE VINTAGE

As far as the big firms are concerned, the peasants no longer link arms and set about converting the grapes into must by treading with their bare feet. Feet have been in very short supply of recent years, many active men having fled the Douro to earn more money elsewhere in Europe.

The big firms, Croft, Cockburn, Sandeman and others, now buy the grapes from the farmers, vinification taking place in new automatically controlled wine-making centres up the Douro fitted with auto-vinificators. Croft, for example, can make up to 3,500 pipes in six weeks with twenty men. To do this by treading would need seven thousand pairs of feet. However a small part of the total produced, remaining in individual proprietors' hands, is still made in the traditional way.

THE ENGLISHMAN'S WINE

As far as we know, George Saintsbury never went to Portugal yet it was that bearded old wine chronicler who first called port 'The Englishman's Wine'*. And rightly! We discovered it, made it, shipped it and were about the only people to drink it.

By 1700, there was in Oporto a small colony of British merchants with Consul-General, chaplain and a club – the famous 'Factory', which has never made anything except its members and their guests convivial. Wines of the Douro had become popular in England and Warre, Croft and Taylor were already established. In 1705 the duty was reduced to one-third of that on French wines and doubtless glasses were raised by many an 'Old Kaspar' before his cottage door to celebrate Marlborough's famous victory. But of course the wine was not yet fortified; otherwise there could have been no boast from Dr. John-

son of drinking three bottles without being the worse for it.

Brandy, it is believed, was first added between 1720–1750, transforming the Douro's rather ordinary dry table wines into sweet, harmonious, full-bodied and – far better travelling wines. The brandy had stopped the fermentation leaving sugar unfermented in the wine.

In 1775 the vintage was so good that the wine was sent to England *unblended* for bottling. With the recent introduction of the cork and the consequent change to bottles shaped so that they could be laid on their sides, 1775 marked the first of a long unending line of vintage ports, for which generations of Englishmen have waited a score of years or, fading away first, handed on to their children.

Concurrently the shippers have learned to select, blend, mature and transform the bulk of their purple wine into mellow ruby, or even more delicate, tawny port, blends perhaps 10 to 15 years old on average, though they can contain wine older than those who drink them.

In spite of wars, revolutions, fire and flood, the English in Oporto multiplied and generally prospered. Napoleon was a help; during the Napoleonic wars British patriots eschewed French wines and when Wellington relieved Oporto in 1809, after two months of enemy rule, the celebrations were decidedly injurious to military efficiency for some weeks. Among the shippers when peace returned, three Scotsmen – Sandeman, Cockburn and Graham – began to make their mark.

In Victoria's reign peace by no means always reigned between growers, shippers and the Portuguese government, yet rows were resolved and the character of the world's greatest dessert wine has now remained unchanged for over a century. The Portuguese laws governing the origins and the production of port (and Madeira) gained the protection of British law under the Anglo-Portuguese Commercial Treaty of 1914.

Port comes only from Portugal and from the Lodges of Vila Nova de Gaia, which lie facing Oporto across the last mile of the Douro.

THE LODGES

Coming from Oporto by road, cross the lower section of the two-tier iron bridge, turning sharp right at the end and continuing along the river for up to half a mile before turning away from it into the narrow streets leading to the Lodges. For the most part these are grouped together and, failing direction signs to the firm required, it is best to ask before getting involved in the maze of narrow streets and commercial vehicles.

From Oporto, the Lodges are not a long walk. Do not, however, take the upper section of the two-tier bridge. For pedestrians, the height, the traffic and the narrow footpath make it a terrifying experience.

Coming from the Lisbon direction by road, turn right just before reaching the Arrábida bridge at the end of the *Auto Estrada*. The route to the Lodges is then clearly indicated by Fonseca port signs.

---

*Saintsbury *Notes on a Cellar Book*.

The new wine from the quintas reaches the Lodges in the spring following the vintage. Wines of similar character are grouped in lots, being given a lot number which is never changed. Each Lodge thus contains many lots of differing ages and qualities. The tasters, aided by their record books, know each lot as a master knows his pupils and are able to follow one shipment to a customer with an identical one, or to match any sample sent to them.

All port spends its first two years in wood. Vintage port is then bottled and continues to mature in bottle. All other ports continue to mature in wood. Thus there are two main types: Vintage port – matured in bottle, and Wood port – subdivided into Ruby and Tawny from the colour assumed in the wood. Port can also be made from white grapes, controlling the fermentation to give a dry, medium dry or sweet wine. Chilled, the dry variety makes a good aperitif. White ports do not improve in bottle.

Though the problem of maturing wines and maintaining consistent blends is the same with port as with sherry, the solera system is not used. The wines vary much more from one year to another in this, a more northerly district, so that blending port wines depends much more on highly experienced tasters. And whereas sherry develops in solera with minimal movement, port needs frequent racking to perfect both its colour and flavour.

Shipping in wood for trade purchasers to bottle under their own brand names is declining. After bringing up their children from birth, the shippers are no longer prepared to send them forth nameless into the world. Understandably *they* wish to do the bottling, obtaining the *selo de garantia*, the Portuguese Government seal, the guarantee of authenticity.

Whether the character of port will be changed because the artificial lakes bring more rain to the Alto-Douro remains to be seen as the years pass. Whatever the answer, it will be of small consequence beside the prospects of the better life that science and engineering should bring to the people of Tras-os-Montes.

VISITS TO THE ALTO-DOURO FROM THE LODGES

Given advance notice, most of the shippers are prepared to make arrangements for parties visiting their Lodges to visit a quinta up the Douro as well. It must however be understood that this requires a second day and a very long one too if the party attempts to return to Oporto for the night. Furthermore, hotel and Pousada accommodation in the Alto-Douro is very limited and providing lunch at either quinta or hotel can be a problem.

Oporto

The British Factory House,

Rua do Infante D. Henrique, Oporto.

The origin of the term 'Factory' lies in the early custom of calling English merchants abroad 'factors', the associations they formed becoming 'Factories'.

In Oporto, when a headquarters for the British Association of merchants was needed, a handsome and very solid Georgian building of grey stone was built between 1786 and 1790. The funds were raised by a levy on all goods exported from Oporto in English ships, the contributions being extorted from the exporters, whether British or Portuguese, much to the understandable annoyance of the latter.

Inside the building, no Rupert Brooke is needed to recognise the Factory as 'a corner ... that is forever England'. A classical staircase leads up to stately reception rooms, ballroom and a dining hall glittering with great crystal chandeliers; eighteenth century and, as nearly always, British domestic architecture at its best.

This then is the headquarters of the British Association in Oporto, where important visitors are entertained and the port shippers bring guests every Wednesday to their informal luncheons.

Administration is in the hands of a Treasurer and four assistants elected for one year. The Factory House is not open to British visitors at any specified date and time. Nevertheless, some travel agents have made arrangements for their tours to be taken over and sometimes the port shippers can arrange a reception or buffet lunch for visiting wine clubs or societies. There is of course a charge.

The Treasurer and House Committee prefer that all visiting arrangements are made through one of the port shipping firms. They regret, however, that lack of resident staff, who do not in any case speak English, makes it impossible to cater for the passing visitor or casual callers.

Vila Nova de Gaia

Cockburn Smithes & Cia. Lda.,

13 Rua das Coradas, Vila Nova de Gaia.
Tel. Oporto 39.40.31.

After the expulsion of the French from Oporto in 1809, Robert Cockburn, younger brother of Lord Cockburn, the Scottish judge, went out to start this famous firm, which was founded in the year of Waterloo, 1815. Cockburns (pronounced Co-burn) are of course Scots, and tolerant too in that they have raised no objections to their quinta at Tua, a Douro tributary, being known as the *Quinta dos Inglezes*.

After 1854, the two Cockburn sons, who had succeeded their father, were joined by two Smithes brothers, of whom one took charge of the London office, which has always handled sales in Britain. Marriage brought in a third family – the Cobbs. Reginald Cobb, in charge in Oporto, recently retired with a C.B.E., while his father before him, having refused to declare a vintage in 1917, 1920 and 1924 when most other firms did, shipped 'Cockburn '27' on the eve of retiring, a wine which even old men will not forget.

Since 1962 the firm has been part of the Harveys of Bristol Group and is combined with another old established port house, Martinez Gassiot.

An English-speaking guide is available at the Lodges. Visitors are shown round, have a tasting and may buy wines.

Hours: Monday to Friday 0930–1200; 1400–1700.

Introduction: Unannounced visitors are welcome. Letters will however be provided on application well in advance to: Harveys of Bristol, Harvey House, Whitchurch Lane, Bristol BS99 7JE. Tel. (0272) 836161.

Croft & Ca. Lda.,
23 Largo Joaquim Magalhaes, Vila Nova de Gaia. Tel. Oporto 39.01.81.

The House of Croft, established in 1678, is the third oldest of the port shippers. Exactly one hundred years later John Croft published a Treatise on the wines of Portugal. He described himself as 'a member of the Factory at Oporto and a wine merchant in York'. His son, Sir John Croft, was created a Baronet after the Peninsular War, having aided Wellington during his campaign, as well as distributing a large sum of money to the Portuguese peasants, donated by the British Parliament. He was also made a Life Peer and given the title, Baron da Estrella by the Portuguese King.

The company owns one of the finest vineyards in the Douro, Quinta da Roeda, near Pinhão, and since the war has added Morgan Brothers and Delaforce to the Croft group. At the Lodges in Vila Nova de Gaia visitors are entertained with a tour, a tasting and a colour film. Samples of Croft ports, which are now exported world-wide, can be tasted and purchased. Should any visitors wish to spend another day going up the Douro, arrangements can often be made to visit Quinta da Roeda so long as notice is given well in advance.

Hours: Monday to Friday 0930–1300; 1430–1630. Closed August.

Introduction: By appointment. Letter or introduction card from agents: Morgan Furze Ltd, City Cellars, Micawber Street, London N1. Tel. 01–253 5263.

W. & J. Graham & Co.,
Rua Rei Ramiro, Vila Nova de Gaia. Tel. Oporto 39.60.65.

After William and John Graham had begun in 1820 as partners in a general trading concern, they received a parcel of port in settlement of a bad debt. Shipped to Scotland, where the Graham family had their main office and business interests, the parcel was well received and a request for more soon followed. From this point the Grahams succeeded very rapidly as port shippers.

Up the Douro they own Malvedos, a beautiful quinta,

superbly remote above Pinhão, at the point where the Tua tributary joins the Douro. From it, meteorological information is sent daily to the Factory House.

Since 1970, Graham has been part of the largest independent port-shipping group in Oporto, controlled by the Symington family, which includes Warre, Dow and the small but old firm, Quarles Harris.

Hours: Monday to Friday 1000–1200; 1400–1730.

Introduction: Visitors are welcome at the Lodges but a letter of introduction is desirable from agents: Luis Gordon & Sons Ltd., 9 Upper Belgrave Street, London SW1X 8BD. Tel. 01–235 5191.

Guimaraens-Vinhos S.A.R.L.,
Quinta Dom Prior, Apartado 13, Vila Nova de Gaia. Tel. Oporto 39.01.80.

Manuel Pedro Guimaraens started this company in 1822 and after a spell in Brazil came to London, setting up as Fonseca, Monteiro and Guimaraens. Having married an Englishwoman and educated their three sons in England, the family has always had close contacts with Britain, the company's head office being in London until 1927.

Fonseca ports have always been highly regarded and from 1950 the London agents, H. Parrot and Co., had Patrick Guimaraens looking after their interests until he retired to Portugal in about 1970.

Over half a million bottles of Fonseca vintage port, which will eventually reach some fifty countries, lie in these cellars, where visitors are offered the usual courtesies, including an English-speaking guide.

Hours: Monday to Friday: June to September inc., 0930–1200; 1400–1730.

Introduction: Open to all but letter of introduction is preferred from agents: Cock Russell Vintners, Seagram Distillers House, 17 Dacre Street, London SW1H 0DR. Tel. 01–222 4343.

Quinta do Noval-Vinhos S.A.R.L.,
Rua Candido dos Reis 575, Vila Nova de Gaia. Tel. Oporto 39.11.61 (Quinta Tel. Pinhão 42,317).

From 1813 to 1974 this great Portuguese company, concerned exclusively with the sale of first class port, was called A. J. da Silva. Better known, however, is its wonderful quinta, Noval, and Fernando and Luiz Van Zeller, great grandsons of the founder José da Silva, decided in 1974 that Noval was the better name to perpetuate. Port usually being a blend from many quintas, single quinta wines are rare and Quinta do Noval is about the best known.

'Like a wilder Tuscany' is Sarah Bradford's description* of the view from the windows of the house, a land-mark of great white walls and wide terraces, on the hill above Pinhão. It looks down on the Douro, often glittering in the cleft of the valley, to great blue-cloud shadows on the hills beyond.

In the Lodges, there are English and French-speaking guides to conduct a tour and tasting.

Hours: Monday to Friday 0930–1300; 1500–1800.

Introduction: None necessary. The quinta is open to all during working hours so that visitors can see the vineyard and the wine-making arrangements. Staff however will only be Portuguese-speaking. For special arrangements write to agents: Rutherford, Osborne and Perkin Ltd., Harlequin Avenue, Great West Road, Brentford, Middlesex TW8 9DY. Tel. 01–560 8351.

Sandeman & Co. Lda.,

Apartado 2, Largo Miguel Bombarda, 3 Vila Nova de Gaia.
Tel. Oporto 39.21.31.

In 1790 George Sandeman of Perth began what is now a large limited company by renting a vault in the City of London with a £300 loan from his father. By 1809 he was in Oporto impressing British generals, including Wellington himself, with his professional knowledge of wine.

When he died in 1841, a nephew succeeded and after the Crimean War, John Sandeman, a colonel who had witnessed the charge of the Light Brigade, came out to look after the Oporto house. (John Sandeman is also recorded as the inventor of the penny-in-the-slot machine, though it never became a penny-a-tot machine.)

1790 was a great year for vintage port and Sandeman 1790 contributed to the company's early reputation, just as the Don figure, used as a trade mark from 1928, was to do later on. Since the war, Robertson Bros. and Offley Forrester (who used to ship that delightful single quinta wine, Boa Vista†, from that celebrated quinta above

---

**Sarah Bradford *The Englishman's Wine*.*

†Enquiries regarding Boa Vista should not be sent to Sandeman but to Laurence Hayward Ltd., 11 Gough Square, London EC4.

Régua) have become part of Sandeman Oporto. Experienced guides speaking English, French, Spanish and German show visitors over the Lodges and there are guides too, speaking English, French and Spanish, at Sandeman's Wine Making Centre at Pacheca, near Régua, where visitors are similarly welcome, the hours for both establishments being those below.

Hours: Monday to Friday 0930–1300; 1400–1700. Saturdays, July, August and September only 0930–1300.

Introduction: By appointment. Letter from: Geo. G. Sandeman, Sons & Co. Ltd., 37 Albert Embankment, London SE1 7UA. Tel. 01–735 7971.

Taylor, Fladgate & Yeatman Vinhos S.A.R.L.,

Rua do Choupelo 250, Vila Nova de Gaia.
Tel. Oporto 39.01.87.

No firm has a greater reputation for its vintage port than Taylor, Fladgate and Yeatman, established in 1692 by Job Bearsley, who had been trading for some years from Viana on the Minho river, north of Oporto. The company is still independent and British, the Yeatmans, who originally came out from Dorset, being the surviving family in control.

In 1744, the company bought a small property at Salgueiral, near Régua in the Douro district, which later became a field hospital during the Peninsular War. The famous Quinta de Vargellas in the heart of the Alto-Douro was bought in 1893 and it is the wines of this quinta, capable of making 200 pipes in a good year, which form the basis of all Taylor vintage ports. The company also has another fine property at Pinhão, where wines are kept before being taken to the Lodges in Vila Nova de Gaia.

At the Lodges visitors are welcome. English and French-speaking guides conduct the brief tour among great casks of wines, some of which will be offered for tastings.

Hours: Monday to Friday 0930–1200; 1400–1730.

Introduction: By appointment; a letter of introduction can be obtained from agents: Deinhard & Co. Ltd., 29 Addington Street, London SE1 7XT. Tel. 01–261 1111.

THE TABLE WINE REGIONS

VINHO VERDE

In an earlier paragraph I left the new arrival ensconced in the Pousada at Valença do Minho looking back across to the river Minho to Spain enjoying his first glass of the local Vinho Verde.

These 'green wines' are so called because the grapes are picked early when a bit green. Growing them high on trees and trellises, keeps them cool, giving more acidity; the result is slightly sparkling, or *pétillant* as the French say. Vinho Verde has a liveliness that makes the white variety a refreshing aperitif and good with shellfish. Drink young, serve cold and don't expect the 'prickle' to last more than a few minutes.

The red wines should also be drunk cold but instead of drinking them within a year, Raymond Postgate recommended four years in bottle to eliminate early hardness*; a counsel of commercial perfection, I fear.

The region was demarcated in 1929 and true bottles have a band label marked *Vinho Verde Selo de Origem.* From the coast it is virtually a rectangle fifty miles wide, bounded by the river Minho at the top and the river Douro at the bottom, although in fact there is quite a bulge south of the Douro. Monção, less than ten miles up the Minho from Valença, makes a very good Vinho Verde, called Cepa Velha, worth trying with lampreys at the Pensão Internacional there. Keep clear at Corpus Christi (late June), unless you relish the town fête when Saint George in armour conquers Santa Cuca, a horrid devil dragged round the town.

After Monção, Lima, Braga, Amarante, Basto and Peñafiel are centres for Vinho Verde and there are a good twenty brands in circulation. Not many are exported; British visitors have probably met Casal Garcia from Avelada (see below), a wine purposely made a little less tart for colder English days, and Lagosta, sold by the Real Companhia do Norte in 'lobster pot' glass bottles.

Local enquiries could be made regarding visits to the

*Raymond Postgate *Portuguese Wine.*

Adega Corporativa and the Sociedade Agricola da Quinta de Santa Maria S.A.R.L., both at Barcelos.

Avintes (near Oporto)
Sogrape – Soc. Com. dos Vinhos de Mesa de Portugal, Lda.
Aldeia Nova-Avintes.
Tel. Oporto 9820801.
Head Office: Rua sá da Bandeira, 819–2°, dt°–Oporto.
Tel. Oporto 20761.

From the winery at Mateus all the famous rosé wine is taken by road to a large, modern bottling plant at Avintes, a village on the left bank of the river Douro, 12 kilometres from Oporto (Route 222). Here it is left in concrete 'balloon' tanks before being cooled and pumped in a series of operations to the bottling lines, which can produce 160,000 bottles a day. And from Avintes they go forth to all parts of the world.

Visitors see the bottling plant, taste the wine free and can buy bottles. English-speaking guides are available in summer.

Hours:	Monday to Friday 0900–1200; 1400–1700.
Introduction:	Though open to tourists without introduction, the company prefer people seriously interested in wine to have a letter of introduction from their agents: Hedges & Butler Ltd., Hedges House, 153 Regent Street, London W1R 8HQ. Tel. 01–734 4444.

Peñafiel
Soc. Agr. da Quinta da Aveleda,
Quinta da Aveleda, Peñafiel.
Tel. (0025) 22041.

The beautiful Quinta da Aveleda is close to the town of

Peñafiel, 42 kilometres from Oporto and 74 kilometres from Vila Real on Route 15. (The head office is in Oporto, Rua Sá da Bandeira, 819–1° Esq. Tel. Oporto 20762).

Centuries old and a charming estate, Aveleda in the Minho province is situated in the heart of the Vinho Verde demarcated district and it is here that the white wines, Casal Garcia and Aveleda are made and bottled. Training the vines high above the ground (which becomes very hot in high summer) gives the 'prickly' Vinho Verde acidity, resulting in freshness and fragrance. Fresh and light, these wines need to be served well chilled; with fish, chicken, veal, or simply as aperitifs. The winery and quinta are shown.

Hours: Monday to Friday 0900–1200; 1430–1800. The quinta, not the winery, may be open at weekends and on national holidays.

Introduction: For the quinta none necessary. For the winery a letter of introduction is preferable from agents:
Hedges & Butler Ltd., Hedges House, 153 Regent Street, London W1R 8HQ. Tel. 01–734 4444.

Vila Real
Mateus,
Vila Real, Douro, Traz-os-Montes.
Tel. (0099) 23074.

Mateus is a small village, three kilometres out of Vila Real on N322 towards Sabrosa, leading down to the river Douro valley. From Oporto (116 kilometres) allow 2 hours by car on Route 15. By train, the journey to Vila Real takes 2½ hours. From the frontier at Quintanilla (Bragança), the distance is 169 kilometres through beautiful country, but slow going.

The place to visit at Mateus is the Palace, a national monument and one of the finest examples of Portuguese baroque architecture. Dating back to 1610, it was built by the Count of Vila Real and still belongs to his family. The grapes for the famous Mateus Rosé are grown in the Palace vineyard and surrounding districts; the label on the bottle shews the elegant architecture and the beauty of the lake and gardens. Inside, visitors can admire the wooden ceilings and old furniture. Before (or after) seeing the Palace, they are recommended to visit the Mateus winery of Sogrape close by. For this however it is best to have an appointment, which visitors in Portugal can obtain from Sogrape Sociedade Commercial dos Vinhos de Mesa de Portugal, Lda., Rua Sá da Bandeira, 819, Oporto. From there the wines go to Oporto for bottling – see previous entry, Sogrape.

Hours: The Palace – Daily 0900–2000 (Winter 1700). Admission 25 esc. Winery (No charge) – Monday to Friday 0900–1200; 1430–1800.

DAO

For Anglo-Saxons, pronunciation of Portuguese has some astonishing surprises, fully reciprocated, I find, by the natives when I bid them what I mean to be 'Good morning' *(Bom dia)*, which unfortunately they understand as Good God! Dão is Da-ong with a nasal twang and Coimbra is 'Queenborough'; sounding to English ears, absolutely right for this charming town, which first established its university in the year 1308.

From Coimbra the wine region stretches north and north-west to Viseu and beyond, almost as far as Guarda. Mostly the vineyards are high up in the valleys of the Mondego and its tributaries, Dão and Alva. Going to Lisbon from the Vila Formoso frontier, or from the Douro at Pinhão or Régua, one passes through, probably without realising it. There are said to be 38,000 growers, spread over 1,200 square miles, each making on average 500 dozen cases a year, so little that the situation cries out for a co-operative movement. In fact over a dozen co-operative wineries now exist, with up to twenty-one planned – one in each of the communes.

For the visitor Dão is disappointing because there are very few single estates; the wines are either Dão Tinto or Dão Branco, good wines – the white being quite dry. The controlling authority *Frederação dos Viniculturos do Dão*, is the sole body permitted to award the *selo de garantia*, which appears on bottles as a thin band label.

Ruby red wines are made by mixing white and red grapes; the red grapes alone, so some experts think, would make better wines.

Local enquiries could be made regarding visits to the following wineries:

PLACE	ESTABLISHMENT
Nelas	Vinicola de Nelas
Oliveirinha	União Commercial da Beira
Povoline	J. P. Ferriera dos Santos
Tondela	Companhia Agricola & Pecuaria das Beiras

Viseu
Vinicola do Vale do Dão,
Campo de Viriato, Viseu.
Tel. (032) 23576.

Viseu is 133 kilometres from Oporto (Routes 1 and 2), 70 kilometres from Lamego (Route 2), 110 kilometres from Vila Real (Route 2) and 135 kilometres from the frontier at Vilar Formoso (Route 16). The winery is on the outskirts of the town close to the railway station and 800 metres north of the Grão Vasco hotel, both of which are marked on the *Michelin Guide* plan of Viseu.

The head office is in Oporto, Rua Sá da Bandeira 819-2° dt°-Oporto. Tel. Oporto 20762.

This region of the rivers, Mondego and Dão, is wild country, hilly and thickly wooded with a granite and sandy soil. Its oldest vineyards – said to be the oldest in the whole Iberian peninsular – are around the agricultural town of Viseu, some as high as 500 metres above sea level.

In the sixteenth century Viseu had a school of painting, Vasco Fernandes (1480–c.1543), who became known as 'Great Vasco', being its leading light. Grão Vasco, a very good Dão wine, has been named after him and this is the winery where it is made. The usual courtesies are offered here, including purchase of wines and an English-speaking guide. The Grão Vasco museum (free Saturday and Sunday, closed Monday) in Viseu is worth a call too.

Hours: Monday to Friday 0900–1200; 1430–1800.

Introduction: Open to all, but letter of introduction is preferred from agents:
Rawlings Voigt Ltd., Waterloo House, 228 Waterloo Station Approach, London SE1 7BE.
Tel. 01-928 4851.

ESTREMADURA

The largest wine-producing province of Portugal is Estremadura, stretching between the Tagus and the sea north of Lisbon for some seventy miles and south for a further twenty as far as Setúbal. To the north, red, white and rosé wines, of good ordinary standards, are made – Alcobaça, Leiria, Batalha and Torres Vedras being the principal towns. Most of the grape harvest goes to branded wines. To the south, a dessert wine, Moscatel de Setúbal is outstanding, while Periquita and Palmela clarete the two table wines of the Arrábida peninsula are good.

The *Michelin Green Guide* suggests a 12-mile circular tour from Sintra, where the attraction is the Royal Summer Palace. Colares, a small town close to the Atlantic forty minutes by car from Lisbon, has a pocket of deep sand on clay, which that devastating insect the phylloxera cannot penetrate. These vineyards, which go back to the twelfth century, therefore retain their native Ramisco vines, whereas in practically all other European vineyards the vines have to be grafted on to American root stock, too tough for the root-destroying insect to attack.

Since the vines' roots must be in the clay, under the sand possibly ten feet down, planting vines is a labour to be avoided. It is kept to a minimum, the vineyards being extended by 'layering' – the process of pegging down shoots, which throw up fresh shoots, well-known to gardeners who increase clematis, for example, in this way.

Colares wine is red and much the same strength as claret. Given ten years' bottle age it can be as lovely as a Rhône Côte Rôtie. None of this is unfortunately commercial these days.

Local enquiries could be made regarding visits to J. Jorge da Silva and the Adega Regional, at Colares.

CARCAVELOS

The encroachment by Estoril is so great that there is now hardly any wine from this coastal region seaward of Lis-bon. The white dessert wine, strong and sweet, is almost certain not to be authentic without a Government unbroken seal, *União Vinicola de Carcavelos* on the bottle.

BUCELAS

Mostly dry white wine, hock style, not for keeping. Bucelas is about fifteen miles up the Tagus from the capital.

Local enquiries regarding visits could be made to João Camilo Alves, Lda., at Bucelas.

SETUBAL

Setúbal, 25 miles south of Lisbon, is reached by crossing the Tagus by the great suspension bridge, completed in 1966 and 23 feet longer than the Forth road bridge. A little unfairly, its name was given to a dessert wine of the highest quality, which should be called Palmela or Azeitão, the two towns of the demarcated wine district, or Arrábida, the peninsula on which they stand.

The wine is made from four or five varieties of Moscatel grape. As with port, fermentation is stopped by adding brandy when sugar still remains in the must. Fresh Moscatel grape skins are then added to macerate with the wine. By the spring, wine, brandy and grape skins may not have harmonised completely but they do so in time, producing a marked, intensely fresh fragrance. Maturing in wood continues for at least six years, the finest qualities remaining in cask for twenty years and more. A fifty-year-old Moscatel de Setúbal is likely to become as dark as an old brown sherry, still very sweet and strong in its bouquet and flavour.

The guide books suggest various tours of the peninsula and the old Palacio do Quinta das Torres at Azeitão is a delightful hotel in which to stay.

Setúbal
José Maria da Fonseca Sucessores Vinhos, S.A.R.L.,
Vila Nogueira de Azeitão.
Tel. 20.80.002.

Conveniently sited 27 kilometres south of the Tagus road bridge, travellers between Lisbon and the Algarve should make the most of this old winery. Since 1834 J. M. da Fonseca has made many of Portugal's best table wines and brandies, and their Muscatel of Setúbal is unrivalled. French and English-speaking guides conduct the tour; wines are tasted free and bottles may be bought. Allow 30 minutes for the visit; if you have more time to spare, ask to see the new winery close by. Completed in 1970, its chief function is exporting rosé wines on a vast scale to America.

Hours: Monday to Friday 0900–1200; 1400–1700. Closed August.

Introduction: None necessary; warning is only required for groups exceeding forty people.

MADEIRA

Introduction

This group of islands, of which Madeira is the largest, is about 500 miles south-west of Lisbon and 200 miles from the coast of Africa. The latitude is 32/33°N and the Gulf Stream gives them an equable climate, with good bathing all the year round. There are however no beaches in Madeira; it is all rock bathing.

Funchal, the capital, with 43,000 inhabitants, is beautifully sited on the south coast, with a port capable of taking large cruise ships alongside and an airport thirty minutes' drive away.

GETTING THERE

By sea: Several companies sail regularly from London or Southampton and there are more sailings by Portuguese ships from Lisbon. Union Castle vessels on the Cape Town run sometimes call and Funchal is a popular stop for cruises.

By air: Daily flights from Lisbon with which British Airways usually connect.

THE LONGEST-LIVED WINE

Discovering the island in 1419, Zarco proceeded to name it Madeira, which means 'wood' in Portuguese. He then set it on fire, destroying all the wood and adding a ton or two of potash to the leaf mould of centuries.

It is sometimes suggested that this rich combination brings about the fine quality of Madeira, but in truth great wine invariably comes from vines struggling for existence in poor soil. Be that as it may, Madeira is a great wine standing in the ranks alongside port and sherry. In more leisurely days a mid-morning piece of cake and glass of Madeira often graced our grandfathers' business calls, which we ourselves now make by telephone. Alas, Time and Motion experts have failed to revive the custom in order to cut out kettles, milk and sugar.

The island –35 miles long by 17 – grows four different grapes making four different wines, all of which compete with *homo sapiens* in longevity; indeed, centenarians are not unknown. Human feet, with a screw-press in support, tread the grapes and human backs carry the wine away in goatskins down precipitous paths to the shippers' Lodges at Funchal. There it is heated in concrete tanks to 100–160°F. for three to four months, a technique arising from the olden practice of sending the casks to the tropics

in sailing ships. This *estufado* process darkens the wine and imparts the flavour of burnt sugar. Then comes fortification with brandy, the blending with older wine and finally long years of maturing in cask.

The improvement in most fortified wines derived from a voyage is well-known but cannot be precisely explained. Warmth and the aeration caused by a rolling ship are the contributing factors. With table wines, decanting and acquisition of room temperature illustrate the same phenomenon.

British seafarers were not very far behind the Portuguese. Though every vine had had to be brought to the island and planted, by 1680 the wine trade was thriving, with no less than ten British shippers out of twenty-six all told. Thenceforward the names, familiar enough today, began to recur –Cossart Gordon and Leacock throughout the story, Blandy from Napoleonic times.

Madeira, notably Malmsey and Bual, is the longest lived of all wines. Like sherry it is matured by a solera system, dates quoted indicating when the solera was first laid down. The average age of a solera is estimated at about eighty years. Madeiras are sold by solera names and brand names; the only place name is Camara de Lobos, noted for the finest Malmsey and, incidentally, as the place where Winston Churchill used to paint his favourite view. Very occasionally, old Madeira of a single year may be met but it has never seemed better than solera wines of comparable date. Madeira improves both in the cask and in the bottle.

Some idea of the longevity of Malmsey may be gained from the cellar of J. Pierpont Morgan, the legendary American banker who collected old Madeiras. Included in it were bottles of vintage Malmsey of 1774 and 1795, which seemed to have lost nothing in fragrance and vinosity, when opened in 1934.

THE FOUR VARIETIES

Sercial pronounced Sair-see-ahl is the dry amber Madeira. An excellent substitute for a dry sherry and good too with a melon.

Verdelho (Ver-dayl-yo) is a medium to rich golden wine with a dry finish. It is the all-purpose Madeira, which can be drunk before or after meals.

Bual is a medium-sweet dessert wine like Malmsey but often cheaper.

Malmsey is the greatest of the four and an excellent substitute for port. Full bodied and fruity.

THE VINEYARDS

About the same size as the Isle of Wight but with a mountainous centre culminating in 6,000-foot peaks, Madeira is an island of fruit and flowers. The vines are trained high on trellises among the fruit trees; there is no great area of concentration, as in the Médoc or Côte d'Or for example. The greatest density is close to the west of Funchal, at the back of Camara de Lobos. The *Michelin Green Guide* of Portugal gives a number of tours on which plenty of vines will be seen along the coastal parts.

Nearly all the exporting companies now belong to the Madeira Wine Association, which is at the disposal of visitors.

RECOMMENDED READING

Madeira Rupert Croft-Cooke Putnam London 1961.

This is the only book devoted wholly to Madeira's

wines: Jan Read's and Raymond Postgate's books recommended on p.178 include them.

Further Information

Casa de Portugal (Portuguese National Office)
20 Lower Regent Street
London SW1.
Tel. 01–930 2455.

Funchal

Madeira Wine Association Lda.,

Rua de S. Francisco, 10-Funchal.
Tel. Funchal 201.21.

The entrance to these offices is by the Tourist office at 28 Aveniga Arriaga, the wide avenue leading westwards out of Funchal.

The Association, representing most of the leading exporters, makes visitors welcome here with a free tasting of Madeiras, which may be bought at favourable prices.

Tours of the exporters' wine Lodges can be organised for small groups.

Hours:	Monday to Friday 1000–1200; 1500–1700.
Introduction:	None necessary – open to all.

ENGLAND

England

'I left open my vineyards to the inspection of the curious.'
– Mr. Vispré, who presented to the Society of Arts in 1784 'A Plan
for Cultivating Vineyards Adapted to this Climate.'

Introduction

ROMAN ORIGINS

Big-hearted though it was of Mr. Vispré not only to present his plan but to open his vineyard for inspection, viticulture in England seems to have come to a halt soon afterwards, after being practised – with varying degrees of intensity – since Emperor Probus in 280 A.D. thought it might have a civilising influence on the northern barbarians.

The slow conversion of the Saxons to Christianity, begun with the arrival of Saint Augustine in 597, created the need for Communion wine and that put the monks in business as vignerons. One of the earliest English vineyards (circa 950), owned by the Abbey of Glastonbury, was at Wedmore in Somerset.

From 1066 to 1089 William the Conqueror and the Norman Barons extended the acreage presumably for their own consumption. There was a 12-acre vineyard at Bisham, Berkshire and later, one of eleven acres at Belchamp Walter in Essex. Altogether, Domesday acreage was 135 in 35 sites.

Soon English vineyards were to be found as far north as Yorkshire and westwards to Shropshire and Herefordshire, a total of one hundred and thirty-nine mediaeval vineyards being recorded, mostly attached to monasteries, in William Younger's *Gods, Men and Wine.**

DISSOLUTION AND DECLINE

Until Elizabethan times, it is believed that the country had a warmer and drier climate than we have now. There were extensive fourteenth-century vineyards along the Meuse in Belgium and wines from Gloucester are mentioned as being as sweet as those of France. Thereafter viticulture declined due to a number of factors.

First, the marriage of Henry II to Eleanor of Aquitaine

---

*William Younger *Gods, Men and Wine* Wine and Food Society 1966 Appendix F.

ENGLAND

LINCOLNSHIRE

Stragglethorpe · Great Yarmouth

SUFFOLK
Sudbury

ESSEX
Felsted · Purleigh

LONDON

AVON
· Bristol

SOMERSET
Shepton Mallet

HAMPSHIRE

DEVON

SUSSEX
Beaulieu · Hambledon · Horam
Brighton

Plymouth

Yarmouth · ISLE OF WIGHT

N

Km 0 20 40 60 80 100
Mi 0   20   40   60

in 1152 brought Bordeaux, and other wines from western France, by sea to Bristol at small cost; secondly, after the Black Death, there may have been a shortage of labour and thirdly, after Henry VIII had dissolved the monasteries, there was probably a shortage of technical skill. Thereafter only a few eccentrics such as John Rose, gardener to Charles II, and Charles Hamilton, a Surrey grower around 1790, demonstrated what could be done until the Marquess of Bute did very well with a vineyard at Coch Castle, near Cardiff, which he planted in 1881.

THE PRESENT REVIVAL
Since World War II there has been such an astonishing revival that there are, once more, over one hundred vineyards in England, with two more in south Wales and one in Dublin. About 500 acres* are planted, individual sites varying from 31 acres down to ½ an acre, the average being 2 to 3 acres. The pioneer in this revival was Major-General Sir Guy Salisbury-Jones, G.C.V.O., C.M.G., C.B.E., M.C., now President of the English Vineyards Association, who had already had two distinguished careers, first as a soldier and secondly as Marshal of the Diplomatic Corps.

*Estimate in 1975.

194

Sir Guy, after taking technical advice on the choice of vines, planted the French hybrid Seyre-Villard 5.276 grafted on 16149, with some Chardonnay and some Pinot Noir as well, both of which are used in the making of champagne and thrive in chalky soil rather like his own. For the rest, turn to the Hambledon vineyard entry on p.197 or join his 12,000 visitors when the vineyard is open for 'the inspection of the curious'.

Chairman of the English Vineyards Association, which has over three hundred and fifty members, is Mr. Jack Ward, joint head of the Merrydown Wine Company (p.199). Though the company's sales have primarily always been of cider and fruit wines, they have their own small vineyards and have given the Association a unique benefit by acting as a Wine Co-operative, vinifying grapes and bottling the wines of members, thus saving them the expense of having their own equipment. In short, Merrydown has the tools and undertakes to finish the job.

Here perhaps I should ensure that there is no misunderstanding between two terms – English Vineyard Wines and British wines.

British wines – the well known VP and RSVP brands for example – though the sales are enormous are not true grape wines and are not recognised as such within the E.E.C. The grape juice is imported, from such countries as Cyprus, Argentine, Spain, Greece and South Africa. A heating process in the country of origin concentrates the juice, reducing it in volume to a condition in which it cannot ferment on the journey. Fermentation takes place in Britain, with a culture of yeasts, adding the same volume of water as was removed before shipment by the heating process.

A DUTY-FREE CONCESSION
Hitherto British and English vineyard wines have had to pay the same excise, to the disgust of English vineyard proprietors. From 1 January 1976 they have, however, been granted a useful concession. They can apply to H.M. Customs and Excise, and subject to certain conditions will receive permission to retain up to 120 gallons a year duty free *for domestic consumption*. For those making over 120 gallons, the allowance is increased by 10 per cent. of the quantity made over 120 gallons. The allowance is for domestic purposes; to sell any part of it would be illegal.

Though fairer, the English growers hope that some further adjustment can be made in their favour because southern European growers naturally produce wines of higher alcoholic strength. In Britain a line has long been drawn at 27° proof (15·4° Absolute Alcohol) dividing High Strength (e.g. Port and Sherry) from Low, less being paid on the latter. If the line could be lowered a little, say to 20·3° proof (11·5° Absolute Alcohol) or a second one drawn thereabouts, the light English vineyard and the white wines of the Rhineland and Loire could be liable for less.

English vineyard wines are nearly all white and not unlike Mosel or Loire wines (neither we nor the Germans have enough sunshine to succeed with red wines). The yield should be about 2,000 bottles an acre. Few individual proprietors have large enough vineyards to do much commercially but Hambledon's wine has been supplied to the Savoy Hotel and to the liner *Queen Elizabeth II* and is usually obtainable retail from Peter Dominic shops in Hampshire and West Sussex. Mr. K. C. Barlow's Adgestone, Isle of Wight, has been on leading hotel wine lists including London's Dorchester. Felstar drew praise at a Brussels E.E.C. tasting from no less an authority than the late Count Matuschka-Greiffenclau, owner of Schloss Vollrads.

COSTLY UNDERTAKING
One last word of advice. Drink English vineyard wines! Visit their vineyards! But do not be tempted into the business without the most careful consideration of costs. Assuming the existence of suitable land and buildings for conversion to press house, winery and cellars, Sir Guy Salisbury-Jones estimated the initial cost of establishing a 3-acre vineyard at £15,200, which included a £6,000 wage bill for the first three years before production started. Thereafter the annual cost he gave as £6,250. And this was in 1973, when petrol was 36p a gallon.

In compiling the entries that follow, Mr. Jack Ward, their Chairman, advised me to approach only a few of the one hundred and fifty commercial growers now thought to exist in England and Wales, believing that the rest had

their hands full without the distraction of visitors. The Adgestone vineyard at Sandown, Isle of Wight, where an excellent white wine is made, is one such example. On the other hand, I regret if this has led to omissions.

To the vineyards, I have been able to add James Hawker's Mayflower House in Plymouth from which the Pilgrim Fathers sailed in the Mayflower in 1620 and – most interesting of all – Harvey's Wine Museum in Bristol.

NATIONAL HOLIDAYS
National holidays in England are given below. Please remember when planning visits that establishments will be closed unless otherwise indicated in the entries that follow: 1 January, Good Friday, Easter Monday, the last Mondays in May and August, 25 and 26 December.

Further Information
Merrydown Wine Co. Ltd.,
Horam Manor,
Horam,
Heathfield,
East Sussex. PM21 0JA.
Tel. (04353) 2254.

AVON
Bristol
Harvey's Wine Museum,
12 Denmark Street, Bristol.
Tel. (0272) 298011 and 27661.

Being 120 miles from London and 90 from Birmingham, motorways M4 and M5 make Bristol quickly accessible from all directions and from Paddington the fast trains are frequent. The Museum (CZ c on the *Michelin Guide* Bristol plan) is close to the city centre.

No. 12 Denmark Street has been the Head Office of John Harvey & Sons since 1796 when William Perry founded the company which eventually became Harvey's of Bristol. Until 1960, all Harvey's sherries were blended, bottled and stored in these ancient cellars, which are now in part this wine museum and in part a splendid restaurant, a wine lover's delight, with a wine list of six hundred items.

Casual visitors cannot normally be accepted. Instead, there are guided tours of the museum, with explanations of the old bottles, decanter labels, early English wine glasses and so forth. There is a film on wine production, followed by a tasting of a selection of wines. The tour takes up to 2½ hours and is not suitable for children under 18 years. Cost is 75p a head but Overseas Visitors are free.

Hours:	Monday to Friday. Usual starting times for the tours are 1430 and 1830.
Introduction:	Tours must be prearranged. Write or telephone in advance to John Harvey & Sons Ltd., (Public Relations) at the above address.

Plymouth
James Hawker & Co. Ltd.,
Mayflower House, Breton side, Plymouth, Devon PL4 0BA.
Tel. (0752) 63144/5.

With mostly a double track road from Exeter, and the M5 motorway coming closer in stages to that cathedral city, the journey to Plymouth becomes progressively easier. By rail the Cornish Riviera does those 226 miles from Paddington in 3½ hours and by sea there is a Plymouth-Roscoff (Brittany) car ferry. The city plan in the *Michelin Great Britain Guide* shews Breton side clearly, close to Sutton Harbour and less than a mile from the city centre.

Every British schoolboy knows of Plymouth as the home town of Drake and Hawkins, Raleigh and Grenville, and every American schoolboy, one presumes, as the port from which the first Virginia settlers sailed in 1558. Four centuries later it was the principal port from which the American armies advanced, through 'Utah' and the bloody 'Omaha' beaches, to the liberation of Europe.

James Hawker, wine shippers and makers of Pedlar Sloe Gin, trace their beginnings to 1620 when the Pilgrim Fathers set sail in the *Mayflower* after spending their last night in the firm's cellars close to the quay. Appropriately in 1957, when *Mayflower II* sailed for the New World as a goodwill gesture, Hawker's Mayflower Cellars were the focal point of final festivity.

Visitors are shewn the duty-paid cellars and the bonded warehouse, which is adjoining, the bottling plant, etc.

Hours:	Monday to Friday 0930–1300; 1430–1630. Closed to visitors in December.
Introduction:	Members of the public interested in Pedlar Sloe Gin and fine wines are welcome in parties, large and small, from January to November inclusive. Write or telephone in advance for appointment.

Felsted
J. G. and I. M. Barrett,
The Vineyards, Cricks Green, Felsted, Essex CM6 3JT.
Tel. (024–534) 504.

Felsted is about eleven miles north of Chelmsford (A130 and B1417) and about two miles south-west of Braintree.

Graham Barrett, an ex-printer, began here in 1966; the first wine was made in 1969, recent vintages winning Silver Awards at *Club Oenologique* competitions.

Now he and his wife have over four hectares of vineyard which – as they point out in an excellent leaflet – is a good commercial size. Indeed, with 14,000 vines, it is much larger than the average piece under single ownership in France or Germany.

Their Felstar wines, stocked by Harrods and a number of Essex hotels, are made from half a dozen varieties of vine and, as professional viticulturists, they supply rooted and unrooted cuttings.

Casual visitors regrettably cannot be accepted but some parties are welcome *by appointment* at weekends and possibly in the long summer evenings of the working week. A charge of about 60p a head should be expected.

Introduction:	Strictly by appointment. A stamped addressed envelope could be sent for the Felstar descriptive leaflet.

Purleigh
S. W. Greenwood Esq.,
New Hall, Purleigh, Essex.
Tel. (062–185) 343.

This vineyard is on B1010, four miles from that popular yachting centre, Burnham-on-Crouch, and therefore very close to the misty North Sea. From Chelmsford (10 miles) take the A414, bearing right on to B1010 at the Woodham Mortimer crossroads.

Planted in 1970, in a hollow among some of the few hills in Essex, the vineyard has produced, from four vine varieties, some excellent medium, dry white wines of Loire and Mosel style, which seem to keep well. The area under vine is now being increased to 6 hectares.

The wine is kept in a cellar under the old farmhouse, where sampling and sales take place. A small admission fee is charged. Visitors in 1977 should be able to sample Mr. Greenwood's latest venture, the 1975 Pinot Noir Rosé.

Hours:	Monday to Friday 1000–1800.
Introduction:	By appointment (write or telephone) only.

Beaulieu
Beaulieu Vineyard,
Beaulieu, Hampshire, SO4 7ZN.
Tel. Beaulieu (0590) 612345.

Beaulieu, pronounced 'Bewly,' is a small village 14 miles south-west of Southampton. From Southampton, take the Fawley road A326, turning right on to B3054. From the west, approach via Lyndhurst. Driving time from London or Bristol is about 2½ hours. The vineyard adjoins the National Motor Museum; those with appointments to see the former meet their escorts at the Museum Information Centre.

In the National Motor Museum, set up by Lord Montagu of Beaulieu, over two hundred veteran and vintage vehicles tell the story of motoring from 1895 to the present day. Palace House, once the Great Gatehouse of Beaulieu Abbey and now the home of Lord and Lady Montagu of Beaulieu, and the ruins of the great Cistercian Abbey founded by King John in 1204, can also be combined with the vineyard visit.

Although there may have been a vineyard at Beaulieu during its monastic days, the present 2·2 hectare vineyard, with over 11,000 vines, was not planted by Mrs. Gore-Browne and her late husband until 1958. Vine varieties include Müller-Thurgau, Seyve-Villard and Siebel. The first pressing was in 1961 and, by 1967, the vineyard was established and the wine of that year sold. The wine, named Beaulieu Wine, is white and medium dry. It is made through the Merrydown Wine Company's co-operative scheme and the 1975 vintage yielded seven tons of grapes, making about 7,500 bottles of wine.

Escorted tours of the vineyard can be arranged on written application to: The General Manager, Beaulieu Manor, John Montagu Building, Beaulieu, Hampshire.

Hambledon
Major General Sir Guy Salisbury-Jones, G.C.V.O., C.M.G., C.B.E., M.C., D.L.,
Mill Down, Hambledon, near Portsmouth, Hampshire PO7 6RY.
Tel. (070132) 475.

The downland village of Hambledon may be described as the cradle of cricket; indeed, on Broadhalfpenny Down in 1777 Hambledon defeated the whole of England. Hambledon is 64 miles from London and 15 miles from Southampton. From London take the Portsmouth road A3, turning right 5 miles after Petersfield, at signpost to Hambledon. Entering the village, 150 yards beyond the first speed limit sign, climb the slope up to the vineyard. From Southampton go via Botley. The nearest station is Petersfield (9 miles). From Portsmouth the bus service is SD139 stopping 7 minutes' walk away.

It was in 1952 soon after Sir Guy retired from his last military job as Head of the British Military Mission to France that he began to plant this vineyard on the slopes of Windmill Down below his house. From it the views include Broadhalfpenny Down, where the eighteenth-century cricketers made their bit of history and, on a clear day, the Isle of Wight. The vineyard makes about 8,000 bottles from its 2 hectares and the dry white wine, estate-bottled and matured, is like a dry Vouvray or a crisp Mosel. Hambledon commands a similar price too.

Visitors see the vineyard and press house, where frequent short talks on wine making are given. Unfortunately production is not sufficient to offer tastings but wine can be bought *on weekdays only*. On all open days, tea and light refreshments are available.

Hours: Open to all from last Sunday in July to the first Sunday in October inclusive, from 1430–1730. On weekdays during this period visitors are received by appointment.

Introduction: Open to all on payment of a modest admission fee. Send a stamped addressed envelope for leaflet with small local sketch map, dates and prices of admission.

ISLE OF WIGHT
Yarmouth
R. H. Gibbons & W. B. N. Poulter,
Cranmore Vineyard, Yarmouth, Isle of Wight PO41 0XY.
Tel. (0983) 760561.

Cranmore, lying to the north of A3054, is a farming area two miles east of Yarmouth. From the A3054 (Yarmouth-Newport), take Cranmore Avenue (2½ miles from Yarmouth) which passes the vineyard entrance in half a mile.

From the mainland, visitors can cross to the Island on the ferries Lymington to Yarmouth, Southampton to West Cowes, or Portsmouth to Ryde or Fishbourne. The nearest of these landing places to Cranmore is Yarmouth.

The vineyard, begun as a part-time venture in 1967, is now a full-time business, the present 2·4 hectares having been extended to four. Production at present is 7,000/10,000 bottles a hectare Mosel style wine, which is all estate-bottled. A flourishing nursery grows young vines for sale to amateur and commercial growers.

Casual visitors cannot be accepted but conducted tours take place, which include tasting.

Hours: Weekends only from about mid-July to mid-September. A small charge a head is made.

Introduction: Send stamped addressed envelope for latest leaflet, which will give open days dates, times, ferry and bus routes, etc., and includes a booking slip.

LINCOLNSHIRE
Stragglethorpe
Major Alan Rook, M.A., F.C.A., F.R.S.L.,
Stragglethorpe Hall, Lincoln.
Tel. Loveden 72308.

Stragglethorpe is a small village near the Nottinghamshire-Lincolnshire border, just off the A17 Newark to Sleaford road, between Beckingham and Leadenham and opposite Brant Broughton. (Nottingham 27 miles, Newark 7 miles, Sleaford 14 miles, Lincoln 15 miles, Grantham 12 miles.)

Stragglethorpe Hall was once an offshoot from the Lincolnshire Priory of Sempringham, founded by a Norman knight, St. Gilbert of Sempringham in 1131. On the Dissolution of the Monasteries, the Hall passed into the possession of the Earle family and on the death of Sir

Richard Earle in 1697 at the age of twenty-four, the property reverted to his mother's family, the Welbys of Denton. It is now the home of Major Alan Rook, who is a well-known Nottingham wine merchant.

In 1964, 0·6 hectares were planted with 2,500 vines in the walled kitchen garden and the first vintage in October 1967 produced some 50 dozen bottles of a dry, white wine somewhat similar to a white Burgundy. Named Lincoln Imperial, by 1970 the output had risen to 170 dozen.

Visitors see the vineyard, the gardens and the winery, where they are received by Major Rook, who describes the making of the wine and answers questions.

Hours: Open to all every Saturday afternoon in August from 1400 to 1730. Admission 15p. Children under 14 – 5p. Parties of 40 and over – 10p each.

SOMERSET
Shepton Mallet
N. de M. Godden Esq.,
The Manor House, Pilton, Shepton Mallet, Somerset BA4 4BE.
Tel. Pilton (074 989) 325.

The Pilton Manor Vineyard lies on the western side of the village off the A361 (Glastonbury to Shepton Mallet) in an open valley, running east and west among the foothills leading up to the Mendips. Distances are Wells 6 miles, Glastonbury 7 miles and Shepton Mallet 2½ miles.

Mr. and Mrs. Godden bought this manor house in 1964, replanting the land with a 1·6 hectare vineyard, which they hope to extend to 6 hectares shortly on the other side of the main road. The manor dates back to the thirteenth century and there were Pilton vineyards associated with the Abbey of Glastonbury earlier. The Riesling-Sylvaner wine made and bottled here won the Gore-Brown trophy for the best English vineyard wine of 1973, Lenz Moser of Krems making the presentation at Christie's in London.

Casual visitors are welcome from 24 August to 6 October, Sundays and Bank Holidays, 1200–1800. Bring a picnic or have tea in the Mediaeval Garden. Vines in pots and books are for sale. Open all the year for the sale of wines and vines, with wines on sale by the glass or bottle from 1200–1400 only (the licensed hours).

Groups *(by appointment only)* (min. 10 people) can be given conducted tours of vineyard and winery lasting about 2 hours as follows:

From 10 June to 4 October, Monday–Friday 1700–dusk.

From September to mid-October some groups can be accepted on Saturdays.

An admission charge is made for both casual visitors and groups.

Since dates, times and charges are liable to change, prospective visitors are advised to send a stamped addressed envelope for leaflet with latest details.

Major C. L. B. Gillespie, M.B.E.,
Wootton Vines, North Wootton, Near Shepton Mallet, Somerset BA4 4AG.
Tel. Pilton (074 989) 359.

North Wootton is a tiny farming community in the foothills of the Mendips, 23 miles south of Bath and 3 miles from the cathedral city of Wells. From Wells, take the A39 road to Glastonbury, turning left to North Wootton after 1½ miles. In the village, at the Crossways Inn, turn left and drive half a mile. The vineyard is on this road in an unspoilt and beautiful part of the country; the lanes are narrow, much as they were in the days of horses. It is not easy to find one's way and there is no public transport.

Less than a hectare was planted in 1971 when Major Gillespie retired from the army. Since then he and his wife have converted the old pantiled barns into a picturesque winery, cellar and wine-tasting room. In addition to the estate wine, wines are made for a number of other vineyard owners, who lack wineries of their own.

A novel stop on any itinerary to Bath, Wells and the West, casual visitors (preferably p.m.), are welcome to buy wine on any day except Tuesday, though they must accept the owner and his wife as they find them – most likely steeped in sprays, and wielding secateurs.

Organised tours with personal attention, can be arranged by appointment. Wine tastings (minimum 15, max 36 people) can also be arranged with tour, discussion, soup and cheese. A small charge a head is made for tours and tastings, the latter being higher.

Hours: Casual visitors see above. Tours at any time by appointment. Wine-tastings by appointment but must be in licensing hours, i.e. 1030–1430 and 1800–2230.

Introduction: Send stamped addressed envelope for leaflet with sketch map and details of charges.

SUFFOLK
Sudbury
B. T. Ambrose Esq.,
Nether Hall, Cavendish, Sudbury, Suffolk.
Tel. (078 73) 73636.

Cavendish, one of the prettiest villages in England, is in the Constable country beside the Stour valley on the Essex/Suffolk border. It is directly between Ipswich and Cambridge, and about 27 miles from each of them, on the A1092 road between Long Melford and Clare.

Nether Hall Manor is a Tudor Elizabethan house mentioned in *Treasures of Britain;* the vineyard adjoins the north side of the village green close to the thirteenth-century church.

Four hectares were planted with Müller-Thurgau vines in 1972.

Visitors are welcome by appointment, an off-licence has been granted and wine can be bought. A modest fee is charged and there are good parking and picnic facilities.

Introduction: Full details, with particulars of Open Days, will be sent on receipt of stamped addressed envelope.

SUSSEX
Horam
Merrydown Wine Co. Ltd.,
Horam Manor, Horam, Heathfield, East Sussex.
Tel. (04353) 2254.

Horam, a small Sussex village, is four miles east of the London-Eastbourne road A22, and twelve miles from Eastbourne on A267, the Tunbridge Wells road. From London allow 1½–2 hours by car; from Gatwick Airport forty minutes.

This enterprising company was formed in 1946, occupying Horam Manor in 1947 and its cider soon became well-known in England. Though specialising in fruit wines, it owns two small vineyards from which a Riesling-Sylvaner wine has been produced since 1967. Since 1969, Merrydown has worked a co-operative scheme in order to help the increasing number of English vineyard owners, who lack the facilities for producing their own wine.

Visitors are conducted in parties round the outside of the winery and are shewn one of the vineyards. English wines and the full selection of Merrydown products may be bought at the shop.

Hours: April–August inclusive: Monday to Thursday 1400–1600.

Introduction: Open to all by appointment only, which should be made well in advance, by letter or telephone, for Merrydown at Horam is a popular destination in summer.

SCOTLAND

Scotland

'A plucky little malt; such a pity it's from the wrong side of the loch'
— *Humorist Anon.*

Introduction

There are three kinds of Scotch whisky:

1. Malt, the original whisky of the Highlands, distilled entirely from malted barley in a pot-still.

2. Grain, a whisky of less character, distilled from a mixed mash of cereal grain, preferably maize, in a 'patent' or continuous still. A little green or malted barley is added to the maize at the cooking stage to accelerate the conversion of starches.

3. Blended, the mixture of the two, 40 per cent. Malt to 60 per cent. Grain approximately, which is now known the world over, earning Britain over £300 million a year in exports, while at home we pay £2.87 a bottle Excise Duty, with at least a further 30p V.A.T. added to the retail price (around £3.80 in mid 1976).

Aqua vitae, 'water of life', to the Romans; *eau de vie* to the French; *uisge beatha* in Gaelic: the three words are synonymous and the art of distilling such spirits has been widely practised in Western Europe at least from the Middle Ages. In the Highlands, east and north from the Cairngorms to the coast, the barley was home-grown and the mountain air fresh as the unpolluted water of the burns. There too was the granite rock, from which the water springs, and the rich, dark peat of the moors. These are the fundamentals that combine to make Highland malt whiskies unique and something of a mystery. Even when the water is taken from the same spring, the products of different distilleries cannot be made to taste the same. Abroad, imitations there have been by the hundred yet none has resembled Highland Malt, even though the same peat and Highland water have been shipped specially to make them.

HISTORY

The evolution of 'Scotch' really began after Culloden, where the 1745 Jacobite rebellion was defeated. The English Government thereupon increased the tax on whisky, introducing an army of excise men to enforce payment. This was a step towards commercialisation of

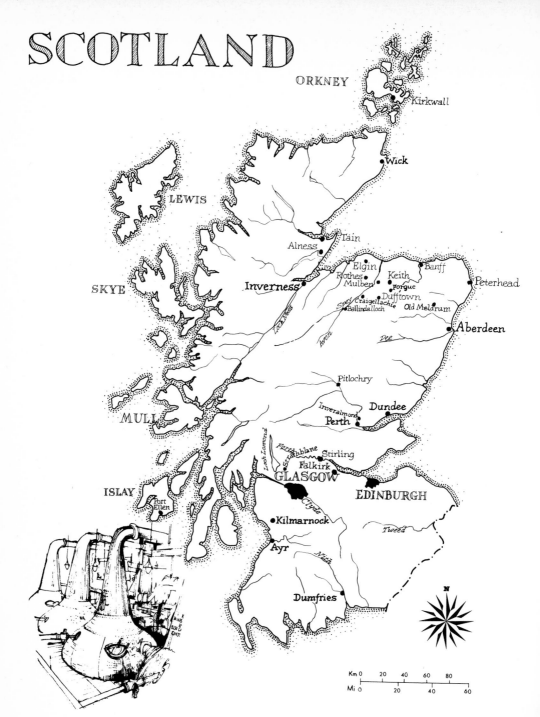

SCOTLAND

ORKNEY

Kirkwall

Wick

LEWIS

Tain

Alness

Elgin   Banff

Rothes   Keith   Peterhead

SKYE   Mulben   Forgue

Inverness   Dufftown

Craigellachie   Old Meldrum

Ballindalloch

Spey   Aberdeen

Avon

Dee

Pitlochry

Dundee

Inveralmond

MULL   Perth

Loch Lomond

Allan   Stirling

ISLAY   Falkirk

Port   GLASGOW

Ellen   EDINBURGH

Clyde

Kilmarnock

Tweed

Ayr

Nith

Dumfries

N

Km 0   20   40   60   80

Mi 0   20   40   60

204

what Burns had called 'the poor man's wine' and the Highlander, having ignored the tax since it was first levied in 1644, set up illicit stills in defiance.

Not until 1823 was the Government persuaded by the Duke of Gordon, the largest landowner in the Central Highlands, to pass a new Act, which banned stills with a capacity of less than 40 gallons and reduced the duty to 2s.3d. a proof gallon from all other stills. It also introduced to Scotland the now familiar system of delaying payment until spirits leave the warehouse.

First Licensed Distillery: The following year George Smith of Glenlivet, architect, distiller and farmer, deciding that illicit stills and contraband whisky could not succeed for ever, took out the £10 licence required by the new Act, rebuilt his distillery and became the first licensed distiller. Though regarded as a blackleg by his fellow farmers on Speyside, who threatened both his life and his property, Smith, a powerful man, stuck to his guns. Gradually others followed his example and the number of illicit stills dwindled from 14,000 in 1823 to 692 in 1834, ceasing slowly thereafter to be a problem.

George Smith died in 1871 at the age of seventy-nine. He had lived to see the railway come to Ballindalloch, seven miles from Glenlivet, saving his horse-drawn carts the thirty-five difficult miles to Moray Firth ports.

The reward for his work came in 1880, when his son John Gordon Smith, secured the right to call his label of whisky 'The Glenlivet' so that thenceforward 'Glenlivet' alone meant 'Smith's Glenlivet' and none other. The honour was deserved. By 1850 Glenlivet had achieved the highest reputation for its malt whisky but the High Court's decision did not ultimately prevent other distillers hyphenating 'Glenlivet' to their names where there was an appropriate generic claim. Many of them did and continue to do so; some far removed from the Glen and its water.

The Patent or 'Continuous' Still: Meanwhile in southern Scotland, a growing number of distilleries were making 'Lowland malt', less strongly flavoured than that of the Highlands though made in precisely the same manner. At the moment when George Smith decided to abide by the law, Robert Stein of the Kilbaigie Distillery, at Kincardine on the Forth, was at work on his new still, which made alcohol in one continuous operation instead of two, as in the pot-still. In 1826 he took out a patent, but Aeneas Coffey of Dublin gave his name to a much better version only six years later.

Coffey's 'patent' or 'continuous' still was revolutionary. A big grain distillery could produce in a week as much as a malt whisky distiller might manage in a year. The process was quicker and independent of locality; whisky moreover could be made largely of unmalted cereals – maize, rye, oats – with only a little malted barley to supply the diastase to convert the unmalted grain during the mashing process.

The working of the patent still is described in all the books on whisky. The grain is usually maize with a little malted barley. What emerges from one continuous operation is a more rectified spirit than that from the pot-still, deriving far less benefit from long years maturing in wood. Though I have tasted grain whisky and there may be one or two brands on the market, it is in no way a substitute for malt. Its great merit is lack of pronounced character, for it is that which makes the blend of the two – grain and malt together – a less pungently flavoured spirit, excellent for daily drinking in climates hot and cold.

Yet there was no immediate revolution. Any sort of whisky shipped to England – it seems impossible to believe now – was usually rectified and turned into gin. Shipment in bottles only became possible under Gladstone's Spirit Act of 1860, after which 'patent still' distilleries sprang up so quickly that six leading companies, John Haig among them, decided collaboration was better than competition. In 1887 they united in The Distillers Company Limited (D.C.L.), destined within fifty years to control the Scotch whisky trade.

The Distillers Company: The rapid increase in demand to the end of the century was primarily caused by the phylloxera, the insect that destroys the roots of vines. The phylloxera epidemic spread across Europe from 1865 and, by the 1880s, Cognac had become so hard hit that Scotch whisky was being used in England to make it go further.

But what *was* Scotch whisky? In Scotland the enterprising new firms had already decided that it had to be a blended whisky; part malt, part grain. How else could a standard, unvarying drink be produced in quantities which would meet the growing demand?

All was not, however, plain sailing. In 1880, shares first offered to the public by the D.C.L. were hardly taken up. It took three years to obtain a quotation for them on the Edinburgh and Glasgow stock exchanges and seven years on the London. Not until the 'What is Whisky?' case did a Royal Commission decide, in 1909, that whisky was 'a spirit obtained by distillation from a mash of cereal grain, saccharified by the diastase of malt'.

Today, great blending and bottling installations have grown up around Glasgow and Edinburgh, where the required number of malt and grain whiskies, of different ages and qualities, are blended and then rested in wood prior to bottling. The 'required number' varies considerably. Some blenders achieve the style they desire with fifteen to twenty makes, others may need as many as fifty.

A law of 1915, confirmed in 1952, requires Scotch whisky to be distilled in Scotland and matured in wooden casks for at least three years. The maturing period now follows in casks of American oak or in former sherry casks. The raw, colourless spirit takes colour from the wood, mellowing as it 'breathes' through the wood. But, since colouring can vary from cask to cask, a little caramelised sugar is permitted at the bottling stage to ensure the spirit in each bottle is of uniform tint. Standard blends e.g. Haig, Bell's, J & B Rare, are likely to exceed four years old when bottled; de luxe blends e.g. Haig's Dimple, Chivas Regal, will exceed twelve years. Bottling on automatic bottling lines (one for bottles, another for halves, etc.) and despatch by sea to all corners of the planet completes the process.

In the D.C.L., the blending companies can now call upon the single whiskies of over forty-five malt and grain distilleries, owned and operated by D.C.L. subsidiary companies. This gives some idea of the huge organisation that has become Britain's leading earner of foreign currency. And the D.C.L., though the largest group, responsible for about six of every ten bottles of 'Scotch', is only 'the larger half' of the trade.

Although purists still regard the admixture of malt and grain whisky as adulteration, there can be little doubt that the blend evolved was a far more acceptable drink than malt whisky. It was cheaper to make in greater quantity and it was easier to sell.

The Great Salesmen: 'The story of its rise from humble beginnings to its present magnificence exemplifies the cardinal virtues of the poor Scot of those days: grit, courage, thrift, plain living, vision, honesty, an immense capacity for hard work and the ability to grasp the golden opportunity when it presented itself.'

So wrote Sir Robert Bruce Lockhart, introducing 'The Whisky Barons' in his book *Scotch* and, as I read the story of those three Scottish Whittingtons (Tommy Dewar, Alec Walker and James Buchanan) coming to London with absolutely no assets but these personal qualities, my mind inevitably goes forward to a fourth, with whom I worked closely for ten years.

Paul Dauthieu, born of French parents and brought up by them in the Highlands, was just such a Scot, sent to London in the twenties when the family fortunes had gone up in smoke. Their hotel near Fort William had been burnt to the ground and Paul slowly progressed from an apprentice in the Savoy kitchens to a commis waiter at Claridge's.

Then, when war came in 1939, scorning an offer of half salary from the wine firm in London which employed him, he started his own shop in Horsham, calling it Peter Dominic. Later, when he was in the R.A.F., his wife ran the shop and after the war for nearly ten years whisky for the home market was so scarce that newcomers to the trade could get very little, particularly of the leading brands. For a retailer with no capital it was a remarkable achievement to build a twenty-branch business by 1962 and sell it for £750,000.

In many ways the whisky barons had it easier, brilliant brand advertising doing so much for them. First to reach London was James Buchanan in 1884, a dissatisfied office boy setting up as an agent for an unknown Glasgow firm. His Black and White horses and delivery vans, so smart and conspicuous, were to delight Londoners for forty years. He was made Baron Woolavington in 1920 and won the Derby twice with Captain Cuttle and Coronach.

The brothers Dewar, John and Tommy, grandsons of a Perthshire crofter, were also made Barons – Forteviot and Dewar respectively. Tommy, the extrovert and great

sporting peer was a patron of motor racing and football; Abbot's Trace, one of his horses, sired a Derby winner.

Alec Walker, born 1869 and still going strong until 1950, only received a knighthood; racing, the sport of Kings, does not seem to have interested him. His father had left the family retail shop in Kilmarnock for London in 1889. Alec, the third son, raised the firm to a leading position, with his brother and Stevenson, another Ayrshire man in partnership. His retirement was largely spent in the club house at Troon, on one occasion, sending for Walter Hagen before the championship to rebuke him for practising from the championship tees.

Leading Brands: Black & White, White Label and Johnnie Walker formed the 'Big 5' with Haig and White Horse. The latter belonged to the Mackie family and in advertising it, every white horse from Napoleon's Marengo to Lord Roberts' Voronel seems to have been made to pull its weight.

The Haigs originally came to Scotland as del Hage, or de Haga, from Normandy and have been distillers in the Lowlands since the eighteenth century. They intermarried with the Steins, Robert Stein being the inventor of the first patent still, and have been the champions of Lowland malt since the Cameron Bridge distillery in Fife was built in 1824. For a hundred years now their main centre has been there and at Markinch, the blending establishment close by. Haig is regarded as the leading brand in England, and Bell's in Scotland.

In the United States two London wine merchants, Justerini & Brooks and Berry Bros. were clever in exploiting the market when Prohibition ended in 1933. J & B Rare and Cutty Sark, high quality blends of a delicate character, had an immediate appeal to a generation which had subsisted for years on a diet of Prohibition hooch.

MAKING MALT WHISKY

Distilling a spirit from grain is more involved than doing it from grapes. Grapes contain sugar and when crushed ferment to make wine, an alcoholic liquid readily distilled into brandy in a pot-still. With grain, the starch in the grain has first to be converted into sugar and a liquid obtained, first to ferment and then to distil.

Making malt whisky, the barley is first cleaned and then 'steeped' for two–three days in tanks of water. On the concrete malting floor the soaked grain is then spread evenly to a depth of 10″ to 14″. In eight to twelve days of warmth and moisture, the grain is worked forward by maltmen with wooden shovels and turned to bring about uniform germination of the seed. The process is laborious, and mechanical means of aeration have been installed at some distilleries.

Further growth is prevented by drying; the fumes from the peat fire used in this drying process pass through the malted barley, imparting the characteristic flavour of Scotch whisky, and out through the pagoda-shaped ventilators of the malt kilns, by which distilleries can be recognised from afar.

Mashing: The dried malt goes to storage. When required it is first ground to a grist and then fed into circular vats known as mash tuns, modern vats of stainless steel holding up to 8,000 gallons, where mixing with hot water converts the soluble starches into fermentable sugars. After half an hour the mixture is drained through the base into the worts receiver, or underback, and the mashing repeated with hotter water and drained again. A third instalment of hot water goes into the mash tun to minimise waste but this time it is drained separately to augment the *next* supply of malt.

Fermentation: The Wort, pronounced *wurt* or *wurts*, must now be cooled to 70°F by refrigeration to permit satisfactory fermentation in the fermentation vats or washbacks. Here yeast is added, bringing about a tumultuous fermentation for 36/40 hours, leaving at last the wash, a fermented brew, mildly alcoholic and suitably named.

So far it is the brewer who has been in charge; now the wash goes to a huge wash receiver, one of the few vessels not locked by the Excise. The starch in the barley has become sugar; the sugar – alcohol. It only requires a distiller to throw away the surplus water and you have the whisky you first thought of.

Distillation: The pot-still can be regarded as a huge copper kettle, in which the wash is boiled. Since alcohol vaporises at a lower temperature than water, the first run of vapour will be alcohol. By leading this vapour to a condenser, which in essentials is a metal coil encased by a water jacket, the raw spirit is obtained.

This alcohol, known as low wines, passes through the spirit safe to the low wines receiver. A gallon wash will make $^1/_5$ to $^1/_6$ proof gallons (roughly a wine bottle) of low wines, which need a second distillation to obtain a potable spirit. The first distillation takes place in a wash-still, the second in a spirit-still, both large copper vessels, side by side, with the fires beneath them.

The aim of the second distillation is to separate the potable portion of the resulting spirit from the first and last flow. The first flow (foreshots) still has too many undesirable elements; the last (feints or tailings) is too weak, which also gives rise to undesirable elements. Yet, indispensable to the character of malt whisky is a small fraction of essential oils (aldehydes, esters and higher alcohols, known collectively as congenerics) and only the experience of the stillman decides the vital moments at which to begin and end the collection of potable spirit.

His work requires considerable skill. The Excise does not permit him to handle or taste his efforts; it only allows him the spirit safe, locked with a glass window through which the liquid can be seen and tested, by hydrometer or by adding drops of distilled water to see whether it remains cloudless.

When distilled, malt whisky is reduced to 63.4 per cent. of alcohol by volume (11° over Proof) by adding water before filling the casks in which it will mature, for up to

fifteen years in a bonded warehouse, usually situated alongside the distillery.

The Distilling Season: The foregoing pages, it is hoped, will give visitors some idea of what they will see on a visit to a malt whisky distillery. The warmer months are not ideal for distilling and July, August and September, or parts of them are 'the silent season' when distilleries refit and staff go on holiday. Visitors, if accepted during these holiday months, may not always be able to see the distilling process actually taking place.

Autumn in Scotland often brings the driest weather so that mid-September to mid-October is a good time to go.

SPEYSIDE
Traditionally malt whiskies are divided into Highland, Lowland, Campbeltown and Islay malts, and of the Highland malts those from Speyside possess a more delicate, less robust character than the remainder.

Getting There: Visitors driving from Glasgow and Edinburgh airports or their rail heads will find the journey to Speyside can be done in 1½ to 2½ hours via Perth. Aberdeen and Inverness also have airports. The two cities are just over 100 miles apart; Keith, where the great concentration of distilleries begins, being roughly halfway between them. The Aberdeen-Inverness rail line also serves Speyside with stations at Huntley, Keith and Elgin.

Motoring north from Perth to Inverness, the Spey, fastest flowing river in Britain, first appears as the road descends from Dalwhinnie and just before Newtonmore. To follow its course to the sea at Speymouth, ten miles north-east of Elgin, turn north-east after Kingussie and Aviemore, taking A95 to Grantown-on-Spey where this road crosses to the east bank. Out of (roughly) a hundred and ten malt distilleries in the whole of Scotland, twenty-eight are to be found in the next thirty-three miles to Elgin, all within a mile or so of the road. And this is not because Spey water is as sought after as Spey salmon. The distilleries do not use Spey water; they take their supplies in nearly every case from a local well.

At Rothes, the central highlands give way to the coastal plain, widening to the north and east as far as the North Sea coast. This was the land of the barley and still is, though nowadays barley is mostly imported to make malt whisky, just as the peat, which imparts the flavour to the maltose, no longer always comes from the Highlands.

Elgin: In the midst of this coastal country is Elgin, where the Georgian houses give an unexpected air of elegance to this small capital town of Scotch whisky. Specialists in bottling and marketing many single malts are two Elgin firms – Campbell, Hope and King, and Gordon and McPhail. The latter bottle and make up those attractive sampling packs of twelve different miniatures, which I first saw in an Edinburgh wine merchant's window. I was determined my firm, Peter Dominic, should stock them and one December, when Cyril Ray recommended the pack in a Sunday *Observer* article only ten days before Christmas, an emergency supply line between Horsham* and Elgin had to be improvised, a young McPhail in his rugger shorts putting a load on the train at Aberdeen one Saturday afternoon five minutes before Kick-Off. This 'Speed Bonnie Malts' operation resulted in several hundred recipients receiving their packs in time for Christmas.

Dufftown: Although Dufftown is only five miles south-east of the Spey at Craigellachie, Professor R. J. S. McDowall in his book *The Whiskies of Scotland* treats Dufftown and nearby malts separately, finding Glenfiddich the driest, Dufftown the most peaty and Mortlach the most fruity, although all are whiskies of the Glenlivet type.

Visitors can go either to Strathmill (p.211) in the village of Keith, which belongs to Justerini & Brooks Ltd., or to Grant's Glenfiddich (p.212) where there are bars, a museum and a gift shop.

NORTHERN MALTS AND OTHERS

Dalmore and Glenmorangie, not unknown to my naval contemporaries when we swung round buoys at Invergordon between the wars, shew the flag for this small group of excellent whiskies mainly sited between Dornoch and Cromarty Firths.

The islands with malt distilleries are the Orkneys (2), Skye (1), Jura (1), with no less than eight on Islay and three on the Mull of Kintyre.

Regarded as essential in small quantities to blended Scotch whisky, the heavy west coast rainfall makes Islay malts powerful and peaty and most of the production is used for blending. Laphroaig, the most pungent, though it has plenty of devotees is not the Highland malt for the novice to try first. Visitors to the island would best begin with one of the others, Bowmore perhaps, working up to Laphroaig.

VISITING PROSPECTS

With 110 malt distilleries it might be supposed that malt whisky is plentiful but nearly all still goes for blending. In fact only some thirty single malts are available to the public and a few vatted malts, a vatted malt being a blend of malt whiskies from several distilleries.

Nevertheless in the past ten years more people have come to realise that a certain French spirit from Cognac is not the only one worth sipping neat, or with a little water, in a balloon glass. The one social error is to add soda, particularly in the presence of a Scot. It could easily convert him to rabid Scottish Nationalism!

Another misapprehension arising from those 110 distilleries is that given time – and considerable stamina – a 110-free-dram holiday tour might be easy. Alas no! Some years ago I recall writing to a distillery in Western Scot-

---

*Headquarters of Peter Dominic 1939–1968.

land asking if a customer could pay them a visit when on holiday. The reply was that visitors were a great distraction to wor-r-k and I gathered that if one should have the temerity to appear at the gate, the Managing Director himself would deliver a kick up the arse, such as Nye Bevan once received at the boot of an infuriated member when descending the steps of a West End London club.

Times change: I do not think my readers' posteriors would now be in jeopardy at any distillery. At one time the D.C.L. hesitated to accept visits from members of the public, chiefly because distilleries are small units, both in terms of the number of visitors who can be accommodated at any one time, and the staff available to look after them, which all too often means the distillery manager.

Doubtless, they would do so again if overwhelmed by a summer invasion but I am assured that visits to their forty-five distilleries can now be made by arrangement with the company operating them: Scottish Malt Distillers Limited, 1 Trinity Road, Elgin, Morayshire. Tel. Elgin 7891.

Since two distilleries at Girvan withdrew from the Guide shortly before going to press, Glenkinchie at Pencaitland, East Lothian could be helpful to those looking for a visit in the Lowlands. However, enough being as good as a feast, the entries that follow should suffice to give interesting whisky visiting in many parts of Scotland.

NON OPERATIONAL

This term, in the entries that follow, indicates that although visitors can be accepted, distilling is not in operation. (See The Distilling Season, p.207.)

RECOMMENDED READING

The Whiskies of Scotland R. J. S. McDowall John Murray paperback 3rd Edition 1975. A copy in the back of the car along with good maps is recommended before setting out.
Scotch Whisky David Daiches André Deutsch 1970. Delightful and complete study by an Edinburgh man who is Professor of English at Sussex University.
Scotch: The Formative Years Ross Wilson Constable 1970. This comprehensive 500-page study covers the first twenty-five years of this century and is for the Trade and those needing specialist knowledge of it.

Further Information
The Scotch Whisky Association,
17 Half Moon Street,
London W1Y 7RB.
Tel. 01–629 4384

GLASGOW
Stepps
James Buchanan & Co. Ltd.,
Cumbernauld Road, Stepps, Glasgow G33 6HR.
Tel. 041–779 2181.

This blending and bottling establishment for 'Black &

White' is about six miles north-east of the city centre on the Stirling road, A80.

Now part of the D.C.L. (Distillers Co. Ltd.), the astonishing career of its founder, James Buchanan, will forever be in the annals of private enterprise. Born 1849, clerk in Glasgow 1863, a job selling whisky in London 1879, found his own firm 1884, buys a Holborn distillery and rebuilds it with borrowed money 1898, wins the St. Leger 1916 and the Derby 1922 and 1926.

Few today will remember the spectacle of those black and white horses and drays which carried the whisky round London, still less the man responsible, James Buchanan, made Baron Woolavington of Lavington, West Sussex, where his house has since become a public school, Seaford College.

D.C.L.'s many blends involve the whiskies of 116 different distilleries throughout Scotland. This establishment carries out the subsequent processes of blending and bottling Black & White and Buchanan's other brands.

Hours: Monday to Friday 0930–1200; 1400–1600.

Introduction: Appointments preferred; best made through the London office: Devonshire House, Piccadilly, London W1A 1BN. Tel. 01–499 5381.

STIRLINGSHIRE
Strathblane

Lang Bros. Ltd.
The Glengoyne Distillery,
Dumgoyne by Strathblane, Stirlingshire.
Tel. Killearn 254.

This distillery is on the A81 Glasgow-Aberfoyle road, 14 miles north-west of Glasgow and three miles north of Strathblane/Blanefield. Professor McDowall* describes it variously as 'a little gem', 'a show piece' and 'nothing like a Speyside distillery'. It was here that the father of Lord Tedder, Marshal of the R.A.F., was an Excise officer and with such a background, the great wartime R.A.F. Marshal became a director of the D.C.L. in his last years.

Lang Bros. bought it in 1876 and although now part of the larger group, Robertson & Baxter Ltd., they see to it that 'the hospitality suite' in a separate chalet is fully used for tasting the product. The product annually is in fact 330,000 proof gallons of fine pot still malt whisky. Most of it, eagerly sought, goes to the blenders; the remainder makes a single eight-year-old malt.

Hours: Monday to Friday 0900–1200; 1400–1700. Non operational mid-July to late August.

*R. J. S. McDowall, The Whiskies of Scotland 1975 edition.

Introduction: By appointment. Write or telephone Head Office (and ask for Mr. Lemkes or Mr. Sinclair). Lang Bros. Ltd., 100 West Nile Street, Glasgow G1 2QT. Tel. 041–332 6361.

ARGYLLSHIRE
Islay

Stanley P. Morrison Ltd.
The Bowmore Distillery,
Bowmore, Isle of Islay, Argyllshire.
Tel. 049–681 225.

Islay, one of the islands of the Inner Hebrides 80 miles west of Glasgow, is served by a British Airways flight daily. By sea, there are regular sailings from Kennacraig (Tarbert) to Port Askaig and to Port Ellen.

The village of Bowmore was begun in 1768 and this distillery – founded in 1779 – is said to be the oldest legal distillery on the island. Sited on the shores of Lochindaal, its very walls being in the sea, *The Tourists' Guide to Islay* of 1881 states, 'This is a feature of the Islay Distilleries, their proximity to the sea being no accident as the operations are favoured by the fresh sea breezes, never wanting at a place like this.'

Bowmore, a traditional distillery, encourages visitors, who come in sufficient numbers to warrant a reception centre being completed for them in 1974. Both Bowmore and Morrison's Glengarioch (p.213) still malt their own barley.

Hours: Monday to Friday 1000–1200; 1400–1600.

Introduction: By appointment. Write or telephone either the Distillery Manager or Stanley P. Morrison Ltd., 13 Royal Crescent, Glasgow, G3 7SU. Tel. 041–332 9011.

Long John International Ltd.
Laphroaig Distillery,
Port Ellen, Isle of Islay, Argyllshire.
Tel. Port Ellen 2418.

Islay, the most southerly of the Inner Hebrides, 80 miles west of Glasgow as the crow flies, is served by a British Airways' flight daily. By sea, there are regular sailings from Kennacraig (Tarbert) to Port Askaig. All Islay's eight distilleries are by the sea, Laphroaig being on the south coast of the 25 × 20 mile island, little more than a mile east of Port Ellen.

Dating from 1815 and now wholly owned by Long John International Ltd., Laphroaig formerly belonged to the Johnston family for many years. Its whisky, the strongest of the Island's malts, is in great demand for blending. The Company's Islay Mist de Luxe contains Laphroaig as well as some of the finest Speyside malts.

Hours:	Monday to Friday 0930–1230; 1430–1600. Saturday 0930–1200. Closed August (Silent Season).
Introduction:	By appointment (1100 or 1500 preferred); please telephone.

ORKNEYS
Kirkwall
Highland Distilleries Co.
Highland Park Distillery,
Kirkwall, Orkney.
Tel. Kirkwall 3107/8.

Founded in 1798, this distillery looks down on Kirkwall, capital town of the island a mile to the north and there are daily air services both from Edinburgh and Glasgow to Grimsetter airport, two miles from the distillery.

Since 1935 the owners have been the Highland Distilleries Company (106 West Nile Street, Glasgow G1 2QY), a group with five distilleries, which includes Famous Grouse among its brands.

Hours:	Monday to Friday 1030–1200; 1430–1630. Closed July and August.
Introduction:	Open to all. Please write or telephone before arrival.

ROSS-SHIRE
Tain
Macdonald & Muir
Glenmorangie Distillery Co.,
Tain, Ross-shire.
Tel. (0862) 2043.

Tain, a small historic town, lies on the A9 road north-east of Alness on the southern shore of Dornoch Firth. The Inverness Airport at Dalcross, fifty miles away by road, has a regular service to Glasgow, with connections to London.

Of the northern group of Highland malts (Old Pulteney, Clynelish, Dalmore, Glenmorangie and Dalbair), Glenmorangie's whisky is the mildest. Established by the Mathiesons in 1843, the Glenmorangie Co. was formed in 1887; Macdonald and Muir of Leith bought the distillery in 1918 and have since become a subsidiary of Macdonald Martin Distilleries.

Hours:	Monday to Friday 1000–1200; 1400–1600. Closed July.
Introduction:	By appointment, best obtained from: Macdonald Martin Distilleries Ltd., Queen's Dock, Leith, Edinburgh EH6 6NN. Tel. 031-554 5111.

Alness
Mackenzie Bros.
Dalmore Ltd.,
Dalmore Distillery, Alness, Ross-shire IV17 0UT.
Tel. Alness 2362/4.

Dalmore, on the Cromarty Firth, with fine views across it, is half a mile beyond Alness, going towards Invergordon and Tain on the B817 road. The Inverness airport at Dalcross, forty miles away by road, has a regular service to Glasgow, with connections to London.

Founded in 1839, the distillery has been in the hands of the Mackenzie family for over a hundred years and is now a subsidiary of Scottish Universal Investments Ltd. It produces the well-known brand of Dalmore Highland Malt – a whisky much in demand by all the best blenders throughout the country.

Visitors are shewn over the distillery and the whole process of the manufacture of Scotch whisky is outlined.

Hours:	Monday to Friday 0930–1300; 1430–1700. Closed mid-July for about five weeks.
Introduction:	By appointment, readily obtainable from Mackenzie Bros. at the address above.

MORAYSHIRE
Elgin
Macdonald & Muir
Glen Moray-Glenlivet Distillery Co.,
Elgin, Morayshire.
Tel. Elgin 2577.

The distillery stands on the right bank of the river Lossie, about a mile west of Elgin city centre, almost where the Inverness-Elgin road, A96 and the B9070 from Dallas meet.

Built in 1897, Glen Moray is a comparatively modern distillery, owned, enlarged and re-equipped by Macdonald & Muir. Almost all the production is required for the firm's Highland Queen, Grand 15 and the Muirhead brands, made by Charles Muirhead & Sons, the retail wine merchants in George Street, Edinburgh, great specialists in Scotch whiskies of all kinds.

Macdonald & Muir is now a subsidiary of Macdonald Martin Distilleries.

Hours:	Monday to Friday 1000–1200; 1400–1600. Closed July.
Introduction:	By appointment, best obtained from: Macdonald Martin Distilleries Ltd., Queen's Dock, Leith, Edinburgh EH6 6NN. Tel. 031-554 5111.

Rothes-on-Spey
The Glenlivet Distillers Ltd.
Glen Grant Distillery,
Rothes-on-Spey, Morayshire.
Tel. Rothes 034–03–243.

This distillery is on the A941 road about 2½ miles out of Rothes towards Elgin.

On Speyside, Grants are as numerous as the bluebells of Scotland in spring, for this is Grant country. The brothers, James and John, established this distillery in 1840 and are not related to William Grant (p.212). In 1953 a merger with Glenlivet brought about The Glenlivet and Glen Grant Distilleries Ltd., which became The Glenlivet Distillers Ltd. on the merger with Hill Thomson in 1970. The Distillers Co. and the Imperial Tobacco Group each have a small interest in the new company.

Close by is their Caperdonich distillery, built in 1897 and completely modernised in 1965, which uses water from the same burn, yet the two whiskies are far from being identical.

Glen Grant single malts of eight, twelve and fifteen years' age have a great reputation. Glenlivet Distillers also market Queen Anne and De Luxe blended whiskies through Hill Thomson & Co.

Hours:	Monday to Friday 1000–1200; 1400–1600. Closed August.
Introduction:	By appointment giving a week's notice to: Mr. A. M. Lawson, P.R.Officer, Innes House, Elgin. Tel. Lhanbryde 251.

BANFFSHIRE
Mulben
International Distillers and Vintners Ltd.
Auchroisk Distillery,
Mulben, Banffshire.
Tel. Mulben 333.

This new establishment, a mile from the Spey, lies seven miles west of Keith and a mile beyond the point where the Elgin road changes on maps from being the red A95 to the yellow B9103.

In 1970, International Distillers and Vintners bought a 90-hectare farm with a spring called Dorie's Well. Having assured themselves that there would always be sufficient water, plans for a new distillery went ahead, the Rt. Hon. George Thomson formally opening it in September 1974 when he was Commissioner, European Economic Community.

Fully operational in 1975, Auchroisk (pronounced *Och-rusk,* which is as near as a Sassenach can get) can make a million gallons of Highland malt annually and has a maturing capacity of fifteen million gallons.

Visitors are shewn the complete mashing, fermentation and distilling process and see over a typical maturing warehouse.

Hours:	Open throughout the year; Monday to Friday 0930–1200; 1400–1600. Non-operational mid July – mid August.
Introduction:	Appointment preferred. Write to: International Distillers and Vintners Export Ltd., 1 York Gate, London NW1 4PU. Alternatively, small parties in the locality can often be accepted if 48 hours' notice is given direct to the Manager of the Distillery.

Craigellachie
Macallan-Glenlivet Ltd.
Macallan Distillery,
Craigellachie, Banffshire.
Tel. Aberlour (0343 935) 471.

Leaving Craigellachie on A941 for Rothes and Elgin, the Macallan Distillery is close to the Telford Bridge, built over the Spey in 1814. This bridge replaced a ford, which was on the route of the herdsmen taking cattle from the flat lands of the Moray coastal plain through the Highlands to southern markets.

The distillery was first licensed in 1824 and is now owned by a public company. No Macallan single malt is less than ten years old, the fifteen-year is unsurpassed and there has been some over twenty-five years old.

Hours:	Monday to Friday 1000–1200; 1400–1630. Closed Silent Season (usually August).
Introduction:	By appointment; write or telephone Mrs. McPherson at the distillery.

Keith
Justerini & Brooks Ltd.
Strathmill Distillery,
Keith, Banffshire.
Tel. 054–22 2531.

Keith lies on the main Aberdeen-Inverness road midway between the two cities. Coming from Aberdeen, cross the bridge in the centre of Keith, the entrance to the distillery is immediately left, marked Justerini & Brooks.

Built in 1823 as an oatmeal mill for the local farming community, Strathmill was rebuilt as a Highland malt distillery in 1891. Since then it has operated continuously, distilling capacity being increased over the years from about 100,000 to 650,000 gallons a year.

Most of its production is now needed for J. & B. Rare. Justerini & Brooks is a subsidiary of International Distillers and Vintners, who now prefer to shew visitors Auchroisk, their new establishment completed in 1974 (p.211) and since awarded a big European architectural prize.

Visitors are shewn the complete mashing, fermentation and distilling process and see over a typical maturing warehouse.

| *Hours:* | Open throughout the year: Monday to Friday 0930–1200; 1400–1600. Non operational mid July – mid August. |

Introduction: Appointment preferred. Write to: International Distillers and Vintners Export Ltd., 1 York Gate, London NW1 4PU.

Alternatively, small parties in the locality can often be accepted if 48 hours' notice is given direct to the Manager of the Distillery.

Dufftown
William Grant & Sons Ltd.
Glenfiddich and Balvenie Distilleries,
Dufftown, Banffshire.
Tel. Dufftown 375.

Dufftown is on the A941, four miles south-east of Craigellachie. Keith, on the main Aberdeen-Inverness line and 11 miles away, is the nearest good rail connection. These two distilleries are close together, just off the A941, to the north of the town and on the river Fiddich, tributary of the Spey.

Behind this company is the romantic story of William Grant who slaved and saved for twenty years in the neighbouring Mortlach distillery. Then, in 1886, with £120 and the help of seven sons, he built the Glenfiddich distillery, following in 1893 with a second one – Balvenie close by, which still contains one of the few traditional floor maltings on Speyside.

Although both were much enlarged in 1955, the process remains virtually unchanged. Glenfiddich Pure Malt Scotch Whisky is the only Highland Malt to be bottled at its own distillery. The company's other brands are Balvenie Pure Malt, Grant's Royal 12-Year-Old (blended) and Grant's (blended) Scotch Whisky.

There is no entry charge and a team of guides shew visitors the whole process, from mashing to bottling, in a 30-minute tour. After that there is a free dram and the Gift Shop where souvenirs are on sale.

Hours: Monday to Friday 1000–1200; 1400–1630. Evening and weekend visits by special arrangement.

Introduction: None necessary. Any special requirement should be arranged with: The Chief Guide, William Grant & Sons Ltd., at the distillery, address above.

Ballindalloch
George and J. G. Smith, now The Glenlivet Distillers Ltd.
The Glenlivet Distillery,
Ballindalloch, Banffshire.
Tel. Glenlivet 202.

From Grantown-on-Spey take the A95 north-east towards Craigellachie for about 12 miles as far as the Dalnashaugh

Inn on the Avon, tributary of the Spey. Turn right on to B9008, which runs south. The distillery, standing on a bare slope where the Livet burn runs down to join the Avon, is reached after about 4 miles. From Aviemore (33 miles) allow about 1¼ hours for the journey.

Smith's Glenlivet is the oldest licensed distillery in Scotland. Already described in the historical paragraphs of the introduction (p.205), George Smith was the first to accept the English Government's new legislation and took out a licence in 1824, surviving many threats to his life and property from fellow farmers, who regarded his as a 'blackleg' for years afterwards.

But Glenlivet's whisky already had a reputation; two years earlier George IV visiting Scotland had insisted upon it. Then, the Court decision of 1880 that only this distillery could sell its malt whisky as 'The Glenlivet' was proof enough that the reputation had been maintained. Smith's Glenlivet has remained a malt whisky of the highest quality, as all those who met there on 31 August 1974 to celebrate the 150th anniversary of George Smith going legal, would be the first to testify.

Visitors see all the processes.

Hours: Monday to Friday 1000–1200; 1400–1600. Closed August.

Introduction: By appointment giving a week's notice to:
Mr. A. M. Lawson, P.R. Officer, Innes House, Elgin.
Tel. Lhanbryde 251.

ABERDEENSHIRE
Forgue, By Huntly
Wm. Teacher & Sons Ltd.
Glendronach Distillery,
Forgue, By Huntly, Aberdeenshire.
Tel. 046–682 202.

From Huntly take the Banff road, A97, turning right on to B9001 after about 5 miles. The distillery, beyond Forgue, is about two miles from the town in the small secluded Dronac glen.

Wm. Teacher, still an independent public company, own two Highland malt distilleries, Ardmore and Glendronach. All Ardmore's malt whisky is needed for blending Teacher's Highland Cream but both eight-year-old and twelve-year-old single malts from Glendronach are sold successfully.

Nineteen-year-old William Teacher founded this company in 1830 and although Glendronach was not bought by them until 1960, it had achieved distinction in 1826 by being second only to Glenlivet in accepting the British Government's new deal and taking out a licence.

Visitors are shewn the complete cycle of pot still distilling from the malting to the final distillation, in a distillery which has not departed from the traditional malting method.

| *Hours:* | Monday to Friday 0900–1200; 1400–1700. Non operational July 24 approx. to 1 September. |
| *Introduction:* | Though callers are welcome without notice, a guide should be present if prior arrangements are made, either with the Manager of the Distillery (telephone above), or with the Tours Director: Wm. Teacher & Sons Ltd., 14 St. Enoch Square, Glasgow G1 4BZ. Tel. 041–204 2633. |

Peterhead

Long John International Ltd.
Glenugie Distillery,
Peterhead, Aberdeenshire.
Tel. Peterhead 2110.

Peterhead is on the coast 32 miles north of Aberdeen and less than a mile north of Peterhead the little river Ugie flows into the North Sea. The distillery, just off the main road A952, immediately after passing Peterhead Prison on the left, is difficult to miss.

Built in 1875, Glenugie is a small distillery taking its water from the Ugie burn. Close by is their local landmark – an ancient stone tower standing apart like that at Château Latour and leaning a little in the Pisa manner.

In the 1930s when gin sales were bad, the London gin company, Seager Evans, acquired Long John Scotch whisky by taking over a firm called W. H. Chaplin. Post-war Long John did so well that Sir Albert Richardson (architect of Bracken House, the *Financial Times* building in Cannon Street) was employed to design Tormore (near Grantown-on-Spey). In 1970, Seager Evans changed their name to Long John International.

Tormore and Glenugie supply some of the malt whisky for Long John's blends; neither has sufficient left to bottle as a single malt.

| *Hours:* | Monday to Friday 0930–1230; 1430–1630. Saturday 0930–1300. Closed mid-July – mid-August (Silent Season). |
| *Introduction:* | By appointment (1100 or 1500 preferred); please telephone. |

Old Meldrum

Stanley P. Morrison Ltd.
The Glengarioch Distillery,
Old Meldrum, Aberdeenshire.
Tel. Old Meldrum 065–12 235.

One of the oldest distilleries in the north-east, Glengarioch, near Old Meldrum on the A947 road to Banff is 17 miles from Aberdeen and less than a dozen from Dyce, airport of 'the granite city'.

Dating from 1788 the distillery takes its name from the Garioch, the granary of Aberdeenshire, in which Robert

the Bruce defeated the Earl of Buchan's army in 1308, just one mile away.

A traditional distillery, still malting its own barley, Glengarioch is built of Aberdeen's greyish-white granite. Visitors are very welcome and a reception centre was completed in 1975 for them.

| *Hours:* | Monday to Friday 1000–1200; 1400–1600. |
| *Introduction:* | By appointment. Write or telephone either the Distillery Manager or Stanley P. Morrison Ltd., 13 Royal Crescent, Glasgow G3 7SU. Tel. 041–332 9011. |

PERTHSHIRE
Pitlochry

Arthur Bell & Sons Ltd.
The Blair Athol Distillery,
Pitlochry, Perthshire.
Tel. Pitlochry 2161.

Coming up the Inverness road, A9, from Perth, this distillery is easily recognised on the southern outskirts of Pitlochry, 30 miles from Perth. The smaller town of Blair Atholl is seven miles further on.

Arthur Bell & Sons Ltd., which began in Perth as a small whisky shop in 1825, is one of the largest independent firms in the Scotch whisky industry. Arthur Bell himself did not appear until 1851, but by the time that he died in 1900, he and his two sons were in sole charge of the business. Since 1942, there have been no Bells on the Board but they are remembered – as are the Dewars – for their philanthropic work for Perth.

Bell's own four distilleries: Inchgower, near Buckie, Dufftown-Glenlivet and Pittyvaich-Glenlivet on the outskirts of Dufftown, the fourth being this Blair Athol distillery, which was first licensed in 1825 and is as picturesque as any in Scotland.

| *Hours:* | Monday to Friday only at 1500. Closed to visitors during the 'silent season' 1 July to 8 August approx. |
| *Introduction:* | By appointment. Please write or telephone giving at least 3 days' notice of intended visit to: Arthur Bell & Sons Ltd., Cherrybank, Perth. Tel. 0738–21111. |

Inveralmond

John Dewar & Sons Ltd.,
Inveralmond, Perth, Perthshire.
Tel. (0738) 21231.

Perth is 'The Gateway to the Highlands', connected by motorail services from London and other English cities. Edinburgh and Glasgow are 1 and 1½ hours by road respectively. Inveralmond lying one mile north of Perth on the Inverness road (A9), is hard to miss.

In 1846, rather late in life, John Dewar started his own business as a wine and spirit merchant in the High Street of Perth. He died in 1880 and was succeeded by his two sons, John and Tommy, destined to become the first whisky barons in 1923. The first distillery was rented in 1887 and the next built in 1896. By the twenties, when the company was known world-wide, there were seven. Long since a member of the D.C.L., the sales of Dewar's White Label and the De Luxe Ancestor have earned The Queen's Award for export achievement no less than five times.

Visitors are taken over the main blending and bottling areas and the cooperage. Whisky is sampled but may not be bought. Explanations are in English; given good notice, guides speaking other languages are possible.

Hours:	Monday to Friday 1030–1215; 1430–1615. Closed second half of July.
Introduction:	By appointment, obtainable either from Inveralmond or from: Dewar House, Haymarket, London SW1. Tel. 01–930 4921.

AUSTRIA

Austria

Introduction

GETTING THERE

By road: Salzburg and Vienna are continuous motorway driving from The Hague, passing close enough to Koblenz, Wiesbaden and Würzburg to make Mosel, Rhine and Main the first part of a wine tour.

By rail: There is a night car sleeper service, Dusseldorf–Salzburg in summer and a day car carrier service, Frankfurt–Salzburg. Normal passenger trains from London via Ostend take about 24 hours. From Paris, the Orient Express takes about 16 hours.

By air: The daily services, London–Vienna and Paris–Vienna, each take about two hours.

By water: From Vienna itself there are Steamship Company services up and down the Danube and a Hydrofoil* service to Budapest.

WHITE WINES

Some seven million Austrians, living in this happy holiday country twice the size of Switzerland, drink a great deal of beer and quite a lot of wine – 42.2 litres a head in 1972–3 so they say – rather more than the country makes itself so some has to be imported. Eighty-five per cent of Austrian production is white wine and annually over 17,000 hectolitres go to Britain, where most good wine merchants stock one or two brands.

In the mountainous western half of Austria, where the skiers go, the wines met are more likely to be Italian – from the Alto Adige, which was Austrian in pre-Great War days until 1919. Austria's own vineyards are all in the flatter eastern half, in the Danube valley and near her borders with Czechoslovakia, Hungary and Yugoslavia. Apart from one or two small patches, they lie within fifty miles of Vienna – 47,000 hectares in all, split into 70,000 holdings, achieving an annual output of two million hectolitres.

Among the grape species, Riesling, Traminer and Syl-

---

*For other Danube services, see Hungary, p. 223.

vaner will be familiar, but Austria's original contribution to drinking is the wine of the Grüner-Veltliner grape, which emerges fresh as a daisy straight from a thousand casks in the taverns of the town.

This grape is also capable of high quality. Prinz Metternich grows it on his Grafenegg estate at Krems and his crisp, dry estate-bottled Schloss Grafenegg, widely sold in Britain, is exclusively from the green or Grüner-Veltliner.

Vienna itself is famed more for wine drinking than making. Nevertheless Klosterneuberg, on the northern outskirts where the Weinviertel district runs north from the Danube to Czechoslovakia, makes good wine and has a wine museum. A dozen miles to the south of the city, in the region of the Vienna woods, Gumpoldskirchen makes a luscious golden wine, from late-picked grapes in good years, that does a little to redress the balance of warm water from the springs of Baden nearby.

WINE TOURS

Burgenland: A trip south-eastwards from Vienna leads to the Burgenland, where good white wines are made from vineyards on the shore of the reedy Lake Neusiedl. Eisenstadt, capital of Burgenland, is the local tasting centre, where sampling can be combined with homage to Haydn either in the great hall of Esterhazy Castle or in his own modest house, open to visitors. Rust, Mödling and Perchtoldsdorf are typical wine villages of the region.

Wachau: The vineyards of this stretch of the Danube, centred on Dürnstein upstream from Krems, are dramatically sited on steep slopes mildly reminiscent of the Mosel. Dürnstein and Langenlois, a village ten miles north-east of Krems, have a baroque charm; the former is where Richard Coeur de Lion was interned returning from the Crusade, the statue of him and Blondin commemorating the troubadour legend. The Co-operative winery is also at Dürnstein.

The tour from Vienna to Dürnstein could pause for a while in Krems and should include a run up the Krems valley amid orchards and vines to Senftenberg.

The *Michelin Green Guide* (English Edition) recommends

AUSTRIA

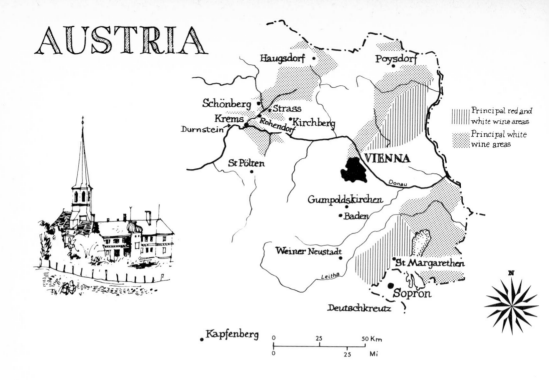

Principal red and white wine areas

Principal white wine areas

Haugsdorf •   Poysdorf

Schönberg • •Strass

Krems • •Rohrendorf •Kirchberg

Durnstein

St Pölten

VIENNA

Donau

Gumpoldskirchen •

• Baden

Weiner Neustadt •

Leitha

•St Margarethen

• Sopron

Deutschkreutz)

• Kapfenberg   0   25   50 Km

0   25   Mi

---

leaving another day to explore the Danube valley from Dürnstein to Enns, pausing at Weissenkirchen, Loiben and Spitz, all three attractive parts of the Wachau.

NATIONAL HOLIDAYS

National holidays in Austria are given below. Please remember when planning your trip that, not only will premises be closed, but that it is not uncommon to remain closed for an extra day (e.g. Friday after a Thursday public holiday) to make a long weekend break.

1 and 6 January, Easter Monday, 1 May, Ascension Day, Whit Monday, Corpus Christi, 15 August, 1 November, 8 and 25 December, and Boxing Day.

RECOMMENDED READING

The Wines of Central and South-eastern Europe R. E. H. Gunyon Duckworth 1971.

Further Information:
Austrian National Tourist Dept.
16 Conduit Street
London W1.
Tel. 01–629 0461.

Krems (near)
Lenz Moser,
3495 Rohrendorf bei Krems.
Tel. (0.27.32) 55.41.

Rohrendorf is on the Danube, 5 kilometres downstream from Krems and about 74 kilometres from Vienna. By motorway and express way it takes about one hour from Vienna to Krems, the autobahn to St. Pölten being the usual route. There is a good train service from the Franz Josefs Bahnhof and in summer it is possible to do the trip to and from Krems on the Danube, the First Danube Steamship Company's fast service taking just under three hours.

Although this Lenz Moser Company was only founded in 1929, the cellars are about a thousand years old and the Mosers have been wine people since 1124. Dr. Lenz Moser is known throughout the world of viticulture as the inventor of the 'High Culture' system, in which the rows are planted at a distance apart of 3.5 metres, the vines in each row being 1.2 metres apart. 3.5 metres allows all agricultural machinery to pass between the rows and the system became the basis of mechanization. The vines themselves grow to a height of 1.2 to 1.3 metres, at which the fruit-bearing laterals are tied to wires running horizontally along the rows. Regarded as revolutionary and disastrous by conventional growers at the time, the system gives a higher crop with fewer vines in a given area.

The company is a major exporter to many countries, the fresh, prickly white Schluck being particularly popular in Britain.

Hours:	Monday to Friday 0900–1600.
Introduction:	None necessary. Lenz Moser receive visitors from Mondays to Fridays *without* appointment and sometimes on Saturdays if enquiry is made by telephone in advance. Any further information from agents: Lawlers of London Ltd., 6A Station Road, Redhill, Surrey. Tel. (0737) 61415.

Gumpoldskirchen
Alois Morandell & Sohn,
Weinkellerei, A-2352 Gumpoldskirchen.
Tel. (02252) 5566.

The vineyards of Gumpoldskirchen, south of Vienna, are 40/60 minutes' drive using the Südautobahn, to which there are plenty of directions in the city. The light, fruity Gumpoldskirchner is as celebrated a wine as any in Europe and there are direction signs to the Morandell Weinkellerei on reaching the town.

Founded in 1920, this is the only Austrian wine company to have established its own company in Britain to act as agents and distributors for their Austrian estate-bottled wines, such as 'Steiner Hund'. They also ship fine wines from the Italian Tyrol, having their own cellars at Caldaro, still called Kaltern by the Austrians (see Italy, Trentino-Alto Adige).*

Visitors can taste and buy wines; English is spoken.

Hours:	Monday to Friday 0800–1200; 1400–1600.
Introduction:	By appointment, arranged through: Morandell (U.K.) Ltd., 6–8 The Highway, London E1 9BQ. Tel. 01–481 9251.

Vienna
Rudolf Kutschera und Söhne,
Weingut, Wein und Sektkellereien, 1195 Wien XIX –Nussdorf, Heiligenstädterstrasse 205.
Tel. (0222) 37.12.50 and 37.22.30.

The village of Nussdorf, now enclosed in the growth of the city of Vienna, lies on the Danube in the nineteenth city district and is reached by the Heiligenstädterstrasse, a main artery running north from the end of the outer ring road (Gürtel) to Nussdorf, where it continues along the river. By public transport, take the D tram which runs round the Ring to its terminus at Nussdorf. Both motor and tram approaches are easily identified on Vienna city plans. There is parking nearby.

Founded in 1876, the Kutschera family specialises in still and sparkling white wines from its own vineyards, in Gumpoldskirchen south of Vienna, and at Krems, north-west of Vienna on the far bank of the Danube. Visitors are welcome at the cellars in Nussdorf to sample wines in the company of English-speaking assistants.

Hours:	Monday to Thursday 0900–1100; 1430–1600. Friday 0800–1200. Closed the week before Easter and the first two weeks in July.
Introduction:	None necessary but 24 hours' notice by letter or telephone is requested.

---

*Visitors are equally welcome at the cellars in Italy. The address is Bahnhofstrasse 10, Kaltern/Sudtirol. Tel. (0471) 53318. Appointments from Morandell (U.K.) Ltd.

HUNGARY

Introduction

Hungary has long been a good wine country, making about 6,000 hectolitres a year (France and Italy about 60,000) rather less than Yugoslavia, much more than Austria and about the same as Algeria and Chile. The quality of her table wines is generally higher than that of these other small countries and in Tokaj she has a truly great wine, even though quality and price nowadays may appeal less to 'the rich man in his castle' and more to the 'poor man at his gate'.

From the west the motorist entering Hungary from Vienna may pass by Sopron, where Haydn is buried and Liszt began his career. Soproni Kékfrankos, a Gamay-style red wine, can be drunk, though it hardly rates as a rhapsody.

Sopron is where the Small Plain begins. To the north the Danube flows east, forming the frontier with Czechoslovakia before making a 90° turn to flow south to Budapest and on across the Yugoslav border to Belgrade. Its course encloses Transdanubia, the western half of Hungary, an excellent wine land, with Lake Balaton, the 'Hungarians' Sea', in the middle.

GETTING THERE

By road: The direct route is as given for Austria. From Vienna the Trans-continental highway reaches the frontier in about 40 miles continuing to Budapest via Györ. Vienna to Budapest is 170 miles. A slower alternative from Vienna is through Burgenland via Sopron, a wine town where the inhabitants voted to be in Hungary after the first World War.

By rail: The Orient express route is Paris – Munich – Vienna – Budapest. Vienna to Budapest takes 4/5 hours.

By air: Almost daily flights to Budapest from London and from other European capitals.

By water: A Danube hydrofoil service (most days May to September) does Vienna to Budapest in 4/5 hours. Motor vessels of the Soviet Danube line and the First Danube

Steamship Co. do the journey as part of a cruise, which extends to the Danube delta from Vienna.

LAKE BALATON

The vineyards lie along the north shore of this huge lake, 50 miles long by 5 miles wide, on the slopes of Csopak, Balatonfüred and the extinct volcano Badacsony.

Today the State viticultural station tries to persuade the peasant growers to eliminate some of their seventeen grape varieties and to concentrate on achieving better quality through Wälschriesling, Kéknyelü and Szürkebarát (Pinot gris), best of the white wine grapes.

MOR

A remarkable dry golden wine comes from Mór, fifty miles west of Budapest. Like that of Colares in Portugal, the sandy soil of Mór is immune from the phylloxera aphid and the vines do not have to be grafted on to the tough American rootstock to ensure freedom from attack. The vine is the Ezerjó, meaning 'A thousand boons' and it was first planted here in the eighteenth century by a number of Bavarian refugees. They too were a boon, siting the vineyards so skilfully that they escape both winter frost and summer shadow.

THE GREAT PLAIN

South and east of Budapest the vineyards, spread over The Great Plain, cover nearly 250,000 acres, well over twice the area of all the rest in Hungary but the wine is very ordinary.

EGER

Interest is aroused again at Eger, a charming baroque town, 85 miles north-east of Budapest, with a red wine district of a dozen villages on the slopes of the hills. Besides the legendary Egri Bikavér (Bull's Blood) from the Kadarka grape, helped a little with Pinot noir and

HUNGARY

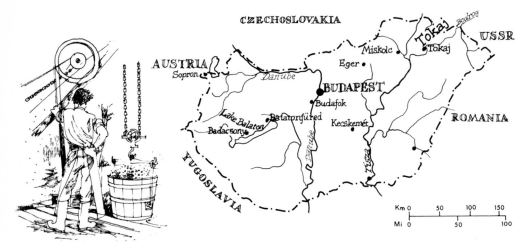

Merlot, there are rich, white dessert wines from Leányka and Mézes Fehér vines.

The legend originates from 1552, when the Turks were besieging Eger. The Magyar defenders, plied with quantities of Egri Bikavér by their womenfolk, fought with such fanaticism that the abstaining infidels, deciding that their enemy had become demented with Bull's Blood, beat a hasty retreat.

Close to the Eger are the foothills of the Matra range where an indigenous Hungarian grape, the Hárslevelü, meaning 'lime leaf', makes Debrői Hárslevelü, a white wine.

TOKAJ

Hárslevelü and Sárgamuskotály, a yellow Muscat, figure in the making of Tokaj but the principal grape that works this miracle of sweetness is the Furmint, which makes much drier wines in the districts already discussed.

Tokaj, 150 miles north-east of Budapest, is a dull town, which happens to be where the small river Bodrog meets the larger Tisza, making it a shipping place that gave the wine its name. Tokaj–Hegyalja is the wine region, comprising some twenty odd hill villages among the foothills of the Carpathians. Given the right conditions, the Furmint and the Hárslevelü react to *botrytis cinerea* in the same way as the Sémillon and Sauvignon grapes react in Sauternes (q.v.).

Carpathian protection from the northerly blast, southerly winds from the plain, morning mist from the Bod-

rog . . . given these in appropriate measure, all should be set for a great *aszú* vintage when picking begins, by tradition, on 28 October (SS Simon and Jude) each year.

These *aszú* grapes, infected with the *botrytis cinerea* or 'noble rot', are gathered and fermented separately in wooden 'hods' called *puttonyos* (capacity 30/35 litres). Any juice that drips of its own accord, without any pressing of the grapes – and there will be only a litre or two from each *puttony* – is collected and kept separately. This is the famous Tokaj Essence, weak in alcohol but full of sugar, so that it will just ferment slowly for years. 'The finest restorative in the world', said Warner Allen.*

Its use, it would seem, was confined to reviving Emperors on their death beds. £220 has been paid in auction in recent years for a half litre, far too expensive for Britain's National Health Service, which keeps us alive too long by other means. Only about sixty barrels of Essence exist and the Hungarians use them for sweetening *Aszú* wines.

After the natural transfusion, the *aszú* berries in their *puttony*, are trodden, or pressed mechanically, into a 'paste'.

Meanwhile, must made from *normally* ripe grapes has been fermenting in the local 130-litre casks called *gönci*.

---

*H. Warner Allen *White Wines and Cognac* 1952.

224

The celebrated Tokaji Aszú as sold to the public is a mixture of the two and its richness is indicated by the number of *puttony* of *aszú* paste added to a *gönci* of normal must. Wines of three, four and five *puttonyos* are those normally met. Capacities being *gönci* 130 litres and *puttony* 35 litres, Tokaji Aszú 5 puttonyos (the richest) is very rich indeed.

A further six years from the vintage with the wine maturing in cask is now required before Tokaji Aszú reaches the market in traditional half-litre bottles.

Other varieties of Tokaji are suitable for drinking with meals, notably Tokaji Furmint and Tokaji Szamorodni, dry to medium dry according to the year, from grapes picked when normally ripe.

Hugh Johnson finds the old cellars of Tokaj among the world's most romantic* and though the wine today is pasteurised he still detects 'a haunting fragrance . . . and the breath of the Bodrog among October vines'.

For visits to Tokaj the motorist should take the M3 to Tokaj as follows: Budapest – Hatvan – Miskolc – Szerencs – Tokaj. In Tokaj the visitor will see the State cellars, which are very old and were taken over from the Imperial or Aristocratic cellars. Other places to see are Tarcal, Tolcsua and Ma'd. There is a museum at Rakosi House (Tel. Sa'toralfaushely 690) which is well worth seeing.

THE STATE ORGANISATION

The Wine Trade in Hungary is State-controlled, not quite so dull as it may sound should one find oneself put 'under the table' by some lovely young blonde in charge of the cellars, as has happened to some journalists.

The State export organisation is known as Monimpex and for members of the Trade the firm of F. & E. May Ltd., 18 Piazza Chambers, Covent Garden, London WC2E 8EN, Tel. 01–836 3012 and 5521, will arrange introductions.

F. and E. May are the sole importers and distributors of all Tokaj and other fine Hungarian wines, mostly estate-bottled.

Visits for members of the public, however, are arranged by the State Travel Agency, IBUSZ. There are a number of other organisations, among them The Tourist Information Service, Budapest VII, Rákóczi út 52, P.O.B.26 BP70, but their staffs do not speak foreign languages – either English or German – and it is therefore best to arrange visits with IBUSZ, which will provide guide/interpreters. IBUSZ has sixteen offices in Budapest and a branch in every Hungarian town at the service of visitors. Correspondence should be sent to:

IBUSZ Management
1053 Budapest
Felszabadulás tér 5.
Tel. 180–860.

*Hugh Johnson *The World Atlas of Wine* 1971.

NATIONAL HOLIDAYS
National holidays in Hungary when establishments are likely to be closed, are: 1 January, Easter Monday, 4 April, 1 May, 20 August, 7 November, 25 December and Boxing Day.

RECOMMENDED READING
The Wines of Central and South Eastern Europe R. E. H. Gunyon Duckworth 1971.

Further Information
Danube Travel Agency Ltd.
6 Conduit Street
London W1.
Tel. 01–493 0263.

Badacsonylábdihegy
Wine Cellars of Badacsonylábdihegy,
8262 Badacsonylábdihegy.

From Balatonfüred the 'Hungarian Riviera' road 71 runs close to the shore, with the hump of the once volcanic Mount Badacsony looming larger until it is reached after 40 kilometres. After the town of Badacsonytomaj, the road, keeping to the shore, circles round the south side of the mount coming to the railhead of Badacsonylábdihegy, which is on the flat ground.

The premises, wines etc. are likely to be similar to those at Balatonfüred and procedure for visiting is the same.

Hours:	Monday to Thursday 0800–1600. Friday 0800–1400. Closed during the vintage 1 September – 15 November approx.
Introduction:	Through IBUSZ, see above.

Balatonfüred
Wine Cellars of Balatonfüred,
8230 Balatonfüred Zrinyi u. 5–7.

Balaton, the largest lake in Europe, 100–200 kilometres south-west of Budapest, is easily seen on the map and easily reached by motorway from the capital. Most of the north shore is a great 'wineyard' and after passing Csopak (one of the best wine villages), Balatonfüred, 134 kms from Budapest, is the first wine town to be reached.

Though grapevines have flourished since Roman times, the scene here is a mixture of old wooden casks and new cement vats. The wines are far more interesting. Those for tasting and purchase include: Badacsonyi Szürkebarát, a full Pinot gris white wine, the green-golden Kéknyelü, Balatonfüredi Olaszrizling, a firm strong white wine from the Italian Riesling grape, and Balatonfüredi Tramini which has that same spicy Traminer flavour so pleasing to Alsace wine enthusiasts.

Hours:	Monday to Thursday 0800–1600. Friday 0800–1400. Closed during the vintage 1 September – 30 October approx.

Introduction: Through IBUSZ, see page 225.

Budafok
Wine Cellars of Budafok,
1221 Budapest, Kossuth Lajos u.84.

In the XXII district 20 kilometres south of Budapest, Budafok is an industrial town, which an English visitor declared made Harlow look like San Francisco.

The limestone hills of Budafok result from deposits of the Pannonian Sea. In the nineteenth century much of the stone was excavated and taken cheaply by water for extensive building in Vienna, Buda and Pest. The resultant galleries, maintaining a uniform temperature and adequate humidity, run for many kilometres and are ideal for keeping wines. The capital city, being close at hand encouraged their use commercially. These cellars, branching hither and thither, in fact cover 20 kilometres.

Visitors walk round part of them, seeing the bottling plant and tasting samples. Wines are for sale to take away. Márka Vermouth is one speciality.

Hours: Monday to Thursday 0800–1600. Friday 0800–1400. Closed during the vintage 1 September – 15 November approx.

Introduction: Through IBUSZ, see page 225.

Eger
Wine Cellars of Eger-Gyöngyös,
3300 Eger, Arnyékszala u. 9–63.

Eger, a charming baroque town, lies at the foot of the Bükk and Mátra hills, 129 kilometres north-east of Budapest. The vines are planted on the slopes of these hills, some twelve hill villages being entitled to use the Eger name.

These are the biggest cellars of Eger, hollowed in the hillside, and the wines are kept, at a steady temperature, in wood and in modern cement cisterns covered by glass plates. There are old wood carvings of peasant origin and

a rare mould on the walls of these caves which put end to end would be 3.5 kilometres long.

The 'bovine' Egri Bikavér and the white Egri Leányka will be to the fore but Eger's vineyards grow the French *cépages*, Pinot Noir, the Bordeaux Merlot and the Cabernet vine as well as the strong native Kadarka so there is a good choice to try and to buy.

Hours: Monday to Thursday 0800–1600. Friday 0800–1400. Closed during the vintage 1 September–15 November approx.

Introduction: Through IBUSZ, see page 225.

Kecskemét
Wine Cellars of Kecskemét,
6000 Kecskemét, Kiskörösi. u.

Kecskemét is an important centre of The Great Plain, 89 kilometres south-east of Budapest close to the main road E5.

These wine cellars, named after the Hungarian wine grower, János Mathias, are in the heart of Hungary's ordinary wine country where the soil of the The Great Plain is sandy and there are a hundred thousand hectares under vine. The modern premises, completed in 1972, process and store, in cement cisterns lined with glass plates some 75,000 hectolitres of wine.

Visitors are shown round and may taste and buy eight kinds of lowland wines and seventy sorts from other districts. Kövidinka is a recommended 'Ordinaire'; likewise Ezerjó, which only achieves quality when made from the Mór vineyards, near Lake Balaton far to the westward.

Hours: Monday to Thursday 0800–1600. Friday 0800–1400. Closed during the vintage 1 September–15 November approx.

Introduction: Through IBUSZ, see page 225.

OTHER COUNTRIES

Other Countries

Due to space and price considerations, the short sections prepared for each of the following countries have had to be reduced to notes as follows:

CYPRUS

Red and white table wines, Cyprus sherry and Cyprus brandy form the back bone of the modernised wine industry on which the economy of Cyprus greatly depends. Of the four big wine making and marketing firms, Keo Ltd. (Tel. Limassol 62053) lay on a good tour with tasting (Monday–Friday, all visitors welcome); Sodap Ltd. (Tel. Limassol 4605) do much the same.

GREECE

In Athens, write or telephone to the Export Department of Andrew P. Cambas S.A., 2 Efpolidos Street. Tel. 32.47.877. The firm's vineyards and winery at Kantza on the inland road to Cape Sounion are worth a visit if you have £5 for the taxi.

In Patras, the Achaia Clauss Visitors' Book contains the names of Liszt, 'Monty' and Gary Cooper. Incomparable film! One wonders whether they got Mr. Deeds 'pixillated' on Mavrodaphne.

ISRAEL

In 1906 Baron Edmond Rothschild helped and financed the growers on condition they formed a co-operative. Today, this co-operative, at Richon-le-Zion, near Tel-Aviv, makes 75 per cent. of Israel's many and varied wines. Visits can be arranged through the Carmel Wine Co. Ltd. in London.

NETHERLANDS

A spirit, not a wine country, famous for gin, Advocaat and Liqueurs. In Amsterdam these may be sampled at Bols Tavern (106 Rozengracht) where 'one on the House' has been arranged for anybody showing this Guide. From Schiedam (near Rotterdam) Johs de Kuyper and Zoon welcome those with introductions from Matthew Clark and Sons, their agents in London.

SWITZERLAND

Swiss wines – red, white and sparkling – are good. Unfortunately, they are apt to cost more than those of their neighbours and are rarely met in Britain. S. Châtenay S.A. of Boudry on Lake Neuchâtel welcomes visitors; their agents in London, Capital Wine and Travers Ltd., arrange introductions.

Le Valais, where the Rhône tumbles down into Lac Leman, is the other region. Local enquiries could lead to one of several firms rejoicing under the name of Bonvin.

YUGOSLAVIA

Excellent wines at low cost, some of them made along the Dalmatian coast, where tourists speaking Croat might trace them to their source. In Ljutomer, the main district nearer to Hungary than the Adriatic, lack of English speakers restricts visits to organised parties. Teltscher Bros., Yugoslav wine distributors in London, are the best advisers.

INDEX

Entries in **bold** refer to establishments open to visitors; entries marked with an * refer to places where there are such establishments; and entries in *italic* refer to names of specific wines and spirits.